Contaminated Soils: From Soil–Chemical Interactions to Ecosystem Management

Other titles from the Society of Environmental Toxicology and Chemistry (SETAC):

Environmental Impacts of Pulp and Paper Waste Streams
Stuthridge, van den Heuvel, Marvin, Slade, Gifford, editors
2003

Porewater Toxicity Testing: Biological, Chemical, and Ecological Considerations
Carr and Nipper, editors
2003

Reevaluation of the State of the Science for Water-Quality Criteria Development
Reiley, Stubblefield, Adams, Di Toro, Erickson, Hodson, Keating Jr, editors
2003

Bioavailability of Metals in Terrestrial Ecosystems: Importance of Partitioning for Bioavailability to Invertebrates, Microbes, and Plants
Allen, editor
2003

Silver in the Environment: Transport, Fate, and Effects
Andren and Bober, editors
2002

Test Methods to Determine Hazards for Sparingly Soluble Metal Compounds in Soils
Fairbrother, Glazebrook, van Straalen, Tararzona, editors
2002

Ecological Variability: Separating Natural from Anthropogenic Causes of Ecosystem Impairment
Baird and Burton, editors
2001

For information about any SETAC publication, including SETAC's international journal, *Environmental Toxicology and Chemistry*, contact the SETAC Office nearest you:

Administrative Offices

SETAC North America
1010 North 12th Avenue
Pensacola, Florida 32501-3367 USA
T 850 469 1500; F 850 469 9778
E setac@setac.org

SETAC Europe
Avenue de la Toison d'Or 67
B-1060 Brussels, Belgium
T 32 2 772 72 81; F 32 2 770 53 86
E setac@setaceu.org

www.setac.org

Environmental Quality Through Science®

Contaminated Soils: From Soil–Chemical Interactions to Ecosystem Management

Edited by

Roman P. Lanno
Ohio State University
Columbus, OH, USA

Proceedings from the Pellston Workshop on
Assessing Contaminated Soils:
From Soil-Contaminant Interactions to Ecosystem Management
23–27 September 1998
Pellston, Michigan, USA

Coordinating Editor of SETAC Books
Andrew Green
International Lead Zinc Research Organization
Department of Environment and Health
Research Triangle Park, North Carolina, USA

Published by the Society of Environmental Toxicology and Chemistry (SETAC)

Cover design by Jason Conder and Michael Kenney Graphic Design and Advertising
Index by Celia McCoy

Library of Congress Cataloging-in-Publication Data

Contaminated soils : from soil-chemical interactions to ecosystem management / edited by Roman P. Lanno
p. cm.
Outgrowth of a workshop held at the University of Michigan Biological Field Station near Pellston, Mich., Sept. 23-27, 1998.
Includes bibliographic references and index.
ISBN 1-880611-31-7 (alk. paper)
1. Soil pollution--Environmental aspects--Congresses. 2. Soil ecology--Congresses. 3. Ecological risk assessment--Congresses. 4. Bioavailability--Congresses. I. Lanno, Roman P. (Roman Peter), 1958- II. SETAC (Society)

TD878 .C66526 2003
628.5'5--dc21 2001054957

This publication was printed on recycled paper using soy ink.
SETAC Press is an imprint of the Society of Environmental Toxicology and Chemistry.
No claim is made to original U.S. Government works.

International Standard Book Number 1-880611-31-7
Printed in the United States of America
10 09 08 07 06 05 04 03 10 9 8 7 6 5 4 3 2 1

♾ The paper used in this publication meets the minimum requirements of the American National Standard for Information Sciences — Permanence of Paper for Printed Library Materials, ANSI Z39.48-1984.

Reference Listing: Lanno RP, editor. 2003. Contaminated soils: From soil–chemical interactions to ecosystem management. Pensacola FL, USA: Society of Environmental Toxicology and Chemistry (SETAC). 445 p.

SETAC Publications

Books published by the Society of Environmental Toxicology and Chemistry (SETAC) provide in-depth reviews and critical appraisals on scientific subjects relevant to understanding the impacts of chemicals and technology on the environment. The books explore topics reviewed and recommended by the Publications Advisory Council and approved by the SETAC North America Board of Directors and the SETAC World Council for their importance, timeliness, and contribution to multidisciplinary approaches to solving environmental problems. The diversity and breadth of subjects covered in the publications reflect the wide range of disciplines encompassed by environmental toxicology, environmental chemistry, and hazard and risk assessment, and life-cycle assessment. SETAC books attempt to present the reader with authoritative coverage of the literature, as well as paradigms, methodologies, and controversies; research needs; and new developments specific to the featured topics. The books are generally peer reviewed for SETAC by acknowledged experts.

SETAC publications, which include Technical Issue Papers (TIPs), workshop summaries, newsletter (*SETAC Globe*), and journal (*Environmental Toxicology and Chemistry*), are useful to environmental scientists in research, research management, chemical manufacturing and regulation, risk assessment, life-cycle assessment, and education, as well as to students considering or preparing for careers in these areas. The publications provide information for keeping abreast of recent developments in familiar subject areas and for rapid introduction to principles and approaches in new subject areas.

SETAC recognizes and thanks the past SETAC books editors:

C.G. Ingersoll, Midwest Science Center
U.S. Geological Survey, Columbia, MO, USA

T.W. La Point, Institute of Applied Sciences
University of North Texas, Denton, TX, USA

B.T. Walton, U.S. Environmental Protection Agency
Research Triangle Park, NC, USA

C.H. Ward, Department of Environmental Sciences and Engineering
Rice University, Houston, TX, USA

Contents

Section II: Chemical–Soil–Organism Interactions

List of Figures

List of Tables

Acknowledgments

This workshop would not have been possible without the support of the SETAC Foundation for Environmental Education and corporate sponsors. The efforts of Linda Longsworth, Cheri Mertins, and Greg Schiefer in preparing for and actually conducting the workshop are greatly appreciated. It would not have run as smoothly without your efforts. The editorial efforts of Amanda Conder and Bonnie Drummond in the preparation of the manuscript are greatly appreciated. The time taken to review the book and the insightful comments provided by Steven Siciliano and Gary Rand helped greatly in the final preparations of the book. The guidance and encouragement of Melissa Winter and Stacey Hagman during the entire process of preparing the book was greatly appreciated. Finally, thanks to Jason Conder for generously contributing his time and artistic talents in the generating the design of the cover.

Roman P. Lanno
Columbus, OH, USA
October 2002

Financial Contributors

American Industrial Health Council

American Petroleum Institute

Chemical Manufacturers' Association

Environment Canada

ExxonMobil

Gas Research Institute

Shell Oil Company

U.S. Environmental Protection Agency

About the Editor

Roman P. Lanno, PhD, is currently an Assistant Professor of Entomology at Ohio State University. Prior to that he was Associate Professor of Zoology and Director of the Ecotoxicology and Water Quality Research Laboratory at Oklahoma State University in Stillwater, OK, USA. He received his PhD in 1991 from the Department of Biology at the University of Waterloo, Canada, and his master's degree in 1984 from the Department of Nutrition at the University of Guelph, Canada.

Current research interests in his laboratory lie in applied and theoretical aspects of chemical exposure assessment and organismal effects assessment in various environmental media. Specifically, research in his laboratory examines the relationship between chemical and biological measures of bioavailability and toxicity endpoints such as lethality, growth, reproduction, or biomarkers. Studies include work with aquatic and terrestrial systems and both metals and organic chemicals. Researchers in his laboratory have developed and applied solid-phase extraction techniques as biomimetic or biological surrogates for estimating the bioavailability of organic chemicals and metals.

Dr. Lanno has been a SETAC member since 1988, serving as a member of the Editorial Board for the journal *Environmental Toxicology & Chemistry* and organizing and presenting short courses in Soil Ecotoxicology and Nutritional Considerations in Toxicity Testing.

Workshop Participants*

Martin Alexander
Cornell University
SCAS Department
Ithaca, New York USA

Herbert E. Allen
University of Delaware
Department of Civil and Environmental Engineering
Newark, Delaware, USA

W. Nelson Beyer
U.S. Geological Survey
Patuxent Wildlife Research Center
Laurel, Maryland, USA

Tina M. Carlsen
Lawrence Livermore National Laboratory
Environment Restoration Division
Livermore, California, USA

Rufus Chaney
USDA-ARS
Euroron Chemistry Laboratory
Beltsville, Maryland, USA

Randy Charbeneau
University of Texas-Austin
Department of Civil Engineering
Austin, Texas, USA

Scott D. Cunningham
E.I. DuPont de Nemours & Company
Department of Environmental Biotechnology
Newark, Delaware, USA

Philip B. Dorn
Shell Development Company
Environmental Technology
Houston, Texas, USA

Clive A. Edwards
Ohio State University
Department of Entomology
Columbus, Ohio, USA

Rebecca A. Efroymson
Oak Ridge National Laboratory
Oak Ridge, Tennessee, USA

Stephen J. Ells
U.S. Environmental Protection Agency
Washington, DC, USA

Carol English
Cytec Industries
Stamford, Connecticut, USA

Anne Fairbrother
ecological planning & toxicology
Corvallis, Oregon, USA

Alan S. Felsot
Washington State University
Department of Crop and Soil Sciences
Richland, Washington, USA

Connie Gaudet
Environment Canada
Guidelines Division
Ottawa, Ontario, Canada

Joop Harmsen
DLO Winard Staring Centre
Wageringen, The Netherlands

Gerry Henningsen
U.S. Environmental Protection Agency
Human and Ecological Assessment
Denver, Colorado, USA

Bruce K. Hope
Oregon Department of Environmental Quality
Waste Management and Cleanup Division
Portland, Oregon, USA

Joseph Hughes
Rice University
Department of Environmental Sciences & Engineering
Houston, Texas, USA

John Jensen
National Environmental Research Institute
Department of Terrestrial Ecology
Sikeborg, Denmark

Larry A. Kapustka
ecological planning & toxicology
Corvallis, Oregon, USA

Roman G. Kuperman
U.S. Army ERDEC
Aberdeen Proving Ground, Maryland, USA

Roman P. Lanno
Oklahoma State University
Department of Zoology
Stillwater, Oklahoma

Greg L. Linder
Heron Works Farm
Brooks, Oregon, USA

David Linz
Linztech
Des Plaines, Illinois, USA

Raymond C. Loehr
University of Texas-Austin
Department of Civil Engineering
Austin, Texas, USA

Warren J. Lyman
Camp, Dresser & McKee, Inc.
Cambridge, Massachusetts, USA

Sara J. McMillen
Chevron Research and Technology Company
Health Environment & Safety Group
Richmond, California, USA

Charlie A. Menzie
Menzie-Cura & Associates
Chelmsford, Massachusetts, USA

G. Ted Nason
Alberta Environment
Edmonton, Alberta, Canada

Charles J. Newell
Groundwater Services, Inc.
Houston, Texas, USA

Susan B. Norton
U.S. Environmental Protection Agency
Washington DC, USA

Stan J. Pauwels
Abt Associates, Inc.
Cambridge, Massachusetts, USA

Kurt D. Pennell
Georgia Institute of Technology
School of Civil & Environmental Engineering
Atlanta, Georgia, USA

Joseph J. Pignatello
Connecticut Agricultural Experiment Station
Soil and Water
New Have, Connecticut, USA

William G. Rixey
University of Houston
Department of Civil Engineering
Houston, Texas, USA

Dennis Rolston
University of California-Davis
Department of Land Air & Water Resources
Davis, California, USA

Jim Ryan
U.S. Environmental Protection Agency
Cincinnati, Ohio, USA

Ann Saterbak
Environmental Directorate
Shell/Equilon
Houston, Texas, USA

Greg Schiefer
SETAC/SETAC Foundation
Pensacola, Florida, USA

*Affiliations were current at the time of the workshop.

Kate Scow
University of California-Davis
Department of Land Air and Water Resources
Davis, California, USA

Gladys L. Stephenson
University of Guelph
Department of Environmental Biology
Guelph, Ontario, Canada

Hans Stroo
Remediation Technologies, Inc. (RETEC)
Carson, California, USA

Henry Tabak
U.S. Environmental Protection Agency-Cincinnati
NRMRL
Cincinnati, Ohio, USA

Mason Tomson
Rice University
Department of Environmental Sciences & Engineering
Houston, Texas, USA

Cornelius van Gestel
Vrije univeriteit Biologie
Department of Ecology and Ecotoxicology
Amsterdam, The Netherlands

Randy Wentsel
U.S. Environmental Protection Agency
Washington, DC, USA

*Affiliations were current at the time of the workshop.

CHAPTER 1

Introduction

Roman P. Lanno, Anne Fairbrother, Connie Gaudet, Ray Loehr, Ann Saterbak, Henry Tabak

Healthy soil systems are essential for sustaining agricultural practices, perpetuating natural resource industries such as timber production, and maintaining the proper functioning of natural ecosystems. Many soil systems worldwide have been contaminated with chemicals and require some form of mitigation or remediation. In order to maintain soil quality and protect human health and the environment from past and current additions of chemicals, it is imperative that we understand the interactions of chemicals with soils, understand the effects of chemicals in soil on organisms, including humans, and develop a sound regulatory strategy for the protection of soil.

For the purposes of this book, soil can be defined as any unconsolidated surface material, substantially of natural mineral origin, other than in-place aquatic sediments. This definition includes native topsoils; agricultural soils; urban, residential, and industrial fill; dredged aquatic sediments; mine tailings; and "manufactured" soils (i.e., blends of natural and/or anthropogenic materials). It becomes readily apparent that this definition deviates substantially from the traditional definition of soils with a well-developed soil profile as described in soil science books. This broader definition is necessary because most contaminated sites are not natural soils and have been subject to drastic anthropogenic modifications, varying greatly in their physical and chemical structure, and containing various amounts and types of minerals and organic matter. They also vary in microbial composition, with a great diversity of species, which are essential to both normal soil functions in nutrient cycling and in the degradation of toxic chemicals. This definition encompasses the tremendous heterogeneity of soils because of their different uses other than for agricultural purposes. Different land uses require different types of soils, and the degree of contamination by chemicals that is acceptable and how it is assessed may vary greatly.

The mere presence of chemicals in a soil does not imply toxicity and may not even imply contamination. Contamination can be defined as making unfit for use by the introduction of undesirable elements. The amount of undesirable elements that constitutes contamination will vary with the specific land-use and the receptor that we wish to protect. In many cases, it is only presumed that a site soil is contaminated, and it is risk characterization that actually determines the relative degree of contamination, and risk managers will then determine the degree of contamination that is acceptable. This emphasizes the need for responsible soil protection using

Contaminated Soils: From Soil–Chemical Interactions to Ecosystem Management. Roman P. Lanno, editor.
 ISBN 1-880611-31-7

the best science to determine the concentration or level of a chemical that can be present in the soil and not result in any significant biological impairment.

In order to begin to understand and integrate the complexities of soil–chemical interactions with biological systems and summarize that information in a format usable by regulators and risk assessors, it was necessary to bring together expertise from diverse backgrounds. By convening a workshop and forcing chemists, soil scientists, toxicologists, engineers, and regulators to work together in a multidisciplinary effort, it was hoped that thinking would extend outside specific areas of expertise and collectively advance concepts in the ecological risk assessment (ERA) of chemicals in soil.

This book addresses the issue of assessing risk to biological receptors from the presence of chemicals in soil, emphasizing ecological receptors rather than humans. It is understood that there are risks to soil structure and function from many other stressors, such as the physical impacts of exploration and remediation processes and the introduction of exotic species into soil ecosystems. These stressors should always be considered when assessing the overall impacts on a soil system but are beyond the scope of this book.

This book is arranged to follow the standard risk assessment paradigm (Figure 1-1; USEPA 1998) with 8 distinct focus areas addressing different aspects of the paradigm. These are grouped into book sections presenting the context for soil management (Section I), the fate and transport of contaminants (Section II), and measures of bioavailability (Section III).

Ecological risk assessment begins with the development of a management goal. Communication between the risk manager and stakeholders is essential to determine what resources are at risk and which ones need to be protected. Once a decision is reached regarding a management goal, the process enters the problem formulation stage where a conceptual model is developed and assessment endpoints are decided upon. During problem formulation, it is possible that development of a conceptual model leads to an alteration in the management goal. At this point, a screening-level assessment can be conducted by comparing soil chemical levels to existing soil-quality guidelines (SQGs) or soil screening levels (SSLs) to see whether they are exceeded. If they are below SQGs or SSLs for the specific chemical in question, then there is no need to proceed any further in the risk assessment of these chemicals because the assessment endpoint is assumed to be protected. If SQGs are exceeded, then risk assessment can proceed to exposure and effects analysis. Information such as the spatial and temporal heterogeneity of chemicals in the soil, bioavailable concentrations of chemicals, and bioaccumulation potential can be assembled and coupled with effects information such as ecological or toxicological dose–response data, toxicity reference values (TRVs), or critical body residues. The relationship between exposure and effects for chemical, physical, and biological conditions present at the site of interest provides some

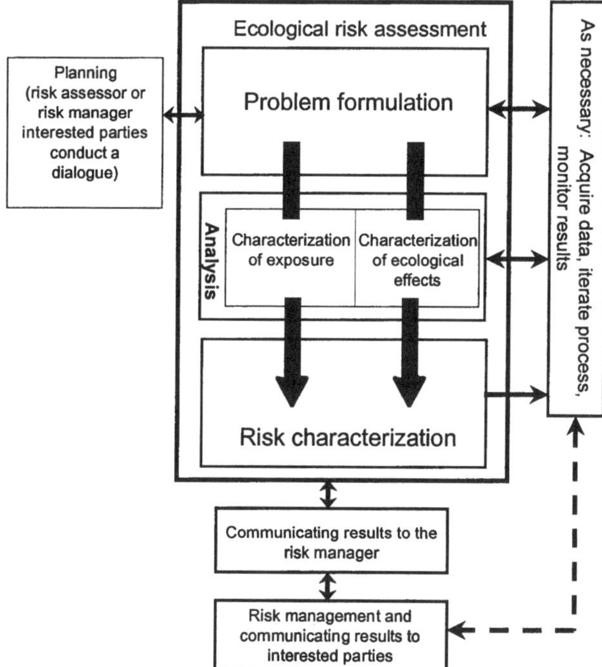

Figure 1-1 The general ecological risk assessment process (USEPA 1998)

relative estimation of risk. This risk characterization may be expressed as a hazard quotient (HQ) or a probabilistic estimate of risk, which is then used by risk managers in deciding whether further assessment of the site is necessary.

One of the key assumptions in estimating exposure in the risk assessment paradigm is that receptors are exposed to the concentration of chemical that is actually measured. In its simplest form, risk is often characterized as an HQ where

$$HQ = C_{total} / \text{reference concentration},$$

where C_{total} = total concentration of a chemical in soil measured using extraction techniques which make available for analytical techniques virtually all the chemical present in the soil; reference concentration = concentration of chemical in soil associated with a toxicological assessment of the chemical; it may be a no-observed-effect level (NOEL), an EC5 value, or a TRV.

If this ratio < 1, then there is assumed to be negligible risk to the receptor from the specific contaminant. If the ratio > 1, then there is assumed to be some risk to the receptor from the chemical in question. As an example, if C_{total} for chemical X at a site is 10 mg/kg and the reference concentration is 100 mg/kg, then the ratio is 0.1

and the risk to the receptor from that chemical is assumed to be acceptable. If C_{total} was 150 mg/kg for the same chemical at the site, then the ratio would be 1.5, and it would be presumed that some risk would be associated with the exposure of the receptor to the chemical and decisions on a course of action would need to be made. One major drawback to using C_{total} is the assumption that all of the chemical (100%) measured in the sample is bioavailable to the receptor. In reality, rarely is a chemical in a specific medium 100% bioavailable. Physical, chemical, and biological modifying factors interact with the chemical to alter its bioavailability, changing the estimated risk. As an example, Conder and Lanno (2000) found that the bioavailable fraction of lead in artificial soil, as estimated in a fraction obtained by 0.1 M $Ca(NO_3)_2$ extraction, was approximately 10% of the total lead measured in the soil. In this case, this information could be incorporated into the risk estimate as follows:

$$HQ = (BF)\ C_{total}\ /\ \text{reference concentration} = (0.1)\ 150\ /\ 100 = 0.15,$$

where BF = bioavailability factor.

In this case, the risk associated with exposure of the receptor to the chemical would now be assumed to be acceptable, due to the consideration of the bioavailability of the chemical. However, this estimate also assumes that the reference concentration was derived for toxicity test data based upon bioavailable chemical measures of exposure, not total chemical levels.

Bioavailability is a term that has different interpretations in different disciplines. Soil chemists may interpret bioavailability as the fraction of chemical in soil that is not sequestered, while microbiologists may interpret bioavailability as the fraction of an organic chemical in soil available for metabolism or mineralization by microbes. For the purposes of this book, the broad definition of bioavailability is "a measure of the potential of a chemical for entry into ecological or human receptors. It is specific to the receptor, the route of entry, time of exposure, and the matrix containing the contaminant" (Anderson et al. 1999). The important concept is that a chemical in soil interacts with various components of the physicochemical environment, rendering some portion of that chemical unavailable for interaction with biological receptors. Figure 1-2 is a graphical depiction of the concept of bioavailability as presented by Landrum et al. (1992). Superimposed on the figure are chapter numbers representing the points where discussions of the various focus area groups of the workshop relate to the concept of bioavailability. When chemicals are present in soil, whether they have been there for an extended duration or were recently applied or spilled, they interact with soil constituents in a dynamic manner. In the case of application or spill, the interaction of the chemical with specific fractions of the soil results in sequestration of a portion of the chemical, which renders it unavailable for interaction with ecological receptors. Sequestration can be defined as "the state in which a contaminant is segregated from and rendered unavailable to a receptor and arises from rate-limiting processes involv-

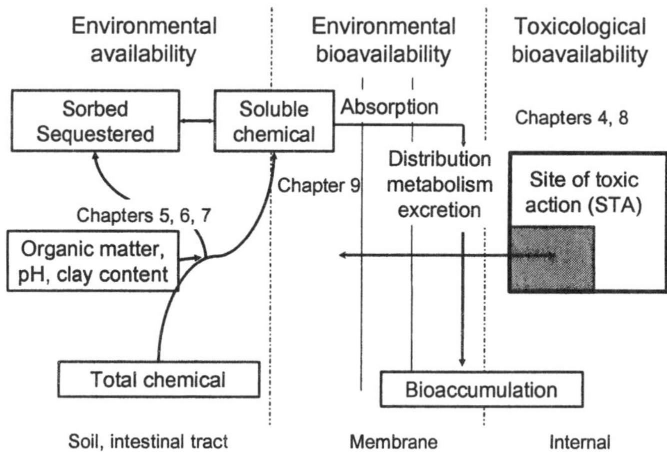

Figure 1-2 Schematic model of bioavailability

ing contaminant interactions with the surrounding matrix, such as phase transfer, complexation, and reversible chemical transformation. Sequestration is specific in relation to the combination of receptor, matrix, spatial and temporal scales, and route of exposure" (J. Pignatello, personal communication). Within the scope of this definition, the proportion of total chemical in soil that is sequestered in relation to bioavailability is dependent upon the organism and the route of exposure. As an example, the amount of a heavy metal that is sequestered from a mole is very much dependent upon the route of exposure. When considering dermal uptake, only the metal present in soluble form in the soil solution may be environmentally bioavailable to the mole, with the rest of the metal sequestered. However, ingested soil will be subjected to the greatly reduced pH of the stomach of the mole, making available a much greater proportion of the total metal present and changing the amount that is sequestered. It should be noted that this definition of sequestration might be extended to include the partitioning of specific chemicals inside an organism into inert forms that are also biologically unavailable to the receptor. An example of biological sequestration would be the sequestration of metals by soil invertebrates into inert granules consisting of an inorganic matrix. The portion of total chemical external to the organism, including in the lumen of the gastrointestinal tract, and not sequestered is referred to as the "environmentally available fraction" (Figure 1-2) and remains available for fate and transport processes. During their movements, soil-dwelling organisms encounter and interact with some portion of the environmentally available fraction. Other wildlife species ingest soil during foraging. The portion of environmentally available chemical that organisms interact with is termed the "environmentally bioavailable fraction." Some portion of the environmentally bioavailable fraction of chemical may cross the external membranes of the organism (e.g., epidermis, gastrointestinal tract) and be absorbed by the organism. Once absorbed, the chemical may be metabolized and excreted, accumulated in tissues, sequestered internally, or transported in the organism to the site of toxic action (STA). Toxicological bioavailability refers to the portion of absorbed chemical that reaches and interacts with the STA. It should be noted that the box representing the STA represents a threshold level. Toxic effects will occur only when the amount of chemical present at the STA completely fills the box and exceeds the threshold. Thus, it is possible to have accumulation of a compound without toxicity. Although this figure is simplistic, it helps in defining the general areas where the focus area groups of the workshop concentrated their efforts. It also provides a starting point for a conceptual model integrating bioavailability into the ERA process at all levels, from soil–chemical interactions to the development of SQGs.

Workshop Organization and Objectives

This workshop was conceived as the first of many addressing a growing concern over the assessment and management of contaminated soils. Although a great amount of information exists regarding the contamination of soils in agricultural settings, the issues related to contaminated soil need to be addressed from a much broader context. That was the impetus in bringing together expertise from diverse backgrounds ranging from soil chemistry to SQG development. The goal of the workshop was to assemble the ideas and discussion of specific focus area groups into a unified body of literature that would allow readers from diverse backgrounds to access information from many areas in 1 place.

The workshop steering committee consisted of Anne Fairbrother, Connie Gaudet, Roman Lanno, Ray Loehr, Ann Saterbak, Greg Schiefer, and Henry Tabak. Following the planning and fundraising efforts of this group and others, coordinated through the Society of Environmental Toxicology and Chemistry (SETAC) Foundation for Environmental Education, the workshop was conducted on 23–27 September 1998, at the University of Michigan Biological Field Station, near Pellston, Michigan.

The specific areas of expertise needed to broadly cover the assessment of contaminated soils were arranged into specific focus area groups. Each focus area group consisted of a chairperson (c) and working group members as follows:

- Risk assessment and risk management (Chapter 2)
  Charlie A. Menzie (c), Rebecca A. Efroymson, Stephen J. Ells, Gerry M. Henningsen, Bruce K. Hope
- Soil-quality guidelines (Chapter 3)
  George E. Nason (c), Tina M. Carlsen, Philip B. Dorn, John Jensen, Susan B. Norton

- Effects of contaminants on soil ecosystem structure and function (Chapter 4)
  Randall S. Wentsel (c), W. Nelson Beyer, Clive A. Edwards, Larry A. Kapustka, Roman G. Kuperman
- Fate of soil contaminants (Chapter 5)
  Dennis E. Rolston (c), Alan S. Felsot, Kurt D. Pennell, Kate M. Scow, Hans F. Stroo
- Transport of contaminants in soil (Chapter 6)
  Randall J. Charbeneau (c), David G. Linz, Charles J. Newell, William G. Rixey, James A. Ryan
- Contaminant–soil interactions (Chapter 7)
  Mason B. Tomson (c), Herbert E. Allen, Carol W. English, Warren J. Lyman, Joseph J. Pignatello
- Biological measures of bioavailability (Chapter 8)
  Sara J. McMillen (c), Cornelius A.M. van Gestel, Roman P. Lanno, Greg L. Linder, Stan J. Pauwels, Gladys L. Stephenson
- Chemical measures of bioavailability (Chapter 9)
  Martin Alexander (c), Scott D. Cunningham, Rufus R. Chaney, Joseph B. Hughes, Joop Harmsen.

The different focus area groups in the workshop addressed topics ranging from how measures of bioavailability can be incorporated into the risk assessment process and SQG development to how the interactions, fate, and transport of chemicals affect bioavailability in soil, and finally, how bioavailability can actually be measured using chemical and biological techniques.

The book begins with a section ("Management Framework") addressing contaminated soils in the broader context of soil management with chapters associated with ERA, SQG development, and effects of contaminants on soil ecosystem structure and function. The second section ("Chemical–Soil–Organism Interactions") provides information specific to understanding and measuring the fate, transport, and interactions of chemicals in soil. The final section ("Measuring Bioavailability") specifically presents biological and chemical techniques for the measurement of bioavailability of chemicals in soil systems.

The primary purpose of this book was to integrate existing information from relevant subdisciplines (e.g., soil chemistry, terrestrial ecology, terrestrial ecotoxicology, risk assessment) into a format that will form a basis for the development of research and management directions. A wealth of information exists on physicochemical interactions of soil with contaminants, toxicity test methods for soil organisms, and ERA, yet this information remains scattered throughout the literature. This book brings together information from these disciplines and organizes it in a manner that we hope facilitates an understanding of the state of the art in each discipline and presents guidance on their interactions and applications in the ERA process. This book is not intended to be a comprehensive compilation of all existing information in the area of contaminated soils, nor will it provide complete step-by-step guidance on conducting ERAs of contaminated soil sites. It is intended to function as a reference framework for all the steps that are needed to conduct a complete risk assessment of sites with contaminated soil. This book will also provide the reader with a window into some of the relevant literature. It is intended to be read in its entirety, especially if some of the subdisciplines are unfamiliar to the reader, because information in all of the chapters is necessary to make informed management decisions regarding contaminated soils.

References

Anderson WC, Loehr RC, Smith BP. 1999. Environmental availability of chlorinated organics, explosives, and metals in soils. Annapolis MD, USA: American Academy of Environmental Engineers. 210 p.

Conder JM, Lanno RP. 2000. Weak-electrolyte extractions and ion-exchange membranes as surrogate measures of cadmium, lead, and zinc bioavailability to *Eisenia fetida* in artificial soils. *Chemosphere* 41:1659-1668.

Landrum PF, Hayton WL, Lee III H, McCarty LS, Mackay D, McKim JM. 1992. Synopsis of discussion on the kinetics behind environmental bioavailability. In: Hamelink JL, Landrum PF, Bergman HL, Benson WH, editors. Bioavailability: Physical, chemical, and biological interactions. Boca Raton FL, USA: Lewis. p 203-219.

[USEPA] U.S. Environmental Protection Agency. 1998. Guidelines for ecological risk assessment. Washington DC, USA: USEPA, Risk Assessment Forum. EPA-630-R-95-002F.

Section I

Management Framework

CHAPTER 2

Risk Assessment and Risk Management

Charlie A. Menzie, Rebecca A. Efroymson, Stephen J. Ells, Gerry M. Henningsen, Bruce K. Hope

Introduction

The discussion presented in this chapter applies to situations where soils are already contaminated (e.g., contaminated sites) as well as to those where contamination might occur in the future (e.g., siting a facility where releases may occur). In this chapter, the term "site" is used to indicate both current and potential future conditions. While most of the examples are drawn from experience with existing contaminated sites, the principles are applicable to situations where future contamination might occur.

Chapter organization

An overarching question raised prior to and during the workshop was "What receptor or function is important to protect?" This chapter begins with a consideration of management goals and how they provide the starting point for answering this question. We then discuss the use of tiered assessment strategies and describe how they may be applied to assessments of contaminated soils. We next present a discussion of the generic risk assessment process as it relates to soils, including problem formulation, analysis (exposure and effects), risk characterization, uncertainty analysis, risk management, and risk communication. The chapter closes with some final thoughts concerning research needs and the dissemination of information. Throughout the chapter we make an effort to identify where tools might be utilized to support exposure and effects assessments and to further refine these as needed.

Several aspects of the process are emphasized:

- the establishment of clear management goals,
- the up-front establishment of criteria by which risks are evaluated and judged,
- the decision framework for using information about risks, and
- the presentation of risk information commensurate with the scale of the problem and the assessment.

Contaminated Soils: From Soil-Chemical Interactions to Ecosystem Management. Roman P. Lanno, editor.
 ISBN 1-880611-31-7

Risk assessment and risk management frameworks and tiers

Risk-based approaches that integrate risk assessment with risk management are useful for organizing information related to evaluating contaminated soils and deciding on appropriate courses of action for protecting and restoring ecological receptors or functions. The types of information that are integrated into an assessment include biological, chemical, and physical data related to evaluating exposure and effects, as well as information drawn from stakeholders concerning the ecological values to be protected. Guidelines are available for the ecological risk assessment (ERA) process (USEPA 1998), as well as for implementing assessments at contaminated sites in the U.S. (USEPA 1997a) and Canada (CCME 1996). Many states and provinces have also developed guidance. The U.S. Environmental Protection Agency's ERA framework outlines a general process (Figure 2-1), which begins with planning and problem formulation and proceeds through analysis of exposure and effects, and characterization of risk. Figure 2-2 illustrates where in the ERA framework information presented in other chapters of this book such as fate, transport, and effects of soil contaminants can be integrated to evaluate and manage risks associated with contaminated soils (Figure 2-2). The terms used in risk assessment are defined in the Glossary and will be discussed further in this chapter.

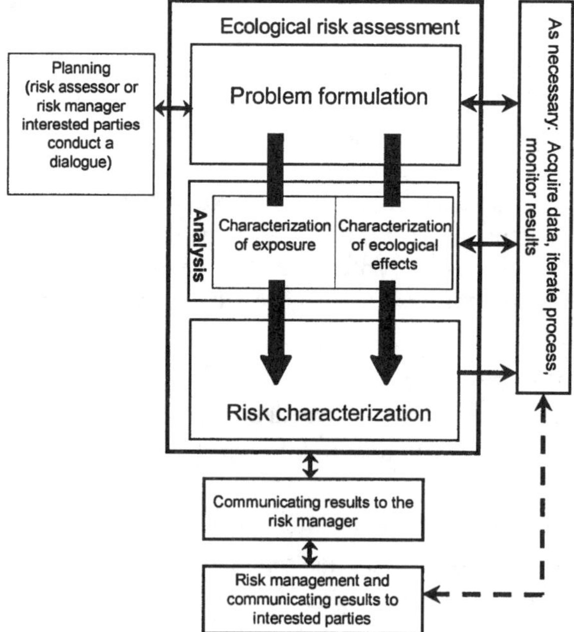

Figure 2-1 The general ecological risk assessment process (USEPA 1998)

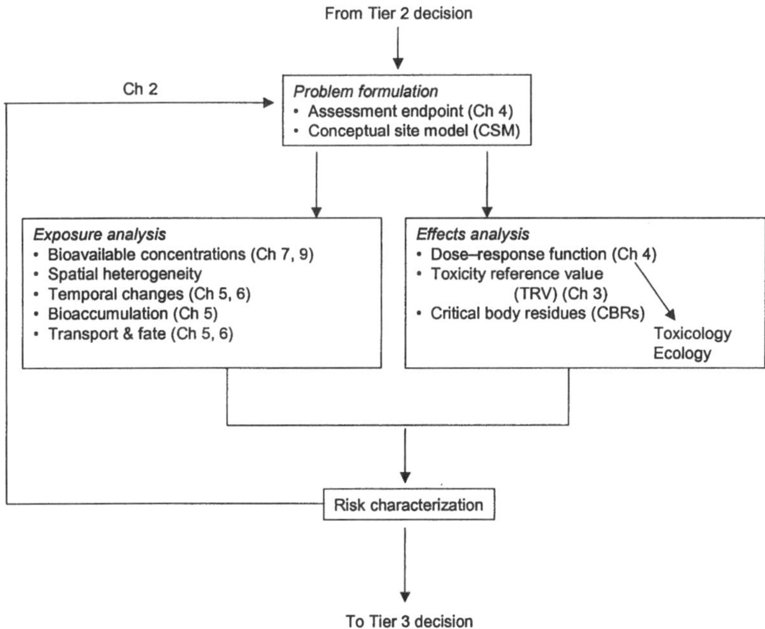

Figure 2-2 Integration of data and concepts from book chapters (e.g., fate, transport, and effects of soil contaminants) into the evaluation and management of risks associated with contaminated soils

While frameworks (Figures 2-1 and 2-2) capture the conceptual features of the risk assessment and management process, they do not identify specific courses of action. As a result, many federal, state, and provincial agencies, as well as professional groups such as the Society of Environmental Toxicology and Chemistry (SETAC) (Ingersoll et al. 1997) and the American Society for Testing and Materials (ASTM 1998), have translated the frameworks into tiered approaches so that they may be implemented in a logical fashion. The USEPA (1997a) has developed an 8-step process that includes 2 or more tiers for ERA at Superfund sites (Figure 2-3). The Canadian Council of Ministers of the Environmental (CCME) and the ASTM risk-based corrective action (RBCA) program have outlined a 3-tier process (Chapter 3).

Contaminated soil issues can be addressed in a number of ways that depend on site conditions, future site use, and site management goals. A "one-size-fits-all" approach for assessing risks at contaminated sites likely will not meet management needs across different scales of site problems. Experience has shown that tiered approaches conserve time and resources, properly focus the scope and extent of the needed studies, and support management decisions commensurate with the scale and nature of the problem. As described in USEPA (1997a), CCME (1996), and

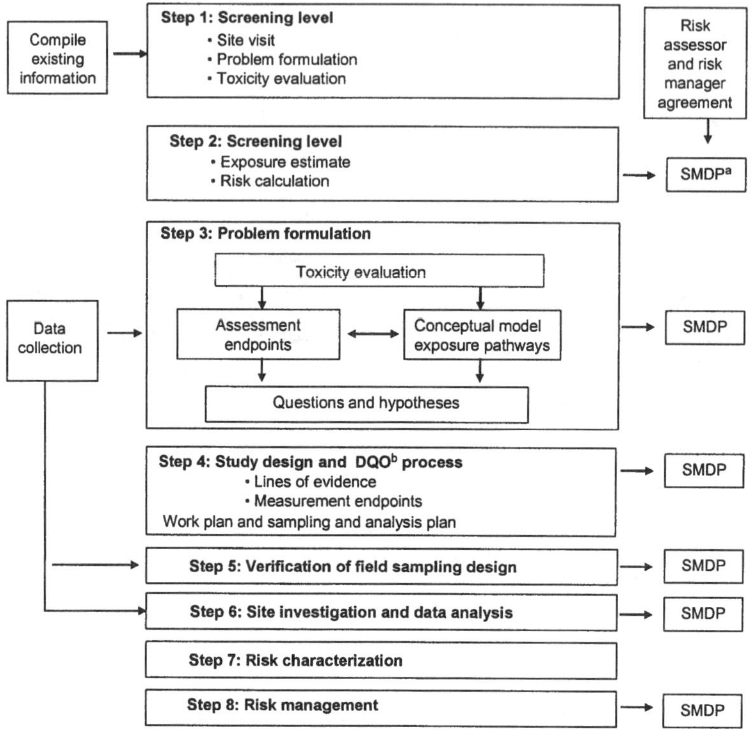

Figure 2-3 Eight-step ecological risk assessment process for Superfund sites (USEPA 1997a)

[a]SDMP = Scientific decision management point.

[b]DQO = Data quality objective.

ASTM (1998), screening (e.g., Tier 1) assessments normally should be performed before designing and conducting more comprehensive assessments. For sites with obvious unacceptable risks, early evaluations are adequate to proceed with a response action. At other sites, screening evaluations can be used to determine if there are any potential adverse risks to the environment. If there are no complete exposure pathways or if soil contaminant concentrations are within accepted screening levels, then there is no need to perform more comprehensive assessments. Screening-level assessments can help focus assessments by identifying chemicals of concern (COCs) at the site, exposure pathways, receptors, and locations needing further evaluation.

The risk assessment and risk management process is often iterative. As new information becomes available, the COC and exposure pathways may change, and new assessment endpoints may be added. Refined information on exposure or effects might be used to calculate Tier 2 values for chemicals in soils that are typically higher than those used in the Tier 1 screening assessment.

A generic, tiered, risk-based approach for evaluating and managing contaminated soils is shown in Figure 2-4 and is used to structure this chapter. The process begins with a preliminary assessment and the establishment of management goals. This critical step is often overlooked in assessments or is not translated adequately into the assessment strategy and management decisions. Assessment and management are closely linked throughout the process. Information that is gathered in early stages is used to make decisions about how to proceed.

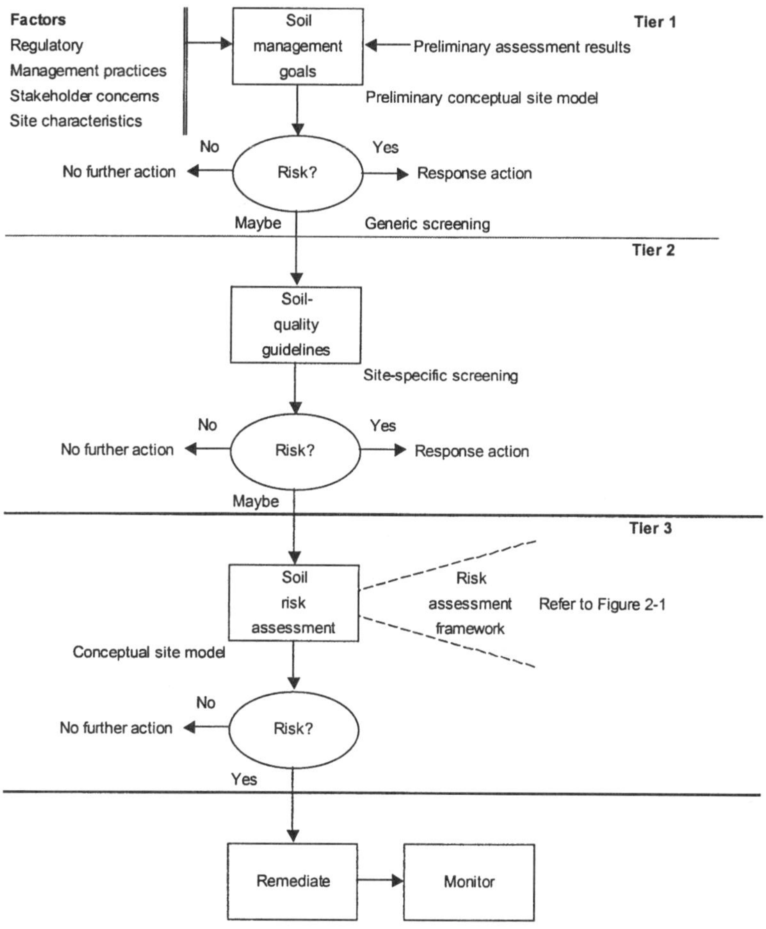

Figure 2-4 Generic, tiered, risk-based approach for evaluation and management of contaminated soils

Assessing exposure and effects of contaminated soils

Most of the chapters in this book touch on soil-specific issues that relate to assessing fate, transport, or effects of contaminants in soils. These include a consideration of spatial and temporal scales as well as soil processes that modify the fate, transport, and bioavailability of chemicals. These are described briefly in this chapter, and reference is frequently made to other chapters for more detailed information. The 3 major scales that are captured to varying degrees when applying a tiered approach are

1) spatial scales (distribution of contaminants and receptors),
2) temporal scales (chemical release, transformation, and sequestration rates; biological processes [e.g., bioaccumulation], location, and life stage of receptor), and
3) effects scales (type and magnitude of effect).

The application of risk assessment to contaminated terrestrial sites must also consider the selection of ecological receptors, the methods for evaluating exposure and effects, and the manner in which risks are characterized and communicated. Uncertainties associated with these aspects of the assessment can be addressed in various ways, and these are customarily included within Tier 2 and 3 assessments.

The exposure of terrestrial organisms to soil contaminants can be influenced markedly by soil–chemical processes that reduce bioavailability. Because bioavailability has emerged as an important exposure variable in risk assessments, it is useful to consider how information on bioavailability can be incorporated into the tiered assessment process.

Establishing Management Goals and Planning

Management goals

Management goals are formulated either at the policy or site-specific level (see Chapter 3 for additional discussion). Technical policy decisions may involve land-use issues, standard practices for addressing particular types of problems, or identification of resources that need to be protected. Management goals related to specific sites may involve similar issues but are applied on a local or site-specific basis. The assessment of risks and decisions concerning what needs to be done flow from stated management goals. Therefore, establishing these goals is critical to successful management of contaminated soils. Examples of management goals with ecological implications include

- protecting, in both the short-term and long-term, ecological receptors utilizing or inhabiting site soils from unacceptable risks resulting from exposure to contaminants present in those soils;
- supporting agricultural use of lands and avoiding activities that would result in ecologically significant effects on grazing wildlife; and
- permitting residential and commercial use of land and supporting fish and wildlife commensurate with that use in associated park and recreational areas.

Each of these examples raises issues of land ownership, land use, future development plans, and ecological values. A management goal sets a framework for discussing these issues and for identifying an appropriate course of action.

When ERA is used as a tool, the risk management goal is translated into assessment endpoints that represent the environmental value to be protected. This section describes some of the factors that should be discussed among risk managers, assessors, and stakeholders (as appropriate) to derive management goals.

A stated management goal charges the risk manager with 2 basic tasks:

1) to decide whether contaminated soils at a site (or associated with a type of activity) present unacceptable risks with respect to the stated goal, and
2) to decide on appropriate actions (i.e., remedies) that are protective of the environment, maintain protection over time, minimize or mitigate collateral injuries to natural resources, and minimize untreated waste.

Meeting these charges for a specific site will require that the risk manager understands the strategic implications of the following key issues embodied in the goals statement and how the definition of these issues will necessarily vary in response to site-specific conditions:

- Short-term versus long-term goals
- Ecological receptors
- Local populations
- Site characteristics
- Unacceptable risks
- Uncertainty and variability
- Exposure
- Contaminants.

These issues influence planning in various ways as discussed briefly in the following text (see also Table 2-1).

Table 2-1 Summary of issues considered in development of site-specific ecological risk management goals

Key issue	Influenced by	Influences
Short-term and long-term goals	• Land use • Physicochemical properties of contaminants and soil	• Magnitude and severity of risk • Period over which risk management decisions must apply • Applicability of natural recovery remedies
Ecological receptors (including local populations)	• Land uses • Physicochemical properties of contaminants and soil • Toxicological properties of contaminants (susceptibility) • Stakeholder issues (relevance)	• Choice of assessment endpoints • Public perception of risk and need for action • Design of the ERA
Site characteristics	• Land use • Legal factors • Spatial extent of contamination	• Magnitude and severity of risk (size) • Design of the ERA
Unacceptable risk	• Existing regulations • Existing site management goals • Public and stakeholder issues (relevance)	• Choice of assessment endpoints • Public perception of risk and need for action • Evaluation and judgment of risks
Uncertainty	• Ecological receptor characteristics and life histories • Contaminant characteristics • Sampling design (data adequacy) • Stakeholder issues	• Possible need for additional sampling and data collection • Evaluation and judgment of risks • Confidence in risk management decisions
Exposure	• Land use • Physicochemical properties of contaminants and soil • Toxicological properties of contaminants • Ecological receptors (susceptibility) • Site characteristics	• Choice of assessment endpoints • Design of the ERA
Contaminants	• Site history and uses • Soil properties • Ecological receptors (bioturbation)	• Choice of assessment endpoints • Identification of exposure pathways and routes • Recovery time

Short-term and long-term goals

Time may be considered in at least 3 ways: 1) "current" versus "future" (often applied to land use), 2) recovery time (i.e., that needed for ecological resources to achieve pre-impact levels or a level acceptable to the stakeholders), and 3) contaminant persistence (resistance to degradation, long half-life, bioaccumulation). "Current" typically considers conditions at a site as they are now (i.e., a time frame of less than 1 year). Evaluating "future" requires the risk manager to project site conditions forward from 1 to 20 or even 50 years. The nature of the management decision will change if it is likely that future uses will be different from current ones. For example, former industrial areas that are being turned over for use as parks or wildlife refuges will require more elaborate risk assessments to ensure protection of future ecological receptors than they would if their use was to remain industrial. In reaching conclusions regarding long-term future risks, a risk manager will need to consider how site conditions influence the rates of both contaminant degradation and ecological receptor recovery (including recovery from the impacts of remedial actions).

Ecological receptors

Ecological receptors may be 1) local animal, plant, or microbial populations that occur or could occur in specific habitats (e.g., wetland, floodplain, grassland, etc.) at sites; 2) individual plants or animals in specific habitats in the case of federal or state threatened and endangered species; or 3) functional features such as soil processes. The ecological receptors selected for assessment define "assessment endpoints" that address the question "What is important to protect?" As such, these receptors need to be, to the greatest extent practicable, both ecologically relevant (i.e., critical to sustaining the ecological structure and function of the populations, communities, and habitats present at the site) and of societal relevance (i.e., reflective of stakeholder concerns). The susceptibility of an ecological receptor also should be considered prior to an assessment endpoint (i.e., they are potentially exposed and sensitive to site-related contaminants).

The perceived relevance of an ecological receptor may diminish as its size (in the case of individual plants or animals) or the extent of its habitat or home range decreases. For example, stakeholders or the public may be more interested in addressing risks to large contiguous areas of forest habitat or to a large bird or mammal species than to a small remnant habitat in an industrial area or to soil invertebrates or soil processes. It may be difficult for risk managers to decide on an appropriate response in cases where the social, political, or economic relevance of the receptor is in question (McDaniel et al. 1995). This perceptual difficulty should not discourage the risk assessor from including such receptors (particularly soil invertebrate fauna or microbial processes) in the risk assessment on the basis of sound science or special circumstances (e.g., vernal pools—small areas of great ecological significance). It should, however, alert the risk manager of the need to educate stakeholders on the special relevance of specific receptors or habitats prior to conducting the risk assessment. The process of how to clearly justify the use of certain ecological receptors or habitats in a risk assessment may be included as a component of the problem formulation stage, and its importance should not be underestimated to the successful communication of ecological risk.

The types of ecological receptors and exposure routes included in ERA will also be influenced by current and future land use. Land-use considerations also influence risk management decisions. For example, it may be decided that ecological risks are not an issue for a site where land-use patterns dictate the current and continued absence (or frequent disruption) of significant ecological features. However, the decision becomes more complicated if the disturbed habitat is used by a threatened or endangered species.

Local populations

The following levels of ecological organization are commonly used in risk assessments:

- individuals of threatened and endangered species,
- local populations of ecological receptors,
- local communities of ecological receptors, and
- local habitats or environments that support ecological receptors, including habitats of special concern (e.g., assessments may include evaluation of soil functions within these habitats and environments).

The "local population" is an operational definition that describes a population of organisms present in the area of a contaminated site and reflects a combination of ecological considerations and risk management judgments. It reflects a recognition that assessments of ecological receptors (other than threatened and endangered species) should be at an organizational level greater than the individual, such as a population (Spromberg et al. 1998). On the other hand, it recognizes that the incremental risk that an individual site poses to the entire breeding population of a species could be small and insignificant even though there may be risks to the subpopulation living at or near the site. This concept has led to the use of local populations of ecological receptors as a basis for judging risks of contaminated soils. The rationale is pragmatic because 1) for most situations it would be extremely difficult to consider all the possible cumulative risks to the larger breeding population, 2) the chosen subpopulation is close to the site and assessment can provide information relevant for site-related decisions, and 3) people typically care about the subpopulations of animals and plants that live in local habitats and environments in their part of the state or province.

The local population has been defined in various ways:

- geographic boundary based on the site—the group of organisms of a species that occupies suitable habitats bounded by the site and/or the contamination released by the site,
- geographic boundary based on contiguous or nearby habitat—the group of organisms of a species that occupies suitable habitats at or near a site as defined by the extent of those habitats, or
- ecological boundary—the area defined by a selected number of home ranges that are judged to be an appropriate scale for defining a small interbreeding subpopulation of a species.

There are cases where the operational definition of a local population will not apply, including 1) the presence of individuals attracted to a particular location from a distance because it represents a favored foraging area or source of water, 2) the periodic presence of migratory species that may be exposed for short periods of time, and 3) assessments focused on risks to breeding populations or metapopulations of a species at regional, national, or global scales.

Site characteristics

The site is typically defined, from a regulatory perspective, as wherever contamination resides or could come to reside. The site may differ and be larger than the legal property boundaries of a facility and is essentially the area of contamination. It is useful to have information regarding the size and spatial pattern of the contaminated area, particularly whether contamination is ubiquitous or localized ("hot spots"). Depending on the types of contaminants and ecological receptors present, it may be possible for the risk manager to decide that the area of contamination is too small to pose an unacceptable risk.

Unacceptable risks

Definitions of acceptable versus unacceptable risk or specific criteria for determining unacceptable risk may have been established by the regulatory agencies or jurisdictions of interest to the risk manager, and the risk assessment will need to be designed to address these criteria. When criteria are not available a priori, they must be established in the context of a given site prior to commencing the risk assessment. For example, if ecological risk at a site was based upon the response of earthworms as an acceptable surrogate for all species of invertebrates at a contaminated site, a risk criterion might be stated as "risk will be acceptable if percent survival of earthworms in a laboratory or in situ bioassay is not significantly less ($\alpha = 0.05$) in site soil samples than in field reference soil samples."

It is critical that decisions regarding risk acceptability criteria are not made unilaterally by risk managers but that they include substantive input from as many stakeholders (industry, public interest groups, regulators, community-at-large, natural resource trustees, etc.) as practicable. Early and continuous involvement of stakeholders in the risk assessment process will favor risk management decisions that are more likely to be supported by all parties.

Uncertainty and variability

Risk management decisions are influenced by both "variability" and "uncertainty," particularly as these affect the ERA process (Hattis and Burmaster 1994). Variability refers to statistical variance derived from random or heterogeneous factors and is determined by sampling procedures and the sample population. It cannot usually be reduced through additional sampling or data collection efforts once a stable variance is reached. Uncertainty, on the other hand, refers to model-specification error, estimation error, or data values that are not known with precision because of limited observations. Unlike variability, uncertainty can be reduced through additional knowledge-gathering activities. The risk manager needs to distinguish between variability and uncertainty, and their sources, in order to direct data-gathering activities toward reducing uncertainty and potentially raising confidence in decisions regarding risk. The effects of variability and uncertainty on a risk estimation may be identified and visualized in a number of ways, including 1) statistical parameters (e.g., standard deviation, variance); 2) ranges, intervals, or lower and upper bounds; or 3) probabilistic methods.

Deterministic and probabilistic methods (e.g., Monte Carlo) provide risk managers with point estimates of risk and multiple descriptors of uncertainty and variability around the risk estimate. This gives more complete information on the likelihood of various risk levels and can be used to identify a range of cleanup values with associated levels of certainty.

Prior to initiating a probabilistic assessment, the risk assessor should discuss the following with the risk manager and stakeholders:

- exposure routes,
- contaminants of greatest potential importance to ecological receptors,
- assessment endpoints for ecological receptors,
- current and future land-use scenarios that are reasonable and appropriate for the specific site, and
- sources and characteristics of the distributions (if any) proposed for use in both the exposure and effects analyses.

Following these discussions and the receipt of concurrence from the risk manager, the risk assessor should prepare a work plan that documents the approach that will be used to perform the probabilistic assessment. This is an important activity, owing to the potential complexity of a probabilistic ERA. Even a simple food web model may involve several ecological receptors, each with its own unique distributions of exposure factors, toxicological response characteristics, and relationships to others in the web. Key factors that should be considered for probabilistic analysis are summarized in Burmaster and Anderson (1994), MacIntosh et al. (1992, 1994), and USEPA (1997b).

While a probabilistic approach can give greater insight into the consequences of uncertainty and variability for risk estimates, it also can be more difficult to use and may require the risk manager to make a clear statement of risk tolerance. There may be some reluctance on behalf of the risk manager to accept personal accountability for a statement of risk tolerance, resulting in a statement of acceptable risk that errs on the conservative side. Deterministic (quotient method) analysis provides point estimates that can be compared directly to risk criteria. In contrast, probabilistic analyses assign probabilities for a range of risk levels. The deterministic estimate provides a clearer basis for judgment, but probabilistic methods provide a more complete picture of risk. As this situation may appear more ambiguous to a risk manager, the impetus to undertake a remedial action may be less or the chosen remedy scaled back substantially, particularly if substantial cost savings could be obtained. This suggests that, with the absence of an equally sophisticated framework for risk management and communication, more sophisticated risk assessment techniques will not necessarily provide an easier path to remedial decisions.

Contaminants

Knowledge of the types or classes of contaminants at a site is typically available either from historical source information or from the results of preliminary assessment testing. Also, data on basic soil characteristics, particularly those relevant to estimating bioavailability, should be obtained as early as possible. Knowledge of soil properties and chemical characteristics is used to estimate bioaccumulation potential, bioavailability, and potential toxicological effects. This information, in turn, is used to identify the exposure pathways as well as susceptible receptors.

Exposure

Movement of contaminants from soils through the food chain to higher trophic-level receptors is a common concern to risk managers. These higher trophic-level receptors are often perceived as having greater societal ("charismatic soil megafauna," e.g., earthworms) or economic (e.g., earthworms as fish bait) value. They may also be located some distance from the soil source area. To evaluate such exposures, consideration must be given to both direct contact with or incidental ingestion of soils, as well as with indirect exposures through ingestion of food items contaminated by contact with soils. The relative importance of these pathways depends on the bioavailability of a contaminant from a soil medium and its biomagnification potential through the food chain. If contaminants that bioaccumulate are found to be present, the risk assessment can be designed to include appropriate measures (e.g., bioaccumulation and biomagnification models, tissue analysis) of potential food web exposure and risk. Also, the risk manager should recognize that land-use patterns (and changes in these patterns) influence both exposure pathways and routes for a given receptor.

Preliminary conceptual site model

Available information on ecological conditions, chemical contaminants, exposure pathways, and receptors should be integrated into a preliminary conceptual site model (p-CSM). The p-CSM is an opportunity for the risk assessor to make an initial evaluation of the potential for site-related risk, given that contaminants, ecological receptors, and complete exposure pathways must all be present for risk to occur. At some simple sites, it may be possible for the risk managers to determine, on the basis of this information alone, that the site is unlikely to present significant ecological risks. If ecological risk is likely, then there may be a number of risk management options available to the risk manager at this point. These options may include, but are not limited to, the following:

- determine that specific soil contaminants are exempt from consideration under the regulations that the risk manager is responding to (examples would be the exclusion of lead from automobile emissions or oil under the Comprehensive Environmental Response Compensation and Liability Act [CERCLA]);
- implement a previously established risk management process, such as a generic or presumptive remedy or one directed at a specific soil contaminant such as lead or total petroleum hydrocarbons;
- conduct a Tier 1 screening-level ERA to further assess the potential for risk at the site (information on land use and habitat, extent and loading of the contaminated area, and types of contaminants would be particularly significant factors in this determination); and
- proceed with a formal site-specific ERA, employing tools, methods, and considerations discussed elsewhere in this volume (it would be expected that information integrated in the p-CSM would be used to define objectives for further ecological assessment activities at the site).

Applying Screening-Level (Tier 1) Assessments

Once a decision has been made to apply a risk-based approach to a site, the tiered process begins with simple screening-level analyses. This occurs within Steps 1 and 2 of the Superfund process (Figure 2-3) and Tier 1 of the ASTM RBCA process. Many states and provinces include a screening-level step in their ERA guidelines. Screening can involve the application of concentration-based screening levels such as the Ecological Soils Screening Levels (EcoSSLs) being developed within a collaborative program led by the USEPA or the soil guidelines developed by CCME. Chapter 3 describes the approaches taken for developing screening levels.

Screening criteria can also involve methods that are not based on chemical concentrations. The following types of screening criteria have been used by various states and provinces and provide a starting point for identifying how sites where chemical releases have occurred might be evaluated:

- environmental "performance criteria" associated with permitted activities,
- proximity of release to relevant ecological receptors and their habitats,
- characteristics of release (size and persistence),
- characteristics of released chemicals (toxicity and bioaccumulative potential),
- identification of conditions that warrant an immediate response action, and
- results of field observations and toxicity tests.

Since screening-level assessments are usually based on limited information, they are designed to be conservative. They usually involve simplified representations of spatial scales (e.g., biased sampling in contaminated areas) and temporal scales (current levels of contamination with no allowance for changes in concentration or availability). Risk magnitudes are usually conveyed as exceedances of the chemical or criteria screening levels in soils.

In a tiered-assessment strategy, if the screening analysis indicates a potential for risk from contaminated soils, a more refined analysis may be undertaken. The transition between a Tier 1 and Tier 2 analysis involves a risk management decision informed by the screening-level risk assessment. The risk manager and risk assessor have several options:

1) additional data can be obtained to better represent exposure,
2) refinements can be made to the screening criteria to make them more site-specific, or
3) a response action can be implemented if this is judged cost-effective.

Refinements of exposure or screening levels could involve the incorporation of information on bioavailability. Chapters 8 and 9 describe some of these methods and Chapter 3 also describes how such information can be used to modify the screening-level analysis. Following a Tier 2 analysis, the risk manager and risk assessor revisit the options mentioned above. If a decision is made to conduct a more comprehensive risk assessment, the analysis proceeds to a third tier (note: some guidance documents refer to this as an iterative assessment process rather that as numerical tiers).

As the analysis proceeds from simple conservative to more sophisticated comprehensive assessments, the level of effort increases, the uncertainties are reduced, and the "scales" and endpoints of the risk are better represented for the site manager. The risk assessor must also seek communication tools (e.g., graphics, tables, narratives) that convey the more complete but complex picture of risk to the manager in a way that can be used to make decisions concerning response actions. The remainder of this chapter describes these components of the risk assessment process.

Problem Formulation (Higher-Tier Assessments)

If a decision is made to proceed to a higher-tier analysis, a formal risk assessment process is initiated, beginning with problem formulation. Some aspects of this process should have been underway during preliminary and screening-level assessments. For example, a p-CSM will have been prepared and information will be available on the nature of the problem.

The USEPA has concluded that the strengths and weaknesses of ERAs appear to originate from decisions made during the problem formulation stage. It is especially important at this stage to identify and contact any stakeholders with responsibilities for the resources being assessed or who may be affected by the site-management decision. If the affected parties do not participate in the early decisions about goals, endpoints, and measurements, the analysis is likely to fail in providing information useful for decision-making. Problem formulation is an iterative process that may be revisited several times during an assessment as stakeholder interests and increases in site-specific information dictate. Problem formulation sets the site-specific questions (hypotheses) that need to be answered (tested) as well as the criteria for judging the answers. Various "tools" (Chapters 4, 8, and 9) provide data on measures of exposure and effects. These data are then analyzed and interpreted in the context of the criteria established during problem formulation to answer the questions (test the hypotheses) posed during problem formulation.

The objectives for problem formulation are to 1) evaluate all available information on ecological conditions, contaminants, pathways, and receptors at a given site; 2) make an initial recommendation as to risks posed by the site; and 3) define objectives for any further ecological assessment (Norton et al. 1992; USEPA 1992a,

1997a, 1998b; CCME 1996). Specific tasks, in the order that they take place during problem formulation, include the following:

1) assemble a site description based on information from site investigations and existing literature, any prior preliminary assessments, and site history, including past and present uses;
2) make an initial identification of contaminants, exposure pathways, and ecological receptors (habitats, local populations, individual threatened and endangered organisms, and soil processes);
3) discuss how the physicochemical and toxicological properties of each contaminant may influence exposure pathways and susceptibility of various receptors;
4) define assessment endpoints appropriate for the site;
5) develop risk hypotheses or testable statements of how site-related contaminants might affect each assessment endpoint;
6) select measures of exposure and effect appropriate for the assessment endpoints that clearly address the question posed by each risk hypothesis; and
7) update the conceptual-site model with the above information so that it can serve as a basis for design of the ERA.

It is important for risk managers to realize that evaluating risk is intended to be an internally consistent process. Or, more succinctly—ask the question, set criteria on how you will differentiate an "unacceptable" from an "acceptable" risk, take some measurements, compare the results to the criteria, and state (along with the degree of confidence in the answer) whether the answer is "unacceptable" or "acceptable" risk. This seemingly obvious point is being made because the connection between questions asked, tests performed, and conclusions reached is not always clear in many ERAs. When assessment endpoints and risk-evaluation criteria have been established through discussions with stakeholders, it is vitally important that the resulting risk assessment clearly connects the questions asked and the answers provided and adheres faithfully to agreed-upon evaluation criteria. The weight-of-evidence approach described by Menzie, Heiger-Bernays et al. (1996) is an example of a formal process that is being applied successfully to Superfund and state sites.

The connection between assessment endpoint, risk hypothesis, measures of exposure or effect, and risk evaluation is illustrated with a simple hypothetical example involving 2 species, a red-tailed hawk that is exposed to soil contaminants indirectly through a food chain (soil→plants→herbivorous rodent→hawk) and a terrestrial invertebrate (earthworm) that is exposed through direct contact with contaminated soils.

Example of Questions and Criteria Posed During Problem Formulation

Assessment endpoints

Assessment endpoint species for our hypothetical example are red-tailed hawk and earthworms. As an assessment endpoint, the earthworm is a surrogate for all soil invertebrates, as well as for the general health of the soil ecosystem. The assessment endpoints could be stated as follows: 1) protect individual red-tailed hawks from acute (mortality) and chronic (reproductive impairment) adverse affects resulting from exposure to contaminants in soils, and 2) protect soil invertebrates from population reduction or reproduction impairment as a result of exposure to contaminants in soils.

Risk hypotheses

For this assessment example, statements of how adverse effects could occur, and the criteria for judging whether these effects constitute the potential for unacceptable risk, are modeled contaminant doses received by red-tailed hawks through ingestion of its prey species and are less than doses reported in the scientific literature as lowest-observed-adverse-effect levels (LOAELs). In this instance, modeling is being used because the birds themselves are not readily available for study or tissue analysis. The potential for risk would be considered unacceptable if modeled doses exceeded the LOAEL. The percent survival of earthworms is not significantly ($\alpha = 0.05$) less in site soil samples than in field reference soil samples. Survival can be ascertained with bioassay tests conducted using soils brought into the laboratory or conducted in situ (i.e., caged field tests).

Measures

For this assessment example, the primary measures of exposure are 1) contaminant concentrations measured in site soils and the tissues of red-tailed hawk prey items, 2) modeled contaminant doses received by red-tailed hawks through ingestion of its prey species and through direct contact with, or incidental ingestion of, contaminated soils, and 3) results of earthworm bioassays that would indicate the potential for exposure by giving indirect evidence of the bioavailability of contaminants.

The measures of effect are LOAELs for the red-tailed hawk or similar bird species obtained from the literature and percent survival of earthworms in an appropriate soil bioassay.

Assessment endpoints

An assessment endpoint is an explicit expression of a specific ecological receptor and an associated function or quality that is to be maintained or protected; an expression that provides a clear connection between regulatory-policy goals and risk assessment results. Assessment endpoints are chosen to reflect environmental values that are protected by law, that provide critical resources, or that provide an ecological function that would be significantly impaired (or that society would perceive as being impaired) if the resource were altered (Norton et al. 1992; USEPA 1992a, 1997a, 1998b). The selection of appropriate assessment endpoints avoids making decisions on the basis of trivial or insignificant effects. Thus, it is conceivable that observable effects could be detected in a system, but because they occur in organisms or processes deemed relatively unimportant, they could be discounted as a cause for remediation.

From the set of ecological receptors identified at the site, specific receptors (habitats, local populations, individual threatened and endangered organisms, soil ecosystem processes and functions) are selected as assessment endpoints. These assessment endpoints are either themselves the object of protection or serve as surrogates for all other ecological receptors requiring protection. A variety of criteria may be used to select assessment endpoints including, but not limited to

- processes considered essential to, or indicative of, healthy functioning of soil ecosystems;
- species that are vital to the structure and function of the food web (e.g., principal prey species or species that are major food items for principal prey species);
- rare, endangered, or threatened species, or those protected under federal or state laws;
- species that exhibit a marked toxicological sensitivity to the contaminant;
- economically important or societally valued species;
- although species for which toxicological data are readily available in the literature may not be relevant to a particular site, they may be justified as a surrogate for species that are relevant to the site;
- species with unique life histories (those that fill unique ecological niches) and/or feeding habits or representatives of a particular guild; and
- species common on or near the site and potentially available for tissue sampling, but their relevance to the site in question must also be justified.

From the standpoint of public acceptance, effects on economically or socially valued populations such as trees, fishes, birds, or mammals are the most understandable. If species that do not meet these criteria are particularly susceptible, then their link to specific policy goals (such as protection of threatened and endangered species) or other environmental attributes (such as aesthetics) should be explicitly described.

On a site-specific basis, assessment endpoints answer the question "What is important to protect?" A surprising number of ERAs have been conducted without a clear understanding of what is supposedly at risk and/or what values are to be protected. The outcome of this failure to clearly connect policy with assessment is misdirected field or laboratory investigations with results that are incapable of supporting risk management decisions. Without clear endpoints, an ecological assessment may easily wander ineffectively through numerous ecological components at a given site without ever reaching a conclusion supportive of risk management decisions.

Establishing assessment endpoints during problem formulation forces all stakeholders (regulatory agencies, natural resource trustees, potentially responsible parties, contractors, etc.) to think through and agree upon a common basis for understanding what is at risk at a given site. Specific requirements for a satisfactory assessment can vary substantially between sites and among individual regulators. Therefore, establishing some form of open communication with regulatory personnel responsible for the site is a necessary, but often overlooked or ignored, aspect of the entire risk assessment process (Moore and Biddinger 1996). Responsible parties often feel that talking with regulators will only open a Pandora's Box of additional out-of-scope requirements. While this is possible, it is equally true that additional work can be levied by an uninformed regulator who feels necessary work is being avoided. A risk assessment project is far more likely to remain within budget and on schedule if there is a clear understanding between the regulated and the regulators as to what constitutes reasonable, acceptable practice (Hope 1995).

Risk hypotheses

Risk hypotheses describe predicted relationships between contaminants, exposure, and assessment endpoint responses. Each hypothesis is a statement of how a given contaminant might affect an assessment endpoint and what the criteria are for judging whether any effect is indicative of risk or not. For example, "...percent survival of earthworms is not significantly ($\alpha = 0.05$) less in site soil samples than in field reference soil samples..." states a position that soil contaminants may cause mortality (adverse effect) in earthworms (an assessment endpoint) and that unacceptable risk will be indicated if site soils are significantly more toxic than reference soils. This position is tested with an appropriate earthworm bioassay. Although such hypotheses may involve statistical tests, they themselves should not be taken as "hypotheses" in a strictly statistical sense (Suter 1996). Other evaluation criteria may involve simple comparisons or ratios, such as "...risk is acceptable if the modeled dose is less than the reference dose..."

Because ecological systems are complex, a number of risk hypotheses may need to be evaluated simultaneously in order to arrive at an estimate of risk. The results of individual hypothesis tests ("lines of evidence") can be combined using weight-of-evidence methods (Menzie, Heiger-Bernays et al. 1996) to make a more holistic evaluation of risk at the site.

Measures

The measurement endpoints must be readily measurable phenomena and appropriate for the exposure pathways, the temporal dynamics of exposure, and the scale of the site being evaluated. Most importantly, they must also correspond to or predict assessment endpoints. There are 3 categories of measurement endpoints (USEPA 1998):

1) Measures of exposure—describes how exposure may occur, including how a contaminant moves through the environment and how it may occur in the same place as the assessment endpoint. Exposure to contaminants in soil may be estimated in a variety of ways, including dietary concentration or dose estimated using food chain exposure modeling, measured contaminant levels in soils, or measured contaminant levels in food items.
2) Measures of effect—evaluates the response of endpoint species when exposed to a contaminant. Typically, these are literature-derived LC50, LOAEL, no-observed-adverse-effect level (NOAEL), or equivalent toxicity reference values (TRVs) for reproductive impairment and other adverse effects in the assessment endpoint.
3) Measures of characteristics—includes site characteristics that influence the behavior and location of assessment endpoints and of contaminant distribution, as well as characteristics of assessment endpoints that may affect exposure or response to the contaminant.

When describing measures, it is useful to address specifically the purpose and rationale of the measure, the approach that will be used to evaluate the measure, and the plan for data collection (Menzie et al. 1992). For example, the measure of effect is toxicity of soils to plant species; the purpose and rationale is that metals in soils can be toxic to plants; the approach is to use controlled laboratory phytotoxicity bioassays, comparing site soils with those from a reference area to determine if effects occurred; and the plan for data collection is to obtain chemical analysis exposure data from site and reference area soil samples and effects data from seed germination and growth tests.

Conceptual site model

Problem formulation concludes with development of a conceptual site model (CSM) that describes and integrates information on ecological receptors, contaminants of interest, exposure routes, potential effects, assessment endpoints, risk hypotheses, and measurement endpoints (measures of exposure, effect, and characteristics) (USEPA 1998). Examples of CSMs for terrestrial receptors that may be exposed to contaminated soils are shown in Figures 2-5 and 2-6. These models include many pathways. However, at a site, a smaller number of pathways would be important. The CSM is used by risk assessors to guide the design of the ERA (specifically, the sampling and analysis plan), as well as by risk managers to convey

Biologically accessible soil zones
Surficial (~ 0 - 6")
Surface (~ 0 - 3')
Subsurface (~ > 3')
Direct exposure
Contact
Ingestion
Inhalation dust or vapors (during burrowing)
Receptors
Plants
Soil invertebrates
Wildlife (mammals, birds)
• Herptiles
• Herbivores
• Insectivores
• Predators
Indirect exposure (ingestion of contaminated foodstuff)

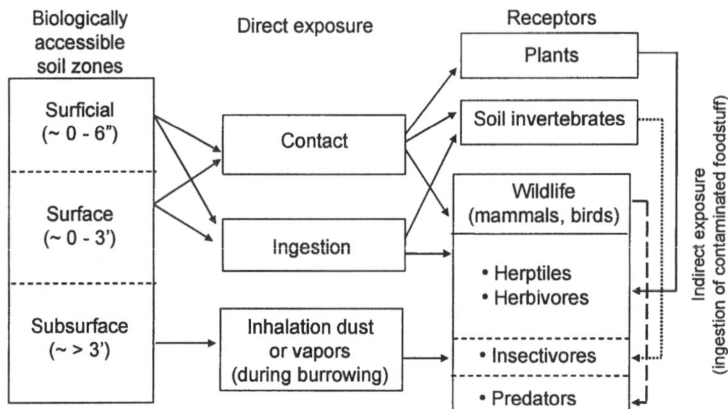

Figure 2-5 Conceptual model for direct and indirect exposure of ecological receptors to soil zones

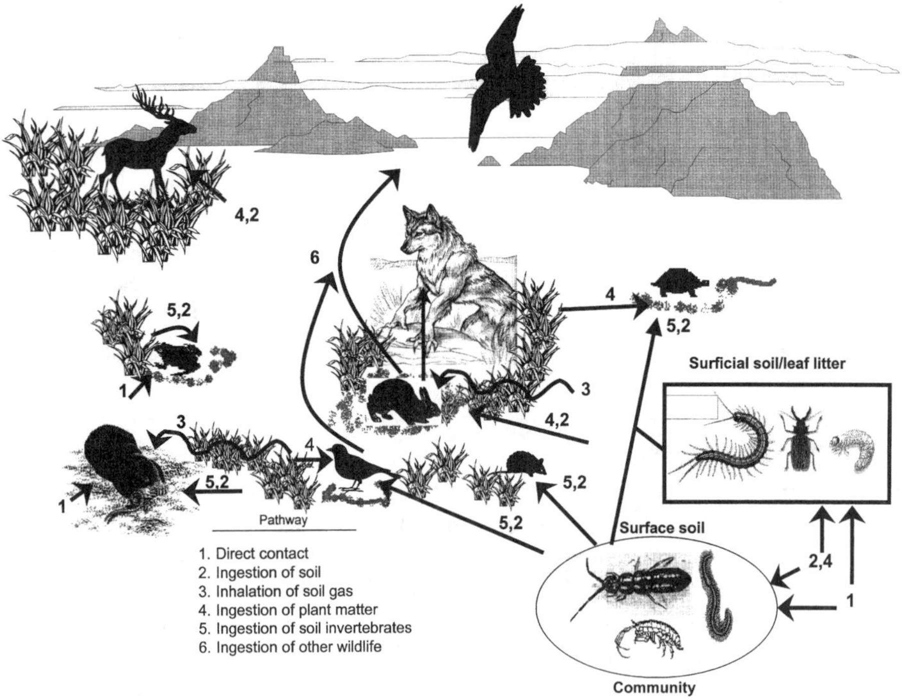

4,2
6
5,2
1
4
3
4,2
Surficial soil/leaf litter
Surface soil
Community
2,4
Pathway
1. Direct contact
2. Ingestion of soil
3. Inhalation of soil gas
4. Ingestion of plant matter
5. Ingestion of soil invertebrates
6. Ingestion of other wildlife

Figure 2-6 Potential pathways for exposure in terrestrial ecosystems

information about potential site-related risks to stakeholders (USEPA 1992a, 1997a).

Good required practices for initiating a risk assessment include

- designing a sampling and analysis plan from the CSM to measure completed pathways;
- tailoring and typing (strategy) to the purpose and objectives of the measure,
  - screening: biased design to look at worse prospective areas,
  - exposure: stratified random in proper media (e.g., surface soils) to obtain a representative concentration term within an exposure unit (home range), may need some purposeful sampling to not overlook heterogeneity of contamination in soils as related to habitat;
- employing relevantly qualified personnel to help design and conduct sampling and analyses;
- pre-specifying method detection limits and other quality assurance aspects of sampling and analysis appropriate for the needs of the risk assessment (these are reflected in the data quality objectives);
- using statistical power analyses to help determine numbers of samples needed to test the risk hypotheses with specified terms (alpha, beta, minimum relative difference, variance, and relative proportion of group sizes); and
- developing an appropriate experimental or statistical design for testing hypotheses (if stated) or answering specific questions. Not all risk assessments will necessarily use analysis of variance (ANOVA) or regression analyses, and other methods of data assessment of summary should be considered (e.g., weight-of-evidence approach).

Analysis: Exposure Assessment

Exposure assessment for contaminated soils involves estimating exposure to define endpoint receptors for the exposure pathways developed within the CSM. Exposure is assessed in a manner that complements information on effects so that risks may be characterized with a consistent body of information. In the context of risk assessments for Superfund and many other programs, both current and future exposure scenarios are considered. In prospective risk assessments, such as those involved in planning future land development or in deciding on the use of a new chemical, potential future scenarios are the focus of evaluation. Exposure estimates also are used to evaluate the relative importance of various contaminant sources or pathways when considering cleanup levels or remediation strategies.

Exposure is typically well characterized in definitive risk assessments (e.g., Tier 3) but less precisely characterized in screening assessments (e.g., Tier 1). Primary routes of exposure to contaminated soils are through the direct ingestion of soil or by the consumption of food organisms that have accumulated chemicals from the soil, and also from direct contact with contaminated soils (plants, some soil invertebrates [e.g., earthworms], dermal exposure of wildlife). Key descriptors in the exposure assessment are bioavailability of the contaminant, spatial scale of the contaminant with respect to the receptor, temporal aspects of the chemical concentration, and activity of the receptor. Exposure parameters may be single values, ranges, or, if possible, distributions with respect to space, time, or organism characteristics. Other chapters in this volume address these issues in some detail, particularly bioavailability.

Measures of exposure available to terrestrial ecological risk assessors include

- concentrations of chemicals in soil that may be adjusted for bioavailability,
- doses to wildlife that are derived from these concentrations or are adjusted for bioavailability using knowledge of species-specific assimilation efficiencies, and
- concentrations of chemicals in tissues of organisms (food items or consumers).

The measured concentration of a chemical in an organism or particular tissue is often a good measure of exposure for persistent, bioaccumulative chemicals such as polychlorinated biphenyls (PCBs) and organochlorine pesticides. For other compounds, such as organophosphate insecticides, which are rapidly metabolized, tissue or body residues may provide little meaningful exposure data. Additionally, trapping vertebrates and sampling plants may not be feasible given the funding level and/or timing of a risk assessment. Because few existing exposure–response models for wildlife relate tissue concentrations to toxicity, the primary focus of this section is on concentrations of chemical contaminants in soil as measures of exposure.

Sampling methods for estimating exposure to soil contaminants

Concentrations of chemicals in soils are most commonly determined using either biased, grid, or stratified-random sampling. Each offers advantages and disadvantages (Table 2-2; Gilbert 1987). Limited investigations during preliminary or Tier 1 analyses typically involve biased sampling. More comprehensive ERAs that account for spatial distribution of contaminants and receptors usually rely on either grid or stratified-random sampling.

It is useful to incorporate receptor-based sampling into the sampling design. This involves considering how receptors might be exposed and the dimensions and extent of areas that the receptors occupy. Samples are taken within the receptors' exposure zones, including contaminated and uncontaminated areas. Stratified-random sample designs are often used to collect samples from these different strata.

Table 2-2 Advantages and disadvantages of various soil sampling strategies

Sampling method	Advantages	Disadvantages
Biased Samples are taken predominantly in areas of obvious soil contamination, as evidenced by past practices, location of production or storage facilities, or soil staining.	Inexpensive and quick. Supports conservative screening analyses.	May not characterize exposure zones relevant to ecological receptors. May yield an overly conservative estimate of exposure. May not be appropriate for comprehensive baseline risk assessments.
Grid Samples are taken at regular intervals on a grid that divides the site into uniform squares, rectangles, or triangles.	Relatively easy to design and implement. Provides even coverage for areas of interest. Can be useful for identifying "hot spots" of contamination. Are well suited for contour and geographic information system (GIS) analyses.	The method can be expensive and time-consuming for tight grids applied to large areas. Certain ecologically important strata may be overrepresented and others underrepresented in the sampling. Finding small "hot spots" may require an unacceptably large number of samples.
Stratified-random Samples are taken within areas ("strata") of the site that have been differentiated on the basis of ecological and or topographic features.	Suitable for baseline risk assessment. Supports area-weighted average methods. The method gives better statistical estimates of population means than other methods. Can be more cost-effective than grid sampling.	The method requires competent ecological and topographic surveys of a site prior to sampling. The method can be labor- and time-intensive.

Total and bioavailable chemical measurements in soils

Total chemical concentrations in soil are the most common measure of exposure. These measures also are typically used to derive bioaccumulation factors or regression relationships with chemical residues in organisms (BJC 1998; Sample et al. 1999). In addition, most of the available exposure–response thresholds, screening benchmarks, and other dose-response models are derived from studies in which only total concentrations of chemicals in soil were measured. Concentrations and doses based upon total chemical measurements are used in both screening and higher-tier risk assessments.

Although methods for adjusting total chemical concentrations for the fraction that is bioavailable to organisms (e.g., plant available nutrients in soil) have been available for some time, the application of these methods to measuring toxicant bioavailability has only been recently developed and their application is not common. A summary of some methods used for adjusting total chemical concentrations to account for bioavailability is presented in Table 2-3.

Current hypotheses suggest that the fraction of total chemical in the soil that is potentially bioavailable to soil organisms such as earthworms, where dermal uptake occurs, is dissolved in soil pore water (Belfroid et al. 1996; Conder and Lanno 2000). Therefore, 1 method for determining the potentially bioavailable fraction of chemicals in soil is to measure toxicant levels in pore water or in a solvent that extracts dissolved and/or weakly sorbed chemicals. Collection of pore water and determination of dissolved chemicals has been applied to both organic compounds and metals. Organic chemicals are often measured in nonpolar or semipolar solvent extracts such as tetrahydrofuran or butanol (Tang et al. 1999; Tang and Alexander 1999), while potentially bioavailable metal levels are often determined in weak salt extracts using $CaCl_2$, $Ca(NO_3)_2$, or ammonium acetate (Conder and Lanno 2000). Simply determining chemical concentrations in soil extracts and assuming that this fraction represents the bioavailable fraction is insufficient, and soil extracts as measures of chemical exposure must be correlated with biological responses to see if they truly represent bioavailability. Soil extracts represent the environmentally available fraction as described in the introduction of this book (Chapter 1) and are representative of point-in-time or instantaneous chemical availability with no temporal integration of exposure. Temporal integration of chemical bioavailability can only truly be achieved by examining the bioaccumulation of chemicals in organisms, and to a lesser extent, using biomimetic sampling devices. Because bioaccumulation studies are expensive and time-consuming, some researchers have examined the relationship between bioaccumulation of organic compounds in organisms (earthworms) and sorption by solid-phase extractants (SPEs) (Tang et al. 1999; Wells and Lanno 2001). SPEs show great promise as biological surrogates with good correlations achieved between sorption of DDT (Tang et al. 1999) and phenanthrene (Wells and Lanno 2001) by SPEs and bioaccumulation in earthworms.

Table 2-3 Measures of exposure to chemical contaminants—alternatives to total concentrations in soil

Method	Assessment endpoint	Test endpoint	Chemical	Reference
Soil porewater and solvent extractions				
Soil porewater measurement	Plant community	Uptake by and toxicity to various crop species	Numerous inorganic chemicals	Efroymson et al. 1997
	Soil invertebrate community	Uptake by earthworm	Metals	Janssen et al. 1997
			Chlorophenols	Van Gestel and Ma 1988
Ammonium acetate extract	Plant community	Uptake by plant	Uranium	Sheppard and Evenden 1992
	Soil invertebrate community	Uptake by earthworm	Uranium	Sheppard and Evenden 1992
Propanol *n*-butanol, ethyl acetate	Plant community	Uptake by plant	Anthracene, fluoranthene, pyrene	Tang and Alexander 1999
	Soil invertebrate community	Uptake by earthworm	Anthracene, fluoranthene, pyrene	Tang and Alexander 1999
Ethanol, butanol	Soil invertebrate community	Uptake by earthworm	Phenanthrene pyrene	Chung and Alexander 1999
Tetrahydrofuran extract	Soil invertebrate community	Uptake by earthworm	DDT, DDE, DDD	Tang et al. 1999
$Ca(NO_3)_2$ extract	Soil invertebrate community	Earthworm lethality and bioaccumulation	Zn, Pb, Cd	Conder and Lanno 2000
	Plant community	Uptake by plant	Zn, Pb, Cd	Basta and Gradwohl 2000
Estimated metal activities in solution	Plant community	Yields of test plants, mostly crops	Cu, Pb	Sauvé et al. 1998
	Microbial processes	Microbial biomass, soil respiration, carbon mineralization, nitrogen mineralization, denitrification, *Rhizobia* survival, etc.	Cu, Pb	Sauvé et al. 1998
	Soil invertebrate community	Nematode community abundance and subsets	Cu, Pb	Sauvé et al. 1998
Solid-phase extractants (biomimetic devices)				
Solid-phase microextraction fiber; semi-permeable membrane device	Soil invertebrate community	Earthworm lethality and bioaccumulation	Phenanthrene	Wells and Lanno 2001
C18 membranes	Soil invertebrate community	Earthworm bioaccumulation	DDT, DDE, DDD	Tang et al. 1999

Experience is being gained on how to use bioavailability measures in ERA. It is advisable that methods for measuring bioavailability continue to be developed and tested for ranges of organisms, soils, and chemicals. In addition, it would be useful for risk assessors if soil chemists and toxicologists state the ranges of uncertainty and limits of applicability for exposure estimates derived from these methods. For example, if a particular organic solvent can be used to estimate the bioavailable fraction of phenanthrene to earthworms, can the same organic solvent be used to estimate the bioavailable fraction of other PAHs in different soils to other soil invertebrates? Can it be used for all nonionic, semivolatile organic compounds?

Many estimates of bioavailability are measures of chemical concentrations in soil pore water, and this may be an appropriate exposure pathway for receptors such as plants and earthworms where direct contact with dissolved contaminants may be an important uptake pathway. However, for exposure pathways involving the incidental ingestion of soil or transfer between solid phases, soil porewater measures may not be correlated with the relevant biologically available fraction.

In predictive risk assessments, estimates of exposures are conducted without the benefit of direct measures of bioavailability. Instead, estimates of bioavailability are made based on chemical properties, soil characteristics, and receptors of concern. It would be useful to have national, or international, databases that relate chemical concentrations in contaminated soil to levels of bioaccumulation and/or toxicity.

Bioavailability is a factor when considering comparisons between site chemical levels and background chemical concentrations. Background chemical levels may be those that occur in a specified area due to natural geological conditions (e.g., elevated metal levels in mining areas) or that are characteristic of a specific area (e.g., deposition of PAHs on soil surface in industrial areas). Regardless, the conditions that define "background" need to be clearly stated prior to comparisons of chemical levels by any methods. Many risk assessments use such comparisons when evaluating exposure conditions, and in some cases, background chemical levels may actually exceed screening levels. The bioavailability of compounds is rarely considered in such situations because comparisons are made strictly on the basis of total chemical levels. The comparison of total chemical levels in soil is meaningless because bioavailability is governed by the form of the chemical present, the physicochemical structure of the soil, and the duration of the soil–chemical interaction period (aging and weathering). For these reasons, background concentrations of chemical may pose very different exposures than chemicals associated with relatively recent additions of chemicals. In such cases, determination of the bioavailable fraction of chemical present would facilitate a true comparison of exposure levels.

The implications of incorporating bioavailability measures into ERA include

- the decreased uncertainty in site-specific exposure estimates for the relevant assessment endpoints;
- the requirement for site-specific soil tests from which to develop exposure–response relationships;
- the increased precision of exposure–response relationships developed from site-specific data (e.g., laboratory toxicity tests, field studies) where the measure of exposure is the bioavailability-adjusted concentration or dose;
- the relegation of generic empirical exposure–response models (where exposure is represented by total contaminant concentrations) to a position of low importance in the weight-of-evidence analysis;
- the possibility of developing empirical bioaccumulation models that relate bioavailability-adjusted contaminant concentrations in soil to concentrations in plants and other biological tissues and wildlife foods;
- the possibility of using estimates of soil porewater concentrations in available exposure–response models where exposure is represented by the concentration in nutrient solution or soil pore water (e.g., the method of Sauvé et al. [1998] estimates exposure from total chemical concentration in soil and soil pH); and
- the increased cost of assessment, to the extent that different extraction methods are required for different chemicals or chemical families and different endpoints.

The use of bioavailability-adjusted measures of exposure would reduce uncertainty but increase the cost of risk assessments. Decisions to proceed with more expensive assessments are usually balanced by the possibility that better estimates of exposure lead to more cost-effective response actions and remedial decisions. Menzie et al. (2000) list 3 questions that a risk assessor or manager should ask when considering the use of information on bioavailability:

1) Do the governing site-management regulations or policies allow for risk-based approaches that can incorporate bioavailability information?
2) Within these risk-based approaches, when is it appropriate to consider information on bioavailability?
3) Will the information provide value for decision-making?

When judging the value of bioavailability information, Menzie et al. (2000) suggest the following be asked:

- Is the chemical in a form that is considered for all practical purposes to be physically and chemically unavailable as supported by regulatory determination, previous studies, or simple tests accepted by the regulatory agency?
- Are risks associated with site soils being driven by chemicals for which the acquisition of bioavailability information is being considered?
- Are other factors associated with the presence of the chemicals (e.g., the presence of a free product or visibly stained soils) likely to determine site management options?
- Will incorporating bioavailability information change the risk estimate by an amount that affects the way in which the overall site risk is managed?

Incidental ingestion of soil by wildlife can result in exposure to contaminants in the soil. The relative importance of this exposure route depends on the relative amount of soil in the diet as well as the degree that food items (plants and animals) accumulate the contaminant in soils (Table 2-4). Assuming that the contaminant in the soil is 100% bioavailable, incidental soil ingestion becomes relatively important (50% or more of the oral dose) at soil ingestion rates of 1% and greater in the diet and where the soil bioaccumulation factor into food items is approximately 0.1 and less. For chemicals that fit this category, information on the bioavailability of the chemical incidentally ingested with soil could result in significant modifications in dose estimates.

Table 2-4 Percent contribution of incidental soil ingestion to oral dose for wildlife at different soil ingestion rates and bioaccumulation factors and a bioavailability of 100%

	Bioaccumulation factor from soil to food source (organism, %)			
% Soil in diet	0.01	0.1	1.0	10
0.01	0.99	0.099	0.01	0.001
0.1	9.1	0.99	0.1	0.01
1.0	50	9	1	0.1
10	92	52.6	10	1

Soil properties

As described in Chapter 7, soil properties can modify exposure of ecological receptors to soil contaminants. Measures of these properties throughout the course of investigations provide insight into their potential importance as well as guide the selection of subsequent measurements of bioavailability. In some cases, information on soil properties can be used to determine if soils are capable of supporting certain plants and animals (i.e., nutrient levels, texture, amount and type of organic matter, physical properties such as compaction). Such information is important for both exposure and effects assessment as well as for identifying other stressors or factors that pose a risk to or influence terrestrial biota.

A short list of soil properties that are considered especially important includes

- total organic carbon and any type of organic carbon—provides an indication of sorption capacity of soils for some organic chemicals and metals (also provides an indication of the amount of organic material available for soil organisms);
- cation exchange capacity—provides a measure of the ability of soil to bind certain metals;
- pH—provides a measure of acidity which affects the availability of some metals and organic chemicals as well as influencing soil biota;
- texture—composition of soil as sand, silt, and clay provides an indication of the physical structure of the soils and is related to the types of organisms that inhabit the soil (also related to the ability of soil to bind chemicals); and
- degree of compaction—provides a measure of the ability of soil to support biota (this is included because this can be a major physical stressor at existing or former industrial sites).

Other measures that can be useful include nutrients (P/K/N), water-holding capacity, and moisture content. Nutrient measurements provide information on the ability of soil to support plant growth, and water-holding capacity is a specific property related to the relative mass of water a soil will retain at a negative pressure, often –0.3 bar. The actual moisture content of a soil is a soil state and is related to the ability of soil to support different types of animals and plants as well as the availability of certain metals and organic chemicals.

Additional measures of soil properties can provide more insight into the availability of chemicals, and these are discussed further in other chapters (especially Chapter 7). However, preliminary and subsequent investigations of contaminated soils would benefit from inclusion of the measures listed above. Depending upon the history of the site and knowledge of the chemicals that are present, a subset of measures could be selected.

Scales of exposure

Risk assessors must define both spatial scales (Table 2-5) and temporal scales (Table 2-6) of exposure appropriate to the particular contaminated site or chemical use. Spatial scales are relevant to particular endpoints and fate and transport processes. Sampling of chemical contaminants in soil and mathematical models of exposure should reflect these scales. Because of the diversity of potential receptors and our low-to-moderate understanding of their activities and locations, there is often a compromise between what is known, what is assumed, and what can

Table 2-5 Spatial scale of processes relevant to exposure

Spatial scale	Fate or transport process	Relevant endpoint
Soil particle to soil aggregate (μm–mm)	Biodegradation, chemical transformation, sorption	Soil processes, invertebrate community
Depth in soil (μm–m)	Contaminant migration	All ecological receptors
Local topographic features (10–1000 m)	Runoff, transport of fugitive dusts	All ecological receptors
Spatial (horizontal) distribution of contamination (variable scale)	Lateral migration	All ecological receptors
Home ranges of ecological receptors (<1 m^2–100,000 ha)	NA	One aspect of characterizing exposure to individuals within a population
Areal dimension for "local populations" of ecological receptors (often operationally defined)	NA	Vertebrate populations, plant community
Spatial scales for other contaminant and noncontaminant environmental stressors (variable scale)	All transport and fate processes	All ecological receptors

Table 2-6 Examples of processes and behavior that may influence risk estimates for contaminated soils with respect to time

Fate process or receptor behavior	Importance
Rates of chemical loss via volatilization, leaching, and/or advection	Reduces chemical exposure with time to contaminants in soil but may influence exposure in other media
Rates of sequestration	Reduces chemical exposure with time
Rates of transformation	Reduces chemical exposure with time for parent compounds but could result in exposure to metabolites
Daily, seasonal, and annual behavior patterns of receptors	Influences when and for how long receptors may come into contact with contaminants
"Recovery times" for ecological receptors (species, communities, systems)	Influences the rates by which an impacted receptor may return to or recover within an area. Such information may be valuable for judging the potential for natural or enhanced restoration of an area.

actually be accomplished. For example, it is not feasible to measure chemical concentrations in soil from the surface to 1 cm, 2 cm, 3 cm, etc. because certain soil invertebrates or plants have their primary exposures at all of these depth intervals. Nor is it feasible to measure these concentrations at 1-m horizontal intervals. It also is not desirable to expend resources on this level of sampling if most assessment endpoints are aggregations of ecological receptors (e.g., communities).

A few or many of these scales may be considered, depending on the needs of the assessment for supporting management decisions. Simple screening-level analyses consider only a few scales (e.g., areas where contamination is suspected), while more comprehensive assessments may evaluate the spatial distribution of populations or receptors at the site in relation to the site-wide exposure field for soil contaminants and other stressors.

The depth of sampling should be related to the assessment endpoint. Sampling depths chosen to reflect exposure to the soil invertebrate community should be based on an average estimate of depth intervals where these organisms reside. Similarly, the sampling depth for the plant community should be an average unless a particular threatened and endangered population is of interest, in which case, the rooting depth of the particular species is applicable, if known. A default sampling depth of 30 cm for plant communities may be identified, based primarily on rooting depths in Jackson et al. (1996). However, this default depth should be used with caution because rooting depth will vary greatly with plant species and climatological conditions. A depth of 30 cm may also be a reasonable default sampling for soil invertebrates although the depth of the organic soil horizon also would be appropriate. Default sampling depths should be adjusted if the vertical gradient of contamination is known. If ideal sampling depths are different for different assessment endpoints, a compromise depth may be required. Where contaminant concentrations are measured (e.g., at Superfund sites), exposure depth is not treated as a variable. Therefore, any uncertainty in the use of a particular default depth should be stated. The choice of a 10-cm sampling depth versus a 1-m sampling depth could mean a difference of an order of magnitude (e.g., plutonium in Litaor et al. 1994).

In selecting a statistical representation of concentrations for organisms exposed to chemicals in soils, a fundamental distinction must be made between receptors that average their exposure over space and time and those that have essentially constant exposure. For plants and soil invertebrates that are immobile or nearly immobile, the exposure concentration is best represented by the maximum detected concentration (Sample et al. 1997). Mobility is irrelevant if the organisms do not move the distance between 2 sampling points. However, for mobile ecological receptors (terrestrial wildlife consuming soil, vegetation, or animal foods) that do not experience their environment on a "point" basis, it is necessary to convert measured data from single sample points into an estimate that represents some relevant spatial area, such as their habitat (Sample et al. 1997).

There are a number of ways to consider spatial area when computing exposure concentrations. The simplest approach is to estimate exposure by assuming that contaminants are evenly distributed within a habitat and that ecological receptors forage randomly with respect to contamination within that habitat. In this case, the exposure concentration can be represented by the average concentration or an upper bound on the mean (e.g., the 95% upper confidence limit [UCL]) calculated using USEPA standard methods (USEPA 1992b).

Consideration can be given to the size and quality of habitat on and near the site utilized by mobile species in relation to species foraging areas and contaminant levels in these habitats (Sample et al. 1997). Area-weighted averaging methods that take into account the relative differences in contamination levels per unit area may be used to estimate exposure concentrations for both individual species and populations (Freshman and Menzie 1996). More sophisticated techniques are available, including bootstrapping and probabilistic methods, to enhance area-weighted approaches by allowing for quantitation of variance in both the area and contaminant concentration estimates (Burmaster and Thompson 1997). Finally, geographic information system (GIS) approaches may be used to simultaneously integrate spatial information on contaminant concentrations, habitat distribution, species foraging area, and topographic features that may affect exposure routes (Clifford et al. 1995).

Temporal scales have received little consideration in the assessment of risks of contaminated soils to ecological receptors (Table 2-6). These factors include seasonal patterns of behavior and contaminant uptake and changes in exposure in future scenarios (either prospective assessments or Superfund). Few organisms are exposed to contamination in the winter. Many species are dormant, others are migratory, and the uptake of contaminants by food items (vegetation, invertebrates) is reduced or halted. Changes in chemical concentrations in soil due to biodegradation or other chemical transformations are addressed in Chapter 5. Actual rates of biodegradation depend on a series of environmental factors, including oxygen and nitrogen availability, organic matter and clay content of soil, age of the contamination (in the case of organics), and form of the product (e.g., whether a nonaqueous-phase liquid is present). Exposure estimates are more accurate if reasonable representations of fate are included in the relevant models.

Exposure models

Models have numerous uses in risk assessment. One example is the estimation of tissue concentrations in predatory or protected species that cannot be sacrificed. A second function is to provide an indication of whether or not proposed detection limits are low enough to encompass media and tissue concentrations of ecotoxicological interest. Models of exposure include

1) bioaccumulation models for plants, earthworms, arthropods, small mammals, etc. (wildlife foods and tissue representations of exposure);
2) dietary exposure models for wildlife, where the output is dose to the organism (mg/kg/day); and
3) models of wildlife movement.

In most retrospective risk assessments, chemical concentrations in soil are measured and are the starting point for parameterizing exposure models.

Bioaccumulation models may be mechanistic or based on empirical data ranging from the estimation of whole-body bioaccumulation factors from known chemical properties to more sophisticated models, such as physiologically based pharmacokinetic models, that account for bioavailability and depuration rates in specific tissues within organisms. Empirical models have been applied broadly for evaluating exposure to contaminated soils because they are relatively straightforward and usually rely on data from the site or from similar sites. Mechanistic models are employed for predicting future conditions when body burdens are influenced by several exposure pathways and/or when the relationships between exposure and accumulation are influenced by biological functions that are dose dependent. A model is appropriate for use at a site if 1) it is empirical and incorporates uptake data from a similar soil, 2) it has the flexibility to incorporate site-specific soil properties such as pH or organic matter content, or 3) it is a general model with specified uncertainty bounds.

Dietary exposure to wildlife may be expressed in the following form:

$$E_{\mathrm{t}} = E_{\mathrm{f}} + E_{\mathrm{w}} + E_{\mathrm{s}} + E_{\mathrm{i}} + E_{\mathrm{d}} \qquad (2\text{-}1),$$

where

E_{t} = total contaminant exposure,

E_{f} = contaminant exposure from food ingestion,

E_{w} = contaminant exposure from water ingestion,

E_{s} = contaminant exposure from soil ingestion,

E_{i} = contaminant exposure from inhalation, and

E_{d} = contaminant exposure from dermal absorption.

However, because dermal and inhalation exposures are usually assumed to be negligible at hazardous waste sites, and because this model is to evaluate risks associated with the consumption of contaminants in soil or food exposed to contaminated soil, it can be simplified:

$$E_{\mathrm{t}} = E_{\mathrm{f}} + E_{\mathrm{s}} \qquad (2\text{-}2).$$

Estimates of E_f and E_s can be obtained from measurements or estimates of concentrations of chemicals in food and soil combined with ingestion rates for food and soil. The concentration of a chemical that is directly ingested may be adjusted for bioavailability, though these methods are not well developed yet. The concentration in wildlife food may be measured at a contaminated site or estimated using bioaccumulation models. It is important for the risk assessor to specify the wildlife diet as accurately as possible since contaminant uptake by plants, invertebrates, and small mammals may be quite different. Information on wildlife diets can be found in various texts and research papers on these species. Some of these factors are summarized in documents such as the *Wildlife Exposure Factors Handbook* (USEPA 1993).

Models of wildlife movement are rarely used in ERAs. Assessors should consider the use of these tools if 1) acute exposures at a certain location may result in a risk, 2) exposures at certain life stages (e.g., staging in certain habitats) may result in a reproductive risk, or 3) threatened and endangered wildlife species are present at the site. Akçakaya (1998) has reviewed some of these models.

Although the literature contains a wide array of exposure models, deciding which one to use and exactly how to use it can be a frustrating and time-consuming exercise for the risk assessment practitioner. Models assembled from various sources without careful consideration as to their underlying assumptions, or even the compatibility of their units of measure, can produce misleading and inaccurate estimates of exposure.

Probabilistic estimates of exposure

Exposure estimates are affected by uncertainty and variability, and these should be presented to the extent possible. The methods used to account for and discuss uncertainty range from qualitative discussions (usually used in screening analyses) to sensitivity, interval analysis, and probabilistic methods. A tiered approach for using such methods is described in *Guiding Principles for Monte Carlo Analysis* (USEPA 1997b).

Probabilistic methods incorporate distributions of exposure parameters such as oral ingestion by wildlife. It is important for the risk assessor to describe the variable that is distributed (e.g., time, space) (Suter 1993). For example, exposure parameters (e.g., chemical concentrations) may be distributed with respect to space or time.

Analysis: Effects Assessment

Effects assessment is undertaken to evaluate potential toxicity of soil contaminants to receptors of concern. To do an adequate job of assessing adverse effects, the risk assessor needs to carefully consider the selection of effects measurement or estimation methods and the scale of evaluations. Key criteria for selecting a measure of effect include

- being related to 1 or more assessment endpoints and clearly identifying the relationship,
- being consistent with selected measures of exposure, and
- being an ecological effect that is related to an agreed-upon ecological function. In the case of assessing risks to populations, this usually includes effects on reproduction, growth, and survival.

All studies and modeling for potential effects should arise from the problem formulation on the basis of exposure pathways shown in the CSM. Those pathways that are possibly "complete" (from source, release, and transport to route of entry) for an ecological receptor are candidates for devising measurement endpoints with testable hypotheses on effects. The measures of effects need to be coordinated with the measures of exposure to help establish a causal relationship between the responses and the amount of contact with soil contaminants and other stressors (refer to Chapter 4 and Chapter 8 for more details on types and uses of various biological tests that have relevance for ERA of contaminated soils).

Measures of effect should

1) help support a risk-based conclusion regarding the assessment endpoint,
2) provide information to the risk manager on the "safe" levels of exposure that do not produce unacceptable risk of adverse effects to local populations or to individuals of threatened and endangered species, and
3) give an estimate of variability and uncertainty in the quantitative risk estimates.

Models used to predict effects should be chosen after discussion and approval by the risk manager and risk assessment team during the problem formulation phase, especially if the use of probabilistic risk analyses is considered. The use of a validated bioaccumulation model as a measure of exposure, instead of effects and population models, should include the collection of site-specific calibration and verification data that will provide confidence in the predictive ability of the model estimates.

Risk managers and stakeholders must recognize that certain measures of effects can only be made at specific times of the year and in places when and where critical exposure occurs for receptors, such as during the nesting season for birds or germination and rapid growth of plants, typically in the spring for North America. Other measures of effects can be conducted on resident receptors that do not have critical susceptibility periods and receive more integrated chronic exposures to soil contaminants. Spatial considerations for measures of effects should include adequate representative sampling in appropriate habitats for the receptors of concern, both on contaminated sites where chemical exposure may occur and in reference areas. Sampling sufficient numbers of home ranges for terrestrial receptors to establish more direct exposures, or foraging ranges for more indirect exposures, enables stronger conclusions to be made regarding the risk of adverse effects on local populations.

Effects can be measured or estimated for exposed receptors of concern using a variety of approaches. Toxicity benchmarks, either soil concentrations for plants and invertebrates or chronic daily average doses for vertebrates, are used to assess potential adverse effects by comparing integrated exposures during a critical period with defined acceptable exposures. Toxicity tests can be performed on species of concern or surrogate species, either in the laboratory or in the field, to estimate the types and extent of adverse effects from contact with soil contaminants. These tests may provide indirect results on the relative bioavailability of chemicals. Field observations can be designed to provide information on population abundance and diversity and used to detect phytotoxicity or direct adverse effects in other species. Validated population and bioaccumulation models also may be employed to estimate risks of effects, but they generate more confident predictions when calibrated and verified at the site level. As with measures of exposure, data adequacy and quality are critical needs for valid effects measurements.

Toxicity benchmarks and criteria

Toxicity benchmarks are used to estimate potential for adverse effects due to exposure of receptors to soil contaminants. Toxicity benchmarks may be defined as dietary doses, concentrations in media such as soil or soil-associated prey, or tissue concentrations of contaminant in exposed receptors. Typically, a NOAEL or a LOAEL is derived from chronic exposures in laboratory studies. These are expressed as mg/kg/day for dietary exposure, where kg refers to body weight. Because body weights and daily integrated food intakes vary considerably among animals, dose information allows the analyst to customize exposure for specific types or guilds of organisms. Laboratory studies typically use controlled feeding regimes with measured doses of contaminants that allow a relatively accurate estimate of dietary dose. In contrast, field studies may have accurate information on environmental concentrations but less information on actual doses associated with the feeding habits or other behaviors of receptors.

Since national standard TRVs for various chemicals are only currently being developed (USEPA 2000), the most common method of deriving valid toxicity benchmarks has been to extrapolate NOAELs and LOAELs from the most well designed toxicity studies that are applicable to the soil contaminant and receptor of interest. The ideal toxicity study is conducted with the species of concern, employs a chronic duration of exposure (or exposure during a critical life stage), and evaluates the relevant endpoint related to sustainability of populations (e.g., reproductive success). It uses sound scientific study designs that include

- proper sample size, controls, and multiple doses within a range of responses;
- appropriate routes of exposure;
- accounting for possible confounding or interactive effects; and
- expressing toxicity as a dose–response relationship.

This study design enables the analyst to extrapolate more closely to actual field conditions.

When deriving TRVs using existing literature, it is common to review the quality and applicability of the available studies and to select those that are strongest technically and most relevant for the receptors at a site. Various approaches are used to derive TRV values from available laboratory or field studies. For example, uncertainty factors are used to adjust for differences in taxonomic relationships, durations of exposure, severity of endpoint, and other factors. This may involve dividing the LOAEL or NOAEL values by 1 or more uncertainty factors, each of which may range between 2 and 10. Another approach includes the use of allometric scaling factors to extrapolate study results, and in some cases, no uncertainty factors have been applied to study endpoints (Sample et al. 1996). Currently, there is no universally preferred method for deriving TRVs from study endpoints. Generally, NOAEL doses are used for screening risks, and appropriate LOAELs or equivalents are used to assess likely risks. The collaborative USEPA EcoSSL will be developing a methodology for deriving TRVs (USEPA 2000).

Toxicity tests

Soil contamination can be evaluated for types and amounts of toxicity to ecological receptors by exposing test organisms to the contaminated soil, either at the site or in the laboratory. These tests are conducted under standard conditions in order to control for confounding factors such as non-site related conditions that contribute to toxicity. A variety of toxicity tests are available and should be selected based on the assessment endpoints being evaluated. Such tests can sometimes provide indirect information on relative bioavailability of soil contaminants. Linder et al. (1992) and Menzie et al. (2000 have reviewed the use of soil toxicity tests for soils, and more detail is provided in Chapters 4 and 8 on the types of toxicity tests available.

Field studies and observations

Zooepidemiological studies are sometimes conducted at larger sites to measure effects in receptor populations during exposure to soil contamination under natural field conditions. Realistic responses to soil contaminants via direct contact or through food webs may be determined. However, the responses of animals in the field to exposure may be difficult to discern from natural variability within the population. There could be many factors that contribute to the success or impact on local populations. Concurrent stressors may confound results and make the determination of causality difficult. Some of these limitations in field studies can be addressed by experimental design. Studies that enclose test organisms on-site and at reference areas and examine toxic endpoints can reduce the uncertainty associated with exposure because the duration and location of exposure can be controlled and quantified. These types of studies are conducted with small mammals, invertebrates, or plants. Bioaccumulation studies often provide some of the most useful measures of exposure in terms of site-specific uptake into tissues. Relative bioavailability is calculated by comparing tissue concentrations from known site exposures of total soil concentrations to tissue concentrations from the same exposures to a reference material:

$$\text{Relative bioavailability} = \frac{\text{tissue concentration on-site}}{\text{tissue concentration in lab}}\text{, at same total chemical doses} \quad (2\text{-}3).$$

Collocated measurements of representative soil concentrations and tissue concentrations within the exposure unit area provide a basis for developing site-specific biota–soil bioaccumulation factors. When measuring tissue residues of ubiquitous nonpolar soil contaminants such as dioxins and PCBs, reference areas must be used to provide background levels in food webs and soils. Surveys of population abundance, species diversity, and reproductive success can be used to evaluate effects if these observations can be reasonably associated with concentrations of soil contaminants. Such associations may be used to identify graded or threshold responses to contaminant levels in soils.

Use of reference areas

Reference areas are habitats or areas near a contaminated site with similar characteristics (e.g., soil physicochemical characteristics, slope) except for the presence of site-related contaminants and are used to estimate local background concentrations of chemicals. They can be used to establish local ecological conditions that are not impacted by site-related contaminants and provide a basis for comparison to contaminated site conditions. The greatest challenge when using a reference area for comparison is locating an area that is as similar as possible in soil physicochemical composition, aspect, and other geochemical characteristics to the contaminated site and contains no contaminants. It is essentially impossible to find a reference site with identical physical and chemical soil characteristics. Selection of multiple reference sites (e.g., 3) will help account for natural spatial variability and provide a more robust basis for comparison. An alternative approach is to examine ecological conditions along a gradient of on-site contamination, but care must be taken to assure that no other parameters that are toxic or could modify toxicity (e.g., pH) also vary with the contaminant gradient.

Risk Characterization

Risk characterization for contaminants in soils involves integrating the information on exposure and effects so that the potential for harm can be estimated, described, and presented in a manner that is understandable by risk managers and others involved in the decisions. The characterization of risks associated with contaminated soil can include qualitative (e.g., the presence of sensitive species) as well as quantitative information.

Relating risks to assessment endpoints

The assessment endpoints established during problem formulation are the basis for characterizing risks. Therefore, the discussion of risks should be organized around the specified assessment endpoints. Other issues or potential risks identified during the evaluation should be included but distinguished from the evaluation of the assessment endpoints. Risks are based on 1 or more lines of evidence that incorporate measures of effects and exposure.

Judging risks

The basis for judging risks associated with contaminated soils should be established prior to the conduct of the assessment. Examples include exceedance of a particular hazard quotient (HQ) value, probability of affecting a local population or habitat in excess of a specified amount, or integration and weighting of multiple lines of evidence. The criteria by which risks will be judged may vary among sites but should be clear from the onset of a particular assessment.

Risk estimation methods for soils

Risk estimation methods should vary depending on the needs of the assessment. In Tier 1, and even Tier 2 assessments, simple HQ methods may be used to compare site concentrations or doses to generic or site-specific benchmarks or TRVs. These simple methods are well known and adequately described elsewhere (Suter 1996; USEPA 1997a). The HQ method involves comparing a concentration such as the maximum or upper bound concentration to a TRV or ecological benchmark from the literature or derived for the site. This is similar to approaches used for human health risk assessments (i.e., comparison of exposure dose to risk reference dose [RfD]). However, while these methods are useful for screening-level assessments or where simple conservative assessments with "yes" or "no" answers are desired, they convey little information about the scales of the risks. Scales are especially important in ERAs because these assessments are typically focused on risks to local populations and the habitats that support them. Thus, it is desirable where possible to express risks in terms of the local population (ODEQ 1998). More comprehensive assessments use analyses that estimate risks at 1 or more scales (space, time, and magnitude).

Spatial analysis is useful because it relates risks to physical dimensions (e.g., specific locations or areas) and as such communicates information that can be used by risk managers to judge the size of a potential problem. Available methods range in sophistication and information content.

The spatial analyses of risks include simple tools such as the average concentration with area curve (Figure 2-7), which relates chemical concentrations in soils to target concentrations where effects may occur and to the foraging areas or home ranges of individual animals (Freshman and Menzie 1996). Analyses also include methods that describe risks to fractions of the local population (Freshman and Menzie 1996). This is accomplished using a spreadsheet model to represent the spatial distribution of contamination and habitats. Exposures to individuals of the local population occur depending on their food habits, their forage, and the size of their foraging area. The model estimates exposures and effects to all members of the local population and determines the number of individuals and fractions of the population that are at risk (Figure 2-8).

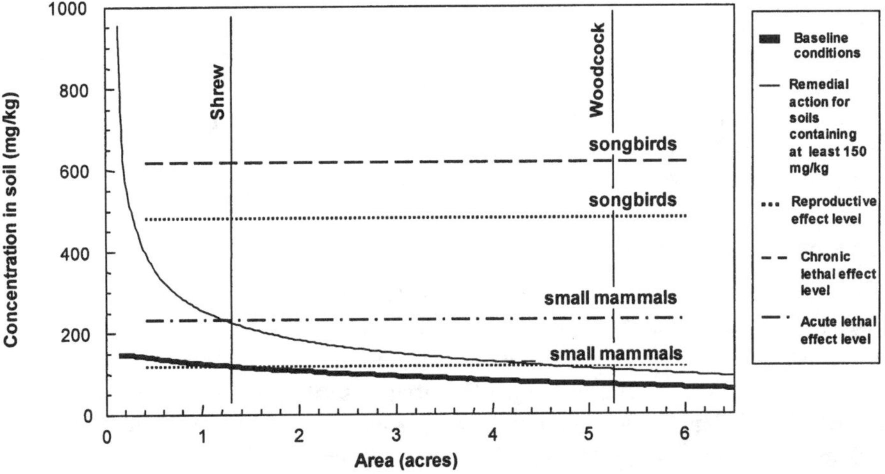

Figure 2-7 Average concentration with area curve exposure model illustrating pre-remedial and post-remedial conditions

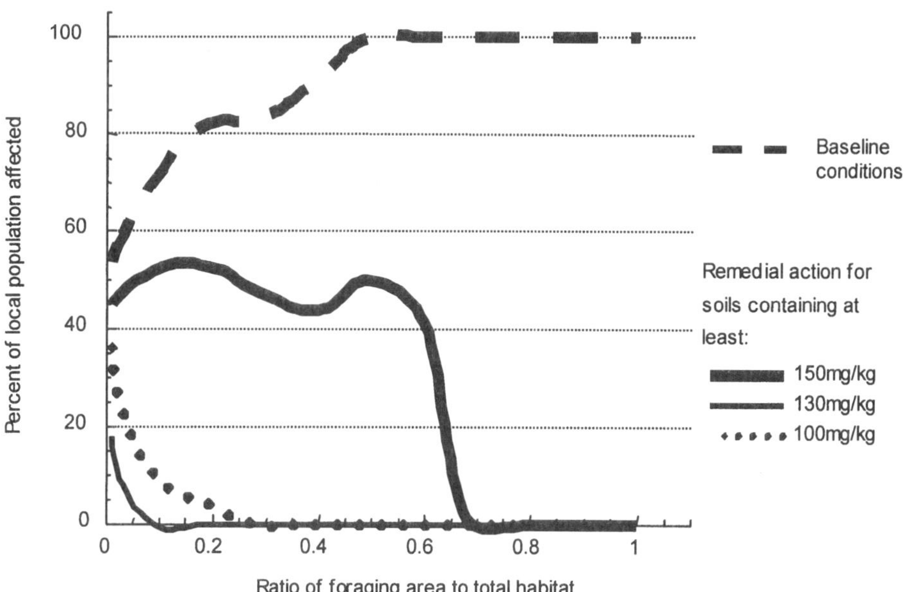

Spatial distribution of chemical "A" in soil mg/kg

60	84	96	97	88	88	91	63	54	87	75	86	88	86	47
49	103	134	146	126	104	70	58	40	145	202	234	98	90	49
40	100	170	177	167	148	80	51	45	136	391	353	100	96	47
39	103	176	377	394	109	88	41	40	110	307	238	146	106	49
45	120	312	956	443	251	89	34	39	109	109	134	197	102	39
35	124	159	208	227	141	85	65	64	104	118	146	132	99	38
38	102	114	150	129	141	82	53	65	72	72	67	69	64	40
27	95	101	110	104	110	64	42	40	41	39	28	16	52	44
27	39	24	24	20	56	48	61	68	58	75	83	73	55	35
26	41	36	32	21	51	43	83	84	84	77	77	71	47	25
18	40	37	32	26	46	57	91	175	144	114	72	77	41	25
16	32	41	44	39	43	50	84	114	180	125	57	75	28	25
15	33	40	55	44	37	53	79	198	267	184	58	78	18	26
14	10	19	35	43	41	60	75	124	181	199	78	69	21	20
9	7	34	40	48	51	56	78	83	77	76	74	63	30	46

Figure 2-8 Example of efficacy of remedial alternatives for multiple sources and soil effect level of 65 mg/kg

Geographic information system technology has also proven useful for analyzing and displaying spatial information. This tool allows the input, storage, management, visualization, manipulation, analysis, and output of spatially referenced data in the form of maps. The technique has been used for sites at the Oak Ridge Reservation in Tennessee, USA (Washington-Allen and Sample 1997). A contaminant map for arsenic was developed and overlaid onto habitat maps for 2 wildlife species—red fox and short-tailed shrew. Exposure models were used to determine where toxic effects might occur.

Akçakaya (1998) reviewed a number of ecological models that consider issues of scale and observed that the spatial structure of habitat has important implications for the dynamics of metapopulations of single species, as well as trophic interactions. Spatially explicit metapopulation models incorporate spatial structure by considering the location, size, and shape of habitat patches and the distances among them. Some models are implemented as software tools, such as risk assessment risk managers audit systems (RAMAS) GIS (Akçakaya 1998). Some theoretical models of trophic interactions consider the effects of spatial structure, but there are no practical tools for modeling trophic interactions in a spatially explicit setting. One method of habitat patch recognition has been implemented in RAMAS GIS and applied to several cases involving threatened species (e.g., Akçakaya et al. 1995; Akçakaya and Atwood 1997). Because this program accepts spatial data in various GIS formats, any landscape model that can predict the relevant habitat variables (such as species and age of dominant trees, average diameter, height, etc.) and export its predictions in 1 of these GIS formats can be integrated into RAMAS GIS.

Information organized in a spatial framework allows managers and risk assessors to ask "what if" questions concerning the environmental costs and benefits of implementing management options. Answers can be displayed in graphs or shown on maps. Not only can these display risks associated with current or future conditions but also potential risk reductions and increases due to implementing management options. These graphics serve as useful communication tools between the risk assessor, manager, and others. Freshman and Menzie (1996) showed how the average concentration with area curve method and a population-level foraging model can be used to illustrate benefits of focused remedial efforts (e.g., Figure 2-7). An example of an approach used to show the risk reduction benefits of remediating mercury-contaminated soil at Oak Ridge is shown in Figure 2-9.

Temporal analysis provides information on the persistence of exposure as well as on the rate that ecological receptors may respond to changes in exposure regime. In general, temporal dimensions of ecological risk have been poorly described for contaminants in soil. However, such information is useful for determining the duration of exposure to chemicals at toxic levels as well as potential rates for natural or enhanced recovery.

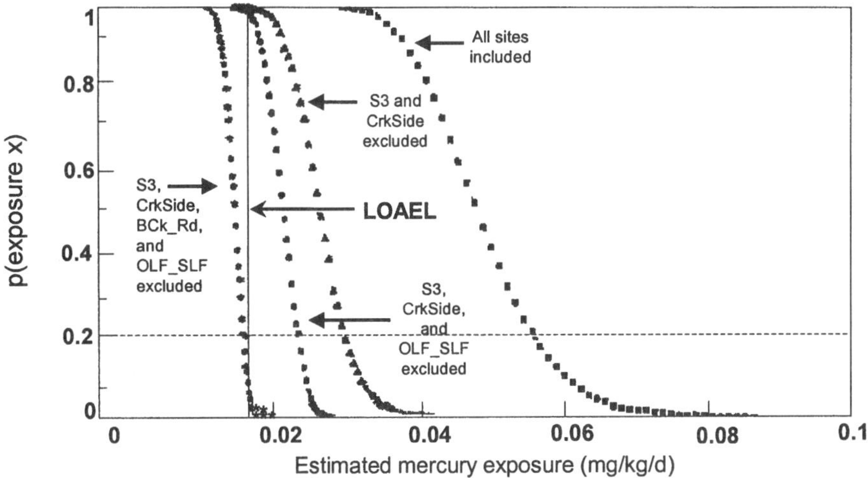

Figure 2-9 Effect of sequential removal of contaminated sites on mean watershed-wide mercury exposure to red fox

Analyses of risk types and magnitudes provide managers with information on the potential severity of a problem. This information can be discussed in a narrative or displayed graphically. The tools described above for spatial analysis can also be used to illustrate the magnitude of the risk by plotting this as a dimension within a spatial context.

Lines-of-evidence methods

One or more risk measures or lines of evidence are used to estimate risks. Because of inherent uncertainty in any measure of effects or exposure, many risk assessments of contaminated soils rely on multiple lines of evidence. Available information from the literature, chemical measurements in soils or tissues, soil toxicity tests, and field observations are integrated.

Lines- or weight-of-evidence approaches have been described for assessing risks to terrestrial receptors. Two methodologies that have been used are Suter (1997) and Menzie, Heiger-Bernays et al. (1996). Each is described briefly below, along with an example.

Suter's (1997) approach relies on an assessment of the logical consistency among various lines of evidence. Each line of evidence is examined to determine whether it is consistent with exceedance of a threshold, inconsistent with exceedance, or ambiguous. The results from all lines of evidence are then examined as a whole to assess whether it is likely or unlikely that the threshold is exceeded. If the results are consistent with respect to the exceedance or nonexceedance of a threshold, then the result of the weighing is clear. However, if there are inconsistencies, the true weighing of evidence must occur and this should involve logical ad hoc expert judgement or consensus. Suter provides an example for a soil invertebrate community (Table 2-7) and notes that, in general, a logical analysis of the data should proceed from more realistic (i.e., site-specific) to more precise and controlled (e.g., single chemical and species toxicity tests). He states that field surveys indicate the actual state of the receiving environment, so other lines of evidence that contradict the field surveys, after allowing for limitations of the field data, are clearly incorrect. For example, the presence of plants that are growing and not visibly injured indicates that lethal and gross pathological effects are not occurring but does not preclude reductions in reproduction or growth rates.

Table 2-7 A hypothetical summary of a risk characterization by weight of evidence for a soil invertebrate community in petroleum-contaminated soil at an industrial site[a]

Evidence	Results	Explanation
Biological surveys	–[b]	Soil microarthropod taxonomic richness is within the range of reference soils of the same type, and is not correlated with concentrations of petroleum components.
Toxicity tests	–	Soil did not reduce survivorship of the earthworm *Eisenia fetida*. Sublethal effects were not determined.
Organism analyses	±[c]	Concentrations of PAHs in depurated earthworms was elevated relative to worms from reference sites, but toxic body burdens are unknown.
Soil analyses	+[d]	If the total hydrocarbon content of the soil is assumed to be composed of benzene, then deaths of earthworms would be expected. Toxicity data for other detected contaminants are unavailable.
Weight of evidence	–	Although earthworm tests may not be sensitive, test results and biological surveys are both negative and are both more reliable than the single chemical toxicity data used with the analytical results for soil.

[a]Suter 1997.

[b] Evidence is inconsistent with the occurrence of a 20% reduction in species richness or abundance of the invertebrate community.

[c] Evidence is too ambiguous to interpret.

[d] Evidence is consistent with the occurrence of a 20% reduction in species richness or abundance of the invertebrate community.

Menzie, Heiger-Bernays et al. (1996) developed a methodology for reconciling and balancing multiple lines of evidence pertaining to an assessment endpoint. Weight of evidence is reflected in 3 characteristics of measurement endpoints:

1) the weight assigned to each measurement endpoint,
2) the magnitude of response observed in the measurement endpoint, and
3) the concurrence among outcomes of multiple measurement endpoints.

First, weights are assigned to measurement endpoints based on attributes related to strength of association between assessment and measurement endpoints, data quality, and study design and execution. Second, the magnitude of response in the measurement endpoint is evaluated with respect to whether the measurement endpoint indicates the presence or absence of harm as well as the magnitude. Third, concurrence among measurement endpoints is evaluated by plotting the findings of the 2 preceding steps on a matrix for each measurement endpoint evaluated. The matrix allows easy visual examination of agreements or divergences among measurement endpoints, facilitating interpretation of the collection of measurement endpoints with respect to the assessment endpoint. A qualitative adaptation of the weight-of-evidence approach is also presented.

Using multiple lines of evidence requires first, that measures of exposure and effects be related to specific assessment endpoints, and second, that criteria are established for judging results. An example of an assessment endpoint and associated measures is given in Table 2-8.

Table 2-8 Example of assessment endpoints and associated measurement endpoints for a New England upland environment

Endpoint	Description
Assessment endpoint terrestrial 1	*Presence and condition of vegetation (trees and shrubs) in upland areas typical of forested areas bordering lakes and ponds*
Measurement endpoint terrestrial 1a	Comparison of composition and condition of vegetation in on-site terrestrial areas to those of comparable reference areas
Measurement endpoint terrestrial 1b	Toxicity of soils to plant species
Assessment endpoint terrestrial 2	*Sustainability of local populations of wildlife typical of forested areas bordering lakes and ponds*
Measurement endpoint terrestrial 2a	Comparison of species composition and habitat use within on-site areas with those in comparable reference areas
Measurement endpoint terrestrial 2b	Assessment of exposure to metals in plants
Measurement endpoint terrestrial 2c	Assessment of exposure via bioaccumulation in invertebrates
Measurement endpoint terrestrial 2d	Assessment of exposure to direct exposure to soil

A summary of the risk characterization for the example introduced in the problem formulation section is given here.

Assessment endpoint A

Protect individual red-tailed hawks from acute (mortality) and chronic (reproductive impairment) adverse affects resulting from exposure to contaminants in soils.

Risk hypothesis A

Modeled contaminant doses received by red-tailed hawks through ingestion of its prey species are less than doses reported in the scientific literature as LOAELs.

Analysis and interpretation

During risk characterization, doses to red-tailed hawks estimated with a food chain model were compared to TRVs (a dose below which no chronic or acute effects are expected) for red-tailed hawks or related bird species, using a quotient methodology. Although the resulting toxicity quotient (TQ) is not a direct measure of risk, it was agreed that TQ values less than or equal to 1 would indicate acceptable risk. The resulting TQ value of 0.22 suggested a negligible potential for an adverse effect in red-tailed hawks. Thus the risk hypothesis associated with this assessment endpoint was accepted.

Assessment endpoint B

Protect the heath of soil invertebrates from reduction or impairment as a result of exposure to contaminants in soils.

Risk hypothesis B

Percent survival of earthworms is not significantly ($\alpha = 0.05$) less in site soil samples than in field reference soil samples.

Analysis and interpretation

The earthworm bioassay indicated no significant difference ($p > 0.05$) in percent survival between control, reference, and site samples, except for 1 sample (T5FD02S01). This sample exhibited significantly lower survivorship when compared to the reference ($t = 1.96$, $p = 0.04$), but not when compared to the control ($t = 0.43$, $p = 0.34$). However, T5FD02S01 had some of the lowest concentrations detected in all samples from the site. Thus there seems to be little connection between contaminant concentrations and toxicity measured in this bioassay. Based on the overall results of the bioassay, which indicated a high degree of survivorship in the vast majority of tests, this hypothesis is accepted.

Conclusions and recommendation

The assessment endpoint for protection of red-tailed hawks from chronic or acute effects is expected to be met. The assessment endpoint for protection of soil invertebrate populations is also expected to be met. Invertebrate populations appear to be unaffected by any contamination on the site, with abundance dictated by availability of suitable habitat rather than contaminant concentrations. Some of the subsurface soil concentrations are higher than surface soil concentrations and may pose a risk to invertebrates and hawks should exposure occur. Although this is unlikely, these soils were identified for removal to eliminate a potential source of ecological risk.

Dealing with multiple stressors

As the size of hazardous waste sites increases, crossing ecological, political, and geological boundaries, ecological risk assessors must adapt their techniques and methodologies to account for the multitude of stressors impacting a site. In order to deal with sites affected by multiple stressors, a number of different methodologies are being developed. Current research trends in multiple stressor analysis focus on watersheds and water quality. However, terrestrial assessments are beginning to include multiple stressor analysis.

Wickwire et al. (1998) have developed a multiple stressor analytical framework for a large terrestrial site. The framework focuses on identifying a range of different stressors and then evaluating vegetation parcels based on the impact of each suite of stressors. The stressor categories include historic factors such as grazing impacts, sulfur dioxide releases and forest harvesting, phytotoxic factors including the release of metals to the soils, and finally, physical factors such as the slope and aspect of each vegetation area. All 3 larger areas interact to define the quality of the site in reference to vegetative habitat. A rank of 1 to 3 is assigned to each factor based on the relative impact of that factor on the quality of the site. An average score for each of the 3 categories is calculated. The results allow a risk assessor to select a vegetation area and predict which of the 3 primary stressors is having the greatest impact on site quality. The risk assessor can appropriate remedial resources to influence the specific stressors at each vegetation area.

The USEPA is initiating a number of assessment projects that focus on the watershed scale and the multiple stressors that influence water quality in each watershed. These concepts are applicable to terrestrial systems. In one study, Lussier et al. (in press) used GIS as an analytical tool for multiple stressor analysis. The GIS not only provides the assessor with a visual understanding of the impact of stressors, but it also serves a predictive role in terms of identifying areas that are threatened or should be monitored closely. Lussier et al. (in press) compiled data on various stressors and land use and overlaid this with area used for various purposes as well as with contaminant source locations. Other examples of ecological models that take into account multiple stressors include those being used to evaluate conditions in the Duwamish River and Elliott Bay (Munger and Toll 2002) and in the Pacific Northwest (Karr 1998).

Presentation and communication

Ultimately the risk information must be conveyed to a risk manager in order to make decisions. Because ecological receptors, their habitats, and soil contamination occur spatially on horizontal and vertical scales, capturing and communicating information on these dimensions can be especially helpful. This is usually best accomplished by graphical displays and maps.

Uncertainties

Environmental decision-making is usually made in the face of uncertainty. As a result, risk managers typically seek information that can help reduce uncertainty as well as help them manage uncertainty. Scientific information contributes to both these objectives. The value of information analysis performed within the context of exposure and effects assessments can point to areas where large gains can be obtained in supporting the technical analyses underpinning management decisions. Because default assumptions are often conservative, important gains in reducing uncertainty can be achieved even when precision in estimates is low (e.g., within a factor of 2 to 10).

When estimating risks to terrestrial wildlife from exposure to soil contaminants, both deterministic and probabilistic (e.g., Monte Carlo) methods may be used to provide risk managers with both point estimates of risk and multiple descriptors of uncertainty and variability around the risk estimate. Discussion of these additional risk descriptors is expected to provide stakeholders with more complete information on the likelihood of various risk levels and risk managers with multiple risk-based soil cleanup goals from which to choose.

Prior to initiating any probabilistic ERA, the risk assessor should discuss with the risk manager and stakeholders such issues as 1) exposure routes, 2) contaminants of greatest potential importance to ecological receptors, 3) assessment endpoints for ecological receptors, 4) current and future land-use scenarios that are reasonable and appropriate for the specific site, and 5) sources and characteristics of the distributions (if any) proposed for use in both the exposure and effects analyses. Following concurrence from the risk manager, the risk assessor should then prepare a work plan that documents the specific approaches, techniques, and information sources that will be used to perform the probabilistic assessment. This is an important activity, owing to the potential complexity of a probabilistic ERA. Key factors that should be taken into consideration are summarized in Burmaster and Anderson (1994) and USEPA (1997b).

Guidance on how to incorporate probabilistic estimates in ERA is being developed by USEPA in Volume 3 of the USEPA Risk Assessment Guidance for Superfund (RAGS) series. Existing guidance given in USEPA (1997a) for Superfund sites identifies 2 general ways to quantitatively evaluate and present uncertainty, and/or qualitative descriptions. Qualitative descriptors include high, medium, or low confidence in certain aspects of the risk inputs and related outputs. These qualitative descriptors might be applied to extrapolation uncertainties and errors in sampling and measuring. Quantitative measures may include calculations over ranges (e.g., interval analysis) or full probabilistic methods. An example is given in Table 2-9.

Table 2-9 Qualitative descriptions of risk inputs and related outputs

Ranges	TRV-NOAEL	TRV-LOAEL
Average exposure factors	Intermediate estimate	Least risk, highest PRG[a]
Upper-end exposure factors	Most risk, lowest PRG	Intermediate estimate

[a]PRG = Preliminary remediation goal.

Risk Management of Contaminated Soils

Site-management decisions stem, in part, from stated site-management goals. These are used to identify the need for and scope of an assessment and to judge the need for and efficacy of remedial options as these relate to protection of ecological receptors and natural resources. However, the risk assessment is one of several tools that the manager relies on. This section of the chapter discusses a number of issues that a risk manager may need to consider during and following completion of the risk assessment.

At some sites, risk management decisions may be made even before a screening assessment has been completed. For example, if there are clearly no potential exposure pathways to ecological receptors, it can be determined that there are no ecological risks and no further actions, including the performance of an ERA. At other sites, it may be obvious that serious ecological risks exist (dead animals, barren soil, etc.) and an immediate response action (an immediate removal action) may be warranted. Such situations normally are addressed early in the site investigation process, before the site even reaches the risk assessment stage.

At most sites, a screening ERA is performed, during which some chemicals normally do not pass the screen. This triggers the performance, or at least the initiation, of a baseline risk assessment. If during the next steps, modifications to the screening values based on site-specific data substantially increase the screening values, the remaining COC could also be potentially screened out. This would also signal an end to the risk assessment at that site.

Upon completion of a properly planned and conducted baseline risk assessment, the risk manager will have answers to the risk hypotheses generated during the problem formulation phase. The risk characterization will have discussed the magnitude of the observed or predicted responses of the receptors, how the receptors are affected (how many and to what extent), the areal extent and period of time over which the effects occur, and the potential for recovery of the affected resources. In most cases, the risk manager will be able to determine whether or not the soil contamination presents an unacceptable risk to the receptors of concern. If there are no unacceptable risks, the risk manager communicates the basis for this decision to the public in a way that is clear to them.

If there are unacceptable risks, the manager must decide what action is appropriate to take at the site to address both short-term and long-term risks. At most sites, there are several alternatives that are considered. The USEPA has developed decision criteria for judging remedial options (USEPA 1991). Other regulatory agencies utilize similar criteria.

U.S. Environmental Protection Agency decision criteria

1: Overall protection of the environment

The risks posed by contaminants must be balanced against habitat destruction resulting from implementation of the remedial action. Factors to consider include the type of community that would be destroyed, the ability to recover or to be restored, and the distribution of this community type in the immediate area (i.e., how rare is this community type in the ecosystem surrounding the site). If the area to be destroyed is small versus the areal extent of the community type in the immediate area, the action may be appropriate. Other factors include the relationship of the measure of effect to assessment endpoints and uncertainty associated with the results of the risk assessment. Remedial objectives may be implemented by site area or community type.

2: Compliance with applicable or relevant and appropriate requirements

Chemical-specific applicable or relevant and appropriate requirements (ARARs) may include state soil criteria that are promulgated as cleanup standards. Location-specific ARARs, such as the Clean Water Act, Section 404, require mitigation when implementation of the remedial objectives impacts wetlands. Action-specific ARARs may be triggered by impacts on species, or their habitats, that are federal or state protected species protected under the Endangered Species Act, Migratory Bird Treaty Act, or equivalent state statutes.

3: Long-term effectiveness and permanence

The ability of a biological community that might be damaged by remedial actions to recover passively or to be restored through active means to a fully functioning status over an extended time frame should be considered in selection of the remedial alternative. For example, the resilience and high productivity of some agricultural land may permit aggressive remediation, whereas the removal of bottomland hardwood forest communities in an area where they cannot be restored due to water management considerations may argue heavily against extensive action in all but the most highly contaminated areas.

4: Reduction in the toxicity, mobility, or volume through treatment

There is a preference for remedies that involve treatment technologies. For example, this could include bioremediation technologies that reduce the bioavailability of PAHs in soil.

5: Short-term effectiveness

The likelihood of the remedial alternative to achieve success and the potential for a biological community to be restored in the short-term should be considered in remedy selection. For example, the ability of an old field community to recover from a remedial action may be 3 to 5 years, whereas a bottomland hardwood forest may require more than 60 years. In addition, if mitigative measures are possible during remedy implementation, their effectiveness and reliability should be considered.

6: Implementability

The technical feasibility of implementing the remedial alternative is considered when choosing the remedy. Owing to the concentrations of contaminants required for protection of a broad array of ecological receptors, technological limitations may preclude achieving such levels in all instances. In addition, actions that result in the recontamination of the area of remediation would not be considered implementable.

7: Cost

Each chosen remedial alternative must be cost-effective, provided that it first achieves adequate protection and meets ARARs. A remedy is cost-effective if its costs are proportional to its overall effectiveness. Ecologically driven remedies may require lowering contaminant concentrations to levels approaching background to be fully protective in all instances. By examining the relationship between contaminant concentration and protectiveness, a cleanup objective may be selected that provides optimal risk reduction per unit cost of implementation.

8: State and/or support agency acceptance

Early and continuous involvement in the risk assessment process of state and other support agencies will favor risk management decisions that have the greatest level of endorsement by all parties.

9: Community acceptance

Communications with the community should be an ongoing process, including explaining the steps in the ERA process and gathering input on the selection of assessment endpoints for the site. Involvement of local citizens in the ERA process will likely favor their greater acceptance of the chosen remedy.

Monitoring

The complex nature of ecosystems and the large number of species potentially affected at a given site may result in a relatively high degree of uncertainty concerning the levels of chemicals deemed to provide overall protection of the environment. At these sites, the risk manager should establish a monitoring plan and a review schedule that is designed to determine if recovery is occurring in an ecologically relevant time frame and whether an additional response action is warranted.

Incorporating bioavailability into ecological risk assessments

Possible approaches for incorporating chemical availability information into tiered assessments are outlined in Table 2-10 for the Superfund process and Table 2-11 for a generic 3-tier approach. The decision to incorporate such information is, in part, a risk management decision informed by risk assessors. As with other aspects of terrestrial ERA for contaminated soils, a tiered approach provides an efficient way to incorporate bioavailability information.

As the tables suggest, information on availability is included after initial screening activities. However, information could be gathered on soil characteristics during all phases of the program to enable the assessor to judge if chemical availability information would be useful. It follows that assessments would benefit from inclusion of measures of selected soil characteristics correlated with bioavailability.

Table 2-10 Example of how and where information on bioavailability could be incorporated into Superfund ecological risk assessments

Steps in Superfund guidance	Activities
Steps 1 and 2	These are screening-level activities that include the development of a conceptual model, selection of conservative toxicity values, and conservative estimates of exposure. Field studies of the health and condition of ecological receptors are usually not performed. *These steps provide a starting point for identifying where information on bioavailability may be useful.*
Step 3 Baseline risk assessment problem formulation	Problem formulation at Step 3 involves further characterization of effects, refining information on fate and transport of contaminants from the source area, selecting assessment endpoints, and developing a conceptual model with working hypotheses or questions. A simple conceptual model from the guidance is shown as Figure 2-1. *At this step, the risk assessor and risk manager could discuss the value of bioavailability information (measures and models) for the assessment of ecological risks. It is valuable to identify the soil and chemical factors that could affect the bioavailability of the chemicals.*
Step 4 Study design and data quality objective process	The conceptual model developed in Step 3 is used to identify measurement endpoints and to establish the scope of the project. Measurement endpoints may include toxicity testing, estimates of body burdens, as well as population/community evaluations. *Measurement endpoints may include bioavailability as a component. The types of information needed to evaluate the assessment endpoints would be identified. Measurement or modeling methods would be selected based on "weight-of- evidence" considerations: 1) soil–chemical relevance, 2) pathway relevance, 3) receptor relevance, and 4) method acceptance. These methods could include laboratory or field measures of bioaccumulation, the use of exposure models that incorporate bioavailability, and the use of toxicity tests, which indirectly relate to bioavailability.*
Step 5 Field verification of sampling design	This involves a check on the scope to insure that it is appropriate and that it can be implemented. *To the extent that measurement endpoints include measures or estimates of bioavailability, these would be checked at this step.*
Step 6 Site investigation and analysis phase	This step implements the sampling plan. Exposure characterization relies heavily on these data and can involve fate and transport modeling. Information on ecological effects comes from the literature and site investigations. *Measures or estimates of bioavailability would be obtained for the site consistent with the study design.*
Step 7 Risk characterization	In risk characterization, data on exposure and effects are integrated into a statement about risk. *Qualitative or quantitative information on bioavailability is incorporated into an overall weight-of-evidence assessment of risk.*
Step 8 Risk management	This step occurs after the assessment. Risk management may involve balancing risk reductions associated with the cleanup of contaminants with potential impacts of the remedial actions themselves. *Qualitative and quantitative information on bioavailability may have a role in guiding decision-making at the site. Such information may be useful for identifying remedial alternatives.*

Table 2-11 Three-tier assessment process for ecological receptors

Three-tier process	
Tier 1 evaluation	This is the initial step in the process.
Inspect site and organize available site information regarding major contaminants, extent of affected environmental media, and potential migration pathways and receptors.	
Implement appropriate initial response action, if necessary. Reevaluate site following initial response action.	
Continued Tier 1 evaluation	The conceptual model is developed in this tier and provides a basis for identifying pathways for which bioavailability information may be important.
Identify potential sources, exposure pathways, and chemicals.	
Develop conceptual site model and locate generic screening criteria.	An initial assessment could be made regarding the value of bioavailability information.
Perform generic screening level evaluation.	Generally, site-specific information on bioavailability is not considered within generic screening levels.
Tier 2 evaluation	Soil and chemical factors that may affect the bioavailability of the chemical should be identified.
Refine and focus Tier 1 evaluation by collecting additional site-specific data as appropriate. Conduct Tier 2 evaluation as per specified procedures.	Simple conservative measures or models of exposure that include bioavailability could be applied. These could include models that utilize published BSAF values, organic matter content of soils, or cation exchange capacity.
Tier 3 evaluation	Information on bioavailability could be used to provide quantitative estimates of risk.
Focus and refine the Tier 2 evaluation. Collect additional site-specific data as needed. Perform detailed site study using quantitative methods. Are unacceptable risks present?	Measures and tests could include more sophisticated models of exposure that incorporate toxicant kinetics as well as measures of bioaccumulation in field or laboratory exposures. Biomimetic techniques (extraction procedures with solid phase extractants) could also be useful at this tier.
Response action program	To the extent that bioavailability information was used to assess risks, this information would subsequently be used to guide response actions.
Identify cost-effective means of achieving final corrective action goals, including combinations of remediation and natural attenuation. Implement the preferred alternative.	
Compliance monitoring	Monitoring could include measures of bioavailability. These could include biological tests, biomimetic procedures, and other methods.
Conduct monitoring program as needed to confirm that corrective action goals are satisfied.	

Integrating human and ecological risk assessment

Human and ecological risk assessments at a site are related in several ways:

1) Initial site investigations should consider the information needs for both human and ecological risk assessment; the conceptual model that guides these investigations should be general enough to consider ecological receptors and pathways along with human receptors and pathways.
2) Information such as exposure modeling that is used for assessing exposure to human receptors is often useful for assessing exposure to ecological receptors as well. Workplans developed for measuring or modeling exposure should consider the needs of both human and ecological risk assessment in order to minimize duplicative efforts.
3) Remedial decisions related to human and ecological receptors are interrelated in that remedial activities can pose risks to receptors or can affect future site use. These interactions can be evaluated through risk balancing discussed in the following section.

The role of natural or facilitated restoration

Ecological systems are dynamic: Receptors migrate into and out of systems, populations reproduce and grow, and dead animals and plants decompose and provide nutrients for subsequent generations. These natural processes can also affect chemical conditions within the soil. Some chemicals are also subject to degradation. In addition, the bioavailability of chemicals can be reduced with time as chemicals become more tightly bound to soils (Chapters 6 and 7). As a result, natural restoration can reduce the risks of released chemicals and may be a desired option if remedial actions are themselves judged to be destructive. The rate of natural restoration is a consideration when judging the role of this process for ameliorating ecological risks. In some cases, these rates may be judged to be too slow. Further, in some cases, there may be concerns that if action is not taken, the risks might increase due to continued release and spreading of chemicals or due to secondary risks such as erosion due to lack of plant cover.

Facilitated restoration may be an option for chemicals released to soils. These processes enhance the rate of degradation processes or reduce the bioavailability of chemicals. A number of remedial actions such as farming, fertilization, soil restoration, in situ bioremediation, and phytoremediation can facilitate the restoration of natural systems. In addition, planting vegetation or introducing soil invertebrates such as earthworms in areas where vegetation or soil invertebrates were lost as a result of the release can enhance restoration processes. Plants and soil invertebrates are vital to the integrity and health of soils. By establishing plants that have been lost, long-term risks associated with soil erosion in areas lacking vegetation are reduced.

Management considerations related to "balancing risks"

Risk balancing involves considering the risks associated with current and future conditions to human and environmental receptors together with the short-term and long-term risks associated with remedial actions. When considering remedial actions to reduce risks to either human or environmental receptors, the risk manager can examine how the actions themselves may affect receptors with respect to potential future uses of the site. For example, if an action being driven on the basis of a human health risk would result in the destruction of a natural resource, the managers may want to examine the merits of the action more closely. This could involve evaluating the technical basis for judging human health risks and alternative methods by which exposure might be reduced (e.g., by limiting access).

Suter et al. (1995) and Suter (1997) have discussed risk balancing for environmental receptors. These authors identified 4 main points:

1) The risks of the remedial actions must be balanced against the risks from the contaminants without remediation.
2) Risks to different ecological assessment endpoints may need to be balanced because remedial actions often damage all components of the ecosystem. Therefore, benefits to some endpoints must be balanced against injury to others.
3) Ecological risks may need to be balanced against human health risks.
4) Risk balancing requires good risk characterizations that provides temporal and spatial information on the nature of the risks.

Summary

Managing sites with contaminated soils requires considerations of a number of technical and nontechnical factors. The site manager relies on various sources of information, including input from risk assessors and stakeholders, to reach decisions concerning appropriate courses of action. Technically sound and well focused assessments of ecological risk are important for insuring that ecological resources are adequately considered and appropriately protected. The impact that this information has on site decisions is also related to how often and well communications occur between the assessors and site managers. It is helpful for risk assessors to understand this process and to be aware of the diverse factors that influence decisions. This helps the assessor develop a broader vision that enables him or her to identify appropriate assessment approaches, information needs, and communication procedures.

References

Akçakaya HR. 1998. RAMAS GIS: Linking landscape data with population viability analysis. Version 3.0. Setauket NY, USA: Applied Biomathematics.

Akçakaya HR, Atwood JL. 1997. A habitat-based metapopulation model of the California gnatcatcher. *Conserv Biol* 11:422-434.

Akçakaya HR, McCarthy MA, Pearce J. 1995. Linking landscape data with population viability analysis: Management options for the helmeted honeyeater. *Conserv Biol* 73:169-176.

[ASTM] American Society for Testing and Materials. 1998. Standard provisional guide for risk-based corrective action. Philadelphia PA, USA: ASTM. PS 104-98. 101 p.

Belfroid AC, Sijm DTHM, Van Gestel CAM. 1996. Bioavailability and toxicokinetics of hydrophobic aromatic compounds in benthic and terrestrial invertebrates. *Environ Rev* 4:276-299.

[BJC] Bechtel Jacobs Company. 1998. Empirical models for the uptake of inorganic chemicals from soil by plants. Oak Ridge TN, USA: U.S. Department of Energy. BJC/OR-133.

Burmaster DE, Anderson PD. 1994. Principles of good practice for the use of Monte Carlo techniques in human health or ecological risk assessments. *Risk Anal* 14:477-481.

Burmaster DE, Thompson KM. 1997. Estimating exposure point concentrations for surface soils for use in deterministic and probabilistic risk assessments. *Human Ecol Risk Assess* 3:363-384.

[CCME] Canadian Council of Ministers of the Environment. 1996. A framework for ecological risk assessment: General guidance. The National Contaminated Sites Remediation Program. Winnipeg MB, Canada: CCME. 108-4/10-1996e.

Clifford PA, Barchers DE, Ludwig DF, Sielken RL, Klingensmith JS, Graham RV, Banton MI. 1995. An approach to quantifying spatial components of exposure for ecological risk assessment. *Environ Toxicol Chem* 14:895-906.

Conder JM, Lanno RP. 2000. Weak-electrolyte extractions and ion-exchange membranes as surrogate measures of cadmium, lead, and zinc bioavailability to *Eisenia fetida* in artificial soils. *Chemosphere* 41:1659-1668.

Freshman JS, Menzie CA. 1996. Two wildlife exposure models to assess impacts at the individual and population levels and the efficacy of remedial actions. *Human Ecol Risk Assess* 2:481-498.

Gilbert RO. 1987. Statistical methods for environmental pollution monitoring. New York NY, USA: Van Nostrand Reinhold Co. 320 p.

Hattis D, Burmaster DE. 1994. Assessment of variability and uncertainty distributions for practical risk analyses. *Risk Anal* 14:713-730.

Hope BK. 1995. Ecological risk assessment in a project management context. *Environ Prof* 17:9-19.

Ingersoll CG, Dillon T, Biddinger GR. 1997. Ecological risk assessment of contaminated sediments. Pensacola FL, USA: SETAC. 389 p.

Jackson RB, Canadell J, Ehrlenger JR, Mooney HA, Sala OE, Schulze ED. 1996. A global analysis of root distributions for terrestrial biomes. *Oecologica* 108:489-411.

Karr JR. 1998. Rivers as sentinels: Using the biology of rivers to guide landscape management. In: Naiman RJ, Bilby RE, editors. River ecology and management: Lessons from the Pacific coastal ecosystem. New York NY, USA: Springer-Verlag. p 502-528.

Linder G, Ingham E, Brandt J, Henderson G. 1992. Evaluation of terrestrial indicators for use in ecological assessments at hazardous waste sites. Corvallis OR, USA: USEPA, Environmental Research Laboratory. EPA-600-R-92-183. 206 p.

Litaor MI, Thompson ML, Barth GR, Molzer PC. 1994. Plutonium-239+240 and Americium-241 in soils east of Rocky Flats, Colorado. *J Environ Qual* 23:1231-1239.

Lussier SM, Walker H, Pesch GG. (in press). Environmental restoration and protection strategies at multiple scales in Rhode Island watersheds. New England, USA: USEPA, Region 1.

MacIntosh DL, Suter II GW, Hoffman FO. 1992. Model of PCB and mercury exposure of mink and great blue heron inhabiting the off-site environment downstream from the U.S. Department of Energy Oak Ridge Reservation. Oak Ridge TN, USA: Oak Ridge National Laboratory, Environmental Sciences Division. ESD Publication 3934, ORNL/ER-90.

MacIntosh DL, Suter II GW, Hoffman FO. 1994. Uses of probabilistic exposure models in ecological risk assessments of contaminated sites. *Risk Anal* 14:405-419.

McDaniel T, Axelrod LT, Slovic P. 1995. Characterizing perception of ecological risk. *Risk Anal* 15:575-588.

Menzie C, Burke AM, Grasso D, Harnois M, Magee B, McDonald D, Montgomery C, Nichols A, Pignatello J, Price B, Price R, Rose J, Shatkin JA, Smets B, Smith J, Svirsky S. 2000. An approach for incorporating information on chemical availability in soils into risk assessment and risk-based decision making. *Human Ecol Risk Assess* 6:479-510.

Menzie C, Heiger-Bernays W, Montgomery C, Edwards D, Pauwels S. 1996. A framework for biological and chemical testing of soil. In: Linz DG, Nakles DV, editors. Environmentally acceptable endpoints in soil. Risk-based approach to contaminated site management based on availability of chemicals in soil. Annapolis MD, USA: American Academy of Environmental Engineers. p 387-465.

Menzie C, Henning MH, Cura J, Finkelstein K, Gentile J, Maughan J, Mitchell D, Petron S, Potocki B, Svirsky S, Tyler P. 1996. Special report of the Massachusetts weight-of-evidence workgroup: A weight-of-evidence approach for evaluating ecological risks. *Human Ecol Risk Assess* 2:277-304.

Menzie CA, Burmaster DE, Freshman JS, Callahan CA. 1992. Assessment of methods for estimating ecological risk in the terrestrial component: A case study at the Baird & McGuire Superfund site in Holbrook, Massachusetts. *Environ Toxicol Chem* 11:245-260.

Moore DRJ, Biddinger GR. 1996. The interaction between risk assessors and risk managers during the problem formulation phase. *Environ Toxicol Chem* 14:2013-2014.

Munger S, Toll J. 2002. Duwamish River and Elliott Bay water quality assessment. A report. The results of a study assessing pollution in the Duwamish River and Elliott Bay. Seattle WA, USA: King County Department of Natural Resources. Available at http://dnr.metrokc.gov/wir/waterres/wqa/wgpage.htm.

Norton SB, Rodier DJ, Gentile JH, van der Schalie WH, Wood WP, Slimak MW. 1992. A framework for ecological risk assessment at the USEPA. *Environ Toxicol Chem* 11:1663-1672.

[ODEQ] Oregon Department of Environmental Quality. 1998. Guidance for ecological risk assessment, level III, appendix A: Procedure for performing a population-level ecological risk assessment. Portland OR, USA: Waste Management and Cleanup Division. 29 p.

Sample BE, Aplin MS, Efroymson RA, Suter II GW, Welsh CJE. 1997. Methods and tools for estimation of the exposure of terrestrial wildlife to contaminants. Oak Ridge TN, USA: Oak Ridge National Laboratory, Environmental Sciences Division. ESD Pub Nr 4650, ORNL/TM-13391.

Sample BE, Opresko DM, Suter II GW. 1996. Toxicological benchmarks for wildlife: 1996 revision. Oak Ridge TN, USA: Oak Ridge National Laboratory. ES/ER/TM-86/R3.

Sample BE, Suter II GW, Beauchamp JJ, Efroymson RA. 1999. Literature-based bioaccumulation models for earthworms: Development and validation. *Environ Toxicol Chem* 18:2110-2120.

Sauvé S, Dumestre A, McBride M, Hendershot W. 1998. Derivation of soil quality criteria using predicted chemical speciation of Pb^{2+} and Cu^{2+}. *Environ Toxicol Chem* 17:1481-1489.

Sheppard SC, Evenden WG. 1992. Bioavailability indices for uranium: Effect of concentration in eleven soils. *Arch Environ Contam Toxicol* 23:117-124.

Spromberg JA, John BM, Landis WG. 1998. Metapopulation dynamics: Indirect effects and multiple distinct outcomes in ecological risk assessment. *Environ Toxicol Chem* 17:1640-1649.

Suter II GW. 1993. Ecological risk assessment. Chelsea MI, USA: Lewis Publishers. 538 p.

Suter II GW. 1996. Abuse of hypothesis testing statistics in ecological risk assessment. *Human Ecol Risk Assess* 2:331-347.

Suter II GW. 1997. A framework for assessing ecological risks of petroleum-derived materials in soil. Oak Ridge TN, USA: Oak Ridge National Laboratory. Report ORNL/TM-13408. 78 p.

Suter II GW, Cornaby BW, Hadden CT, Hull RN, Stack M, Zafran FA. 1995. An approach for balancing health and ecological risks at hazardous waste sites. *Risk Anal* 15:221-231.

Tang WC, Alexander M. 1999. Mild extractability and bioavailability of polycyclic aromatic hydrocarbons in soil. *Environ Toxicol Chem* 18:2711-2714.

Tang WC, Robertson BK, Alexander M. 1999. Chemical extraction methods to estimate the bioavailability of DDT, DDE, and DDD in soil. *Environ Sci Technol* 33:4346-4361.

[USEPA] U.S. Environmental Protection Agency. 1991. Risk assessment guidance for Superfund: Volume 1 – Human health evaluation manual (Part C, Risk evaluation of remedial alternatives), Interim. Washington DC, USA: USEPA, Office of Research and Development. EPA-540-R-92-004.

[USEPA] U.S. Environmental Protection Agency. 1992a. Developing a work scope for ecological assessments. Washington DC, USA: USEPA, Office of Solid Waste and Emergency Response. Publication 9345.0-05I. Eco Update Intermittent Bulletin 1(4).

[USEPA] U.S. Environmental Protection Agency. 1992b. Supplemental guidance to RAGS: Calculating the concentration term. Washington DC, USA: USEPA, Office of Solid Waste and Emergency Response. Publication Directive 9285.7-08I.

[USEPA] U.S. Environmental Protection Agency. 1993. Wildlife exposure factors handbook. Volumes I and II. Washington DC, USA: USEPA. EPA-600-R-93-187a.

[USEPA] U.S. Environmental Protection Agency. 1997a. Ecological risk management guidance for Superfund: Process for designing and conducting ecological risk assessments, Interim Final. Edison NJ, USA: USEPA, Environmental Response Team. EPA-540-R-97-006.

[USEPA] U.S. Environmental Protection Agency. 1997b. Guiding principles for Monte Carlo analysis. Washington DC, USA: USEPA, Risk Assessment Forum. EPA-630-R-97-001.

[USEPA] U.S. Environmental Protection Agency. 1998. Guidelines for ecological risk assessment. Washington DC, USA: USEPA, Risk Assessment Forum. EPA-630-R-95-002F.

[USEPA] U.S. Environmental Protection Agency. 2000. Ecological soil screening level guidance. Draft. Washington DC, USA: USEPA, Office of Emergency and Remedial Response.

Van Gestel CAM, Ma W-C. 1988. Toxicity and bioaccumulation of chlorophenols in earthworms in relation to bioavailability in soil. *Ecotoxicol Environ Saf* 15:289-297.

Washington-Allen RA, Sample BE. 1997. Determination of the spatial risk to wildlife from dispersed contaminants on the Oak Ridge reservation. Oak Ridge TN, USA: Oak Ridge National Laboratory. ES/ER/TM-229.

Wells JB, Lanno RP. 2001. Passive sampling services (PSDs) as biological surrogates for estimating the bioavailability of organic chemicals in soil. In: Greenberg BM, Hull RN, Roberts Jr MH, Gensemer RW, editors. Environmental toxicology and risk assessment: Science, policy, and standardization-implications for environmental decisions: 10th Volume. West Conshohocken PA, USA: ASTM. STP 1403. p 363.

CHAPTER 3

Soil-Quality Guidelines

George E. Nason, Tina M. Carlsen, Philip B. Dorn, John Jensen, Susan B. Norton

Introduction

Contaminated soil is most often evaluated and managed using a risk assessment–risk management (RA–RM) approach (Chapter 2). An appropriate suite of scientific tools to be applied within an RA–RM framework is an important element in a societal response to contaminated soil. Many countries have now developed or adopted numerical soil-quality guidelines (NSQGs) as one such risk-based tool. NSQGs have been developed on the basis of human or ecological health protection or both, although we do not address human health protection aspects in this chapter. Specific applications or roles for NSQG vary, but all participating jurisdictions support their use in the early evaluation of contaminant concentration data and screening or prioritization of site-management actions. Some jurisdictions extend the use to support development of remediation goals in a tiered site-assessment approach. This chapter provides an overview of recent NSQG efforts and discusses considerations in the design, construction, and application of soil-quality guidelines (SQGs) within an overall site assessment framework. We describe challenges and opportunities in both the development and application of NSQGs to the management of soil contamination. Owing to space restrictions, we consider only the development and use of guidelines in the context of existing contamination.

Labels, definitions, and intended uses for NSQGs vary from jurisdiction to jurisdiction. Common labels include criteria, guidelines, screening levels, action or alert levels, standards, and benchmarks. Since there is no intention here to endorse a particular system or jurisdiction, we have chosen to define a neutral working definition that is not, to our knowledge, in common use in any jurisdiction. As used in this chapter, NSQGs are substance-specific benchmarks intended to indicate a particular level of environmental protection, risk, or suitability for use. With this working definition, our intention is to explore challenges and potentials for these risk-based tools. The working definition chosen is not intended to imply or recommend legal obligation or enforceability within any regulatory context.

Contaminated Soils: From Soil–Chemical Interactions to Ecosystem Management. Roman P. Lanno, editor.

History of development and use

Numerical guidelines for gauging the quality of water have been in use for many decades. In contrast, SQGs for contaminants were first introduced by the Netherlands in 1983 ("Dutch List"). The Dutch List was prepared in response to the discovery of drinking water contamination caused by chemicals leaching from buried industrial wastes. Guidelines for a relatively small number of contaminants were presented as a system of 3 levels. "A-values" described upper limits for geochemical background or measurement thresholds for compounds not normally found in soil. Concentrations below the A-value were interpreted as non-contaminated. "B-values" were designed to indicate a contamination level that required further investigation, whereas concentrations between A and B were contaminated but probably not of concern. "C-values" were triggers for remedial action, which could be to remediate back to the A-value (Visser 1993).

Within a few years, the Dutch List had been adopted by other countries where the interpretations and numerical values were altered to meet the needs of adopting countries (Siegrist 1989). As well, other countries began new efforts to develop SQGs under various policy and management premises. By the early 1990s, several countries and jurisdictions within countries had developed, adopted, or adapted NSQGs as one means of evaluating and managing soil contamination (Visser 1993). The predominant theme in these early efforts was to design guidelines to express an acceptable level of environmental quality within a defined set of land-use expectations.

These early NSQGs tended to have sparse documentation describing their development or recommended application. To a large extent, the numerical values were based on professional judgment. Governments, industry, and other stakeholders were keen to improve the scientific basis for NSQGs and support them with explicit documentation on their development and written guidance for their application. Many jurisdictions now have released NSQGs with a basis in risk analysis (Table 3-1), meaning that values were derived through nomination of a tolerable exposure to organisms or other ecosystem components through direct contact with the contamination or along defined transport and exposure pathways. The process essentially involves a reverse risk assessment wherein the risk characterization is developed and fixed and acceptable media concentrations are back-calculated through relevant exposure pathways.

Features of current and emerging numerical soil-quality guidelines

Development of such risk-based NSQGs requires the elaboration of 1 or more exposure scenarios and accompanying sets of endpoints that together define expectations for the environmental "performance" of the site. This performance is tied, consciously or otherwise, to present and future land-use expectations. Recently, these performance and expectation issues have been described under the notion of "services" provided to ecosystems and society (Daily et al. 1997). The

Table 3-1 Overview of ecologically based contaminated site guidance that includes NSQGs

Jurisdiction	Use tiered approach	Use risk-based approach	Use checklists[a]
California	✓	✓	
Illinois	✓	✓	
Louisiana	✓	✓	✓
Texas	✓	✓	✓
USEPA[b] Region 4	✓	✓	✓
USEPA Region 5	✓	✓	
USEPA Region 6	✓	✓	
USEPA HQ 1997	✓	✓	✓
CCME 1997	✓	✓	✓
OMEE 1996	✓	✓	
BCE 1997	✓	✓	
Dutch A-C	✓	✓	
Denmark		✓	
EcoSSLs[c]	✓	✓	

[a] See "Additional Approaches to Numeric Soil-Quality Guidelines", p 109.

[b] USEPA = U.S. Environmental Protection Agency.

[c] EcoSSL = Ecological soil screening level (USEPA).

linkage of these values to the NSQGs development process is shown in Figure 3-1 and discussed further in "Developing Exposure Scenarios for Soil-Quality Guidelines" (p 86).

Risk-based NSQGs are often developed and presented for individual categories based on land use (e.g., OMEE 1996; CCME 1997) or implications for action (VROM 1994). In the former case, a unique exposure scenario is required for each land use supported, whereas the latter case demands well-defined differences in risk level or assessment endpoint. In both approaches, attempts are usually made to integrate information on toxic response across species and levels of organization. Development of exposure scenarios and considerations in support of assessment endpoints are presented in "Developing Exposure Scenarios for Numerical Soil-quality Guidelines" (p 86) and "Development of Numerical Soil-quality Guidelines" (p 94).

The utility and strength of NSQGs depend in large measure on the presence of an assessment framework that puts the tool into the context of a site investigation. Most jurisdictions provide an assessment framework of some type (Table 3-1). A desirable framework guides the user through the assessment process in an efficient way, presenting appropriate options as information from the contaminated site is collected and interpreted. Such a framework will assist in deciding whether a NSQG-based approach is appropriate, a rigorous site-specific risk assessment is justified, or a further action is needed or can be deferred.

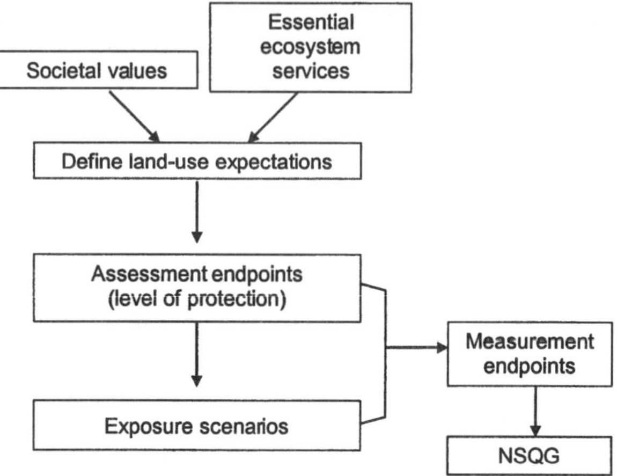

Figure 3-1 Influence of societal and ecosystem values on design of NSQGs

Tiering (Figure 3-2) is a key concept within the NSQG-based approach, which is consistent with an overall RA–RM paradigm (Chapter 2). In Tier 1, a user compares site concentrations to generic NSQGs to provide an indication of the potential for environmental risks. If no exceedances are found and certain other conditions are satisfied, the assessment process is exited. Where exceedances of NSQGs occur, a user may remediate or ascend to Tier 2 through collection and application of certain site-specific information that influences exposure and risk. Once adjustments have been made to NSQGs, the evaluation process is repeated to identify an appropriate choice among no further action, risk management actions including remediation, and further evaluation using site-specific ecological risk assessment (ERA). An overview of present and potential adjustments is furnished in "Adjusting Benchmark Values Based on Site-Specific Data" (p 99). A brief discussion of framework options for other assessment and management strategies and how these interrelate with NSQG-based approaches is presented in the section "Additional Approaches to Numerical Soil-Quality Guidelines" (p 109).

Soils and Society

Humans are almost entirely dependent on soils for their existence. We are physically supported by soil; it is the source of most of our food, and it ultimately provides many of the materials used for our clothing and shelter. Protection of the soil resource is thus not an option or frill, it is a societal imperative. Contamination by toxic substances is 1 of several stresses on the soil resource to which societies must respond.

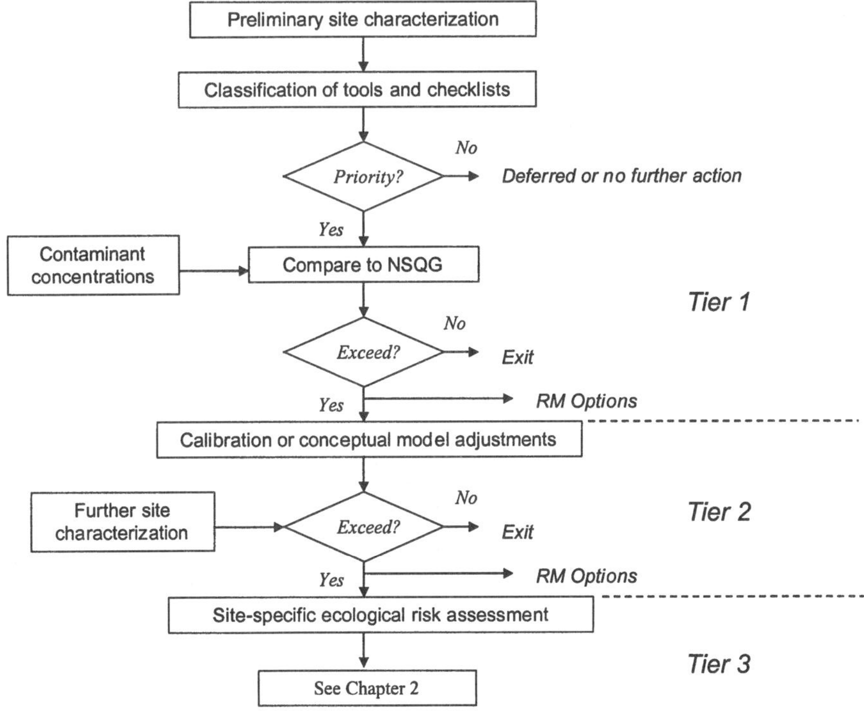

Figure 3-2 Generalized tiered framework for application of NSQGs

Protection of the soil resource, and mitigation where protection has failed, is complicated by the fact that land is bought, sold, and used in many ways (i.e., in many respects, it is handled as a commodity). Systems of land tenure are used to organize use patterns and define rights and responsibilities of parties tied to the land. These rights and responsibilities are expressed and seen in various ways, depending on the form of tenure (e.g., owned, leased, held in public trust). In addition, differing patterns of use, and potential or actual conflicts among these uses, have compelled societies to establish zoning systems, within which the range of uses and activities is restricted.

Land stewardship is a concept that, when followed, serves to minimize impacts on soils. Revered also is the concept of property rights that holds that individuals should be free to conduct whatever activities they see fit, so long as adjacent property owners are not adversely affected. This concept extends to care and control of contaminated lands and restricts the extent that management and regulation apply. Practically, land transfer, redevelopment, and off-site movement of contamination are key drivers for contaminated site assessment because

- future land owners expect to acquire lands in suitable condition to carry out the range of activities permitted under the relevant zoning,
- changing of land use to a more sensitive use may render tolerable contamination problematic, and
- contamination originating on 1 property that impacts an adjacent property indicates a failure of care and control that may require risk management activity.

The critical influence of land use and expectations of the soil resource on development and application of NSQGs is explored below.

Soil services

In western Europe and North America, the management of soils is primarily the responsibility of individuals and private entities. Despite an overall appreciation for the importance of soils in sustaining human societies, many people cannot readily articulate the specific benefits and services provided by or associated with soils. Soils are created by processes that operate on geological time scales, yet they can be depleted and irreversibly changed in a short amount of time (Daily et al. 1997). Remediation of soils can be expensive and ecologically disruptive (Milloy 1995). Because of the large implications in terms of money and sustainability, and the dispersed responsibility for soil management, it is particularly important to discuss and articulate the benefits and expectations associated with soils.

A clear understanding of soil services can direct the development, application, and underlying science supporting NSQGs. By articulating soil services and the expectations associated with soils in the context of different land uses, risk assessors and managers ensure that the right questions are being addressed and that NSQGs are applied appropriately. It is hoped that this discussion will help the scientific community identify and design lines of inquiry that will improve our methods for establishing NSQGs. The best science cannot contribute to guidance for protecting soil if we are not sure what it is about soil we are trying to protect and why.

Articulating soil services provides the basis for deriving management goals that can provide a strong foundation for risk assessment activities. Management goals are desired characteristics of ecological values that the public wants to protect (USEPA 1998). In the development of NSQGs, management goals are defined in generic ways that are intended to encompass most scenarios, yet must be specific enough to identify clear objectives for protection. Objectives more specific than "protect the environment" require considerable thought and discussion. Management goals are identified by considering the regulatory context (e.g., legislative mandates) and engaging interested parties (USEPA 1997). It is hoped the soil services discussed in this book will be useful for these broader policy discussions.

Although there may be others, major services provided by soils and soil ecosystems include

1) support of plant communities,
2) support of wildlife communities,
3) support of biodiversity,
4) buffering and modulation of the hydrological cycle,
5) protection of groundwater and surfacewater quality,
6) decomposition,
7) regulation of element or nutrient cycling, and
8) physical support.

As with any list like this, the soil services are highly interconnected and overlap substantially. We focus on the services of soils and soil ecosystems that may be affected by anthropogenic chemical contamination or different remedial alternatives and are of the most relevance to the issues discussed at this workshop.

Soils support plant communities, wildlife communities, and biodiversity. They support plant communities by providing suitable structural support by providing efficient and safe mechanisms for delivering nutrients, essential elements, and water; and by renewing soil fertility (Daily et al. 1997). Roots keep soil in place, thereby maintaining the presence of the most productive organic layers and also reducing siltation effects on water bodies. Soils provide habitat for many types of wildlife and shelter many components of the food web that wildlife depend upon. In addition, direct ingestion of soil can provide essential elements (e.g., salt licks) and the physical and biological components needed for food processing (e.g., gravel for bird crops, intestinal microorganisms for ruminants). Moen (1988) articulated expectations of soil, suggesting that soils of good quality should pose no harm to any normal use by humans, plants, or animals.

Soils also provide services that are more related to ecosystem functions. They buffer and modulate the hydrological cycle by holding rainwater and then releasing it over time to water bodies. This holding capacity is essential to plant growth and maintaining the structure of streams and rivers. When holding capacity of soils in a watershed is compromised, a chain of events may begin—flow extremes (flooding and low base flows) become more frequent, surface channels erode, sediments are released, and stream biota are impacted.

Moen (1988) suggested that soil of good quality does not contaminate other components of the ecosystem. Soils can also protect ground water and surface water from chemical spills by attenuating chemical movement through the sorption processes. Soil provides the habitat for microorganisms that perform a large variety of transformation processes.

Soil ecosystems play large roles in recycling organic wastes and in the closely related service of regulating element and nutrient cycling. In the process of decomposition, soil organisms render harmless many potential human pathogens in waste and in the remains of dead organisms, and degrade the metabolic waste products of terrestrial animals. Soils and soil ecosystems are important reservoirs of nutrients and elements. For example, the amount of C and N stored in soils exceeds that in vegetation by factors of 1.8 and 18, respectively (Daily et al. 1997). The importance of these reservoirs and the exchange rates between soils and other media is best appreciated when they are altered. For example, changes in the nitrogen cycle have been related to acid precipitation and the pollution of freshwater, estuarine, and marine ecosystems. Changes in the C cycle contribute to global climate change.

Finally, soils physically support plants, animals, and buildings. As with many of the soil services, physical support is most appreciated when it is lacking (e.g., earthquakes or sinkholes) or needs to be replaced (e.g., hydroponic systems).

In summary, soil services provide benefits to society by providing the physical structure for life; supporting the habitat and food for microorganisms, plants, and animals; regulating the cycling of organic material, nutrients, and other elements; and modulating hydrological cycles. Explicit consideration of these soil services can support the development of the most effective guidance associated with soil contamination by providing a focus for both societal dialogue and scientific inquiry.

Land uses and soil services

Management goals are usually identified within the context of a particular situation or group of situations. Soils can be managed in many contexts, ranging from almost total human influence (e.g., urban and industrial areas) to near total lack of influence (e.g., nature preserves). Conversely, the use of land for human endeavors is also associated with different expectations for soil services. The expectations associated with soils and the extent to which they provide different services can vary with the context and so will influence the management goals. These different management goals can then be reflected when NSQGs are developed and applied. In this section, these different management contexts for soil are grouped using land-use categories.

It has long been recognized that the characteristics of soil will influence the most appropriate use of different land parcels. In the U.S. and Canada, we do not expect all types of land to be equally important contributors to and beneficiaries of different soil services, although a multifunctional approach has been adopted by Denmark and the Netherlands (Visser 1993).

Table 3-2 summarizes a view of the general expectations of the soil services associated with different land uses. Table 3-2 is not comprehensive, and alternative groupings may be more useful for a particular numerical guideline application. It

Table 3-2 Expectations of the relative importance of the ability of land uses to support soil services

Beneficial use	Natural	Managed forests	Agricultural	Recreational or park	Residential	Commercial	Industrial
Physical support	+++[a]	+++	+++	+++	+++	+++	+++
Support of plant community	+++	+++	+++	+++	+++	+[b]	
Support of wildlife community	+++	++[c]	++	+	+		
Support of biodiversity	+++	++	+	+	+		
Buffering and modulation of the hydrological cycle	+++	+++	+++	+++	++	+	+
Protection of groundwater and surfacewater quality	+++	+++	+++	+++	+++	+++	+++
Decomposition	+++	+++	++	++	++	++	++
Regulation of elemental and nutrient cycling	+++	++	++	++	++	+	+

[a]+++ Very strong link between land use and service is expected.

[b]+ Some link between land use and service is expected.

[c]++ Strong link between land use and service is expected.

is intended to provide a starting point for discussions that take place at the early stages of developing NSQGs between the risk assessors, the management community, and the community of interested parties.

The land uses along the columns in Table 3-2 are arranged in roughly decreasing order of the strength of connection with soil services and increasing order of human influence. Natural areas such as wilderness areas and wildlife refuges are expected to support all of the soil services. We also have high expectations that managed forests will support most categories. Expectations regarding biodiversity and wildlife communities are somewhat lower because these areas are often managed for a particular plant species or community structure.

There are also high expectations for the existence of many soil services in agricultural lands and our managed park and recreational areas. It is expected to have the capability to grow plants. Also, it is expected that the soils will help buffer off-site impacts associated with hydrology (e.g., runoff) and on-site chemical contamination. While nutrient and elemental cycling and decomposition remain important, we also recognize that modern agricultural methods involve significant management of element cycles and sole reliance on natural soil processes is no longer the norm. The use of these lands to support wildlife and biodiversity likewise takes a lower priority, although we note that recent trends in agriculture point to the benefits derived by supporting biodiversity on these lands.

The expectations for residential land and managed recreational areas fall in the middle of the range of soil services. It is expected that these areas will support plant communities, but supporting biodiversity or wildlife areas will not be the principal land use. The increased impervious surface areas associated with these residential land (e.g., roofs, driveways, and parking lots) limit their ability to provide a buffer for hydrological or chemical impacts. The increased interest in designing best management practices (e.g., drainage swales, constructed wetlands) provides evidence that we still expect that these services will be provided.

Expectations for soil services within the most intensive categories of commercial and industrial land uses are the lowest. Still, it is expected that soils will be able to physically support human activities and that off-site impacts of any chemical spills will be buffered. Expectations for plant communities fall more into aesthetic uses such as landscaping. The benefits associated with plant communities in industrial areas are tempered by the potential risks associated with attracting wildlife to contaminated areas.

The identification of which soil services are most appropriate for a given site should consider potential future land uses as well as current land uses, and it is not always a straightforward task. Coordination with zoning bodies and stakeholder groups can help identify what future land uses are most likely. In addition, assessors should consider the land uses of adjoining parcels and the landscape context of the site. A commercial site located within a forested area may have very different expectations regarding its support of plant or wildlife communities than a similar site within an urbanized or industrial area.

The expectations of soil services are also influenced by the size of the area, and the loss of a soil service from a large area, no matter the designation of land use, is an issue of concern. The loss of services from a small area may be of less concern. However, even small areas may perform important roles in the landscape. For example, the loss of vegetation from a small riparian area may have large effects on water quality. In addition, the creation of sink areas or attractive nuisances for wildlife may have a larger effect than the absolute size of a parcel may suggest.

Another point to consider is whether application of NSQGs for commercial or industrial land use will result in ongoing impairment of soil services that may be difficult or technically impossible to relieve. One might expect that, over a sufficient period of time, a full suite of soil services can be returned to any commercial or industrial site. However, examples do exist where contaminated soil has been left as unusable for the foreseeable future (e.g., Times Beach, Missouri, USA). Whereas societies may be willing to see a certain diminution in soil services in commercial and industrial areas, they likely would not wilfully support a policy of irreversible impairment. Thus, there is a serious responsibility for developers of NSQGs to ensure that endpoints chosen for various land uses do not foreclose options for reuse by future generations. Principles of environmental stewardship are relevant in the development of NSQGs.

Levels of protection and assessment endpoints

Numerical soil-quality guidelines are developed by applying the principles of risk assessment and are improved by clearly identifying assessment endpoints. As with management goals, assessment endpoints are developed in a generic way for NSQGs, and they must apply to a variety of situations. In addition to assessment endpoints, the establishment of NSQGs requires that the level of protection be set generically prior to the application of the guideline values. This section discusses both of these issues.

Levels of protection

The level of protection associated with an NSQG identifies the probability, magnitude, and type of effects that are considered significant in a particular policy context. Levels of protection fall within the range of a very low probability that any effect will be associated with exposure to a stressor to a high probability that biologically significant effects will be observed. Selection of appropriate levels of protection is part of the risk management process and often occurs after the risks associated with a situation have been assessed. In the case of NSQGs, however, a level of protection must be selected a priori so that a specific number can be generated.

The levels of protection associated with NSQGs have been selected by considering their purpose (e.g., screening out low risk sites, identifying high risk sites) and, in some countries, by the type of land use. In Europe, NSQGs have been associated with several levels of protection (Visser 1993; Ferguson et al. 1998). The guideline calculated on the basis of the highest level of protection is used to identify sites that do not require further investigation. The guideline associated with a lower level of protection (i.e., a greater probability of effects) is used to identify sites that require intervention.

In Europe, land parcels are often expected to support a variety of land uses (i.e., multifunctional land-use policy in the Netherlands and Denmark), so NSQGs do not vary by land use. However, larger European countries such as Germany and France do provide for different land uses within their NSQG frameworks.

In the U.S. and Canada, most NSQGs are used to identify sites or chemicals that can be eliminated from further analysis. They are intended to be used with the limited data available at the early tiers of risk assessment. The uncertainty associated with limited data availability is partially addressed through attempts to bias sampling toward contaminated areas, but it is also addressed by using a high level of protection. The high level of protection minimizes the chances that sites or chemicals that truly pose a significant risk are not eliminated. In Canada, the services to be protected vary according to land use, and so the NSQGs vary (BCE 1996; OMEE 1996; CCME 1997).

In overview, land-use management goals vary among jurisdictions; smaller countries with high population densities favor multifunctional land-use scenarios, whereas larger, less populous countries, such as Canada, have developed NSQGs for various land uses.

In theory, a system could be developed where NSQGs vary by both the level of protection and land use. For example, values could identify concentrations corresponding to low or high levels of concern within different land-use categories (e.g., industrial, agricultural, or natural preserves). At the time of writing this book, however, we are not aware of any such applications.

Assessment endpoints

Assessment endpoints are explicit expressions of the environmental value that is to be protected. They are the highest values that can be assessed formally and provide the objective for the measurements and calculations used to estimate risks. Identifying assessment endpoints during the development of NSQGs can ensure that the final values comprehensively address soil services and are focused on the issues most relevant for managing risks. They provide a way to make the assumptions and objectives underlying the NSQGs explicit for users who may or may not have been involved in their development.

Three criteria are commonly used to identify assessment endpoints: ecological relevance, organism susceptibility, and relevance to management goals (Suter 1993). During NSQG development, these criteria must be considered in a way that will be applicable to a broad array of chemicals, regions, and scenarios. The criterion of relevance to management goals is addressed through soil services and land use, as discussed above. Considering the criteria of ecological relevance and susceptibility in a generic way also poses challenges. Soil systems vary considerably in different areas, for example, from the arid plains of central North America to the rain forests of the Pacific Northwest. In addition, different components of soil systems have differential sensitivity and exposure to different types of chemicals. Because NSQGs are applied across many different systems and are developed for many different types of chemicals, they may incorporate a broader range of variability and assessment endpoints than is appropriate for a particular site. The use of site-specific information to refine the application of NSQGs is discussed in "Adjusting Benchmark Values Based on Site-Specific Data" (p 99).

Table 3-3 provides a list of example assessment endpoints that may be useful for developing NSQGs, linked with the soil services discussed earlier. Listed are example measures that have been used by the European Union and Canadian NSQG efforts to illustrate how the assessment endpoints are addressed using different measures and models. The linkages between chemical contamination and these assessment endpoints can be complex, and each assessment endpoint may be addressed by several measures or models. Chapter 4 describes the different measurement options that are available and how they are linked with assessment endpoints and soil services.

Arraying soil services, assessment endpoints, and measures, as in Table 3-3, can provide important insights into areas that are addressed by NSQGs and areas that may require further development. For example, groundwater and surfacewater quality, decomposition rates, and many of the plant population and community endpoints are addressed through several different types of measurements. The relationship between chemicals and the ability of soils to maintain physical structure and mitigate extreme flow events appear to be areas that could benefit from further development.

Together with the levels of protection, the assessment endpoints provide the foundation for identifying and using measures of effect and exposure models to calculate NSQGs. NSQGs are developed separately from their application. Explicit statements of the intended level of protection and the assessment endpoints used during development will help ensure that NSQGs are applied and communicated appropriately.

Table 3-3 Example assessment endpoints associated with soil services

Soil service	Assessment endpoint	Example measures and models used to develop NSQGs
Physical support	Soil physical and chemical structure	Partially addressed by invertebrate and plant tests
Support biodiversity	Presence of taxonomic groups	Acute and sublethal invertebrate tests (e.g., earthworms, springtails, enchytraeids); also tests associated with wildlife and plant populations and communities
Support wildlife populations and communities	Abundance and age or size structure of wildlife populations	Toxicity tests with mammals and avian species and plants; modeling of steady-state partitioning in food web compartments
Support plant populations and communities	Abundance and age or size structure of plant populations	Emergence and growth tests with plants
		Foliar injury in plants
	Crop productivity	Acute and sublethal invertebrate tests (e.g., earthworms, springtails, enchytraeids)
	Growth and appearance of landscaping plants	
Regulation of element and nutrient cycles	Size and location of available and unavailable nutrient and element reservoirs	Functional tests (e.g., nitrification)
		Acute and sublethal invertebrate tests (e.g., earthworms, springtails, enchytraeids)
Decomposition	Decomposition rates	Cotton strip, litter bag, and carbon mineralization tests
		Acute and sublethal invertebrate tests (e.g., earthworms, springtails, enchytraeids)
Buffering of hydrologic cycle	Frequency of extreme flow events	Partially addressed by invertebrate and plant tests
Protection of water resources	Groundwater and surfacewater quality	Biodegradation tests, modeling to ground and surface waters

Developing Exposure Scenarios for Numerical Soil-Quality Guidelines

Identifying the exposure scenarios to be used in the development of NSQGs requires numerous assumptions concerning the conceptual model of the ecosystem, land use, and expected receptors within the ecosystem, as well as considerations of spatial and temporal scales. Because NSQGs will be used by site managers to potentially eliminate sites from consideration for future remedial action, or to downgrade the priority of these sites, conservative assumptions are necessarily built into the exposure scenarios used in NSQGs. However, making these assumptions explicit and transparent to the user of NSQGs may provide an opportunity to make adjustments to the CSM using readily obtained site-specific data to render the screening process more realistic for a given site without proceeding to a full-scale ERA.

Ecosystem conceptual models

Risk assessments on contaminants in ecosystems begin with a conceptual model of how an ecosystem works in order to identify potential pathways for exposure to contaminants. Such a conceptual model is also required during the development of NSQGs. This model is usually formulated in terms of the macro-spatial scale.

However, often what is not recognized is that these models involve an implicit assumption concerning ecosystem stability at the macro-temporal scale. When most people look out upon a landscape, they assume they observe a representation of what the ecosystem should look like, what it has always looked like, and what it should look like in the future. Such a view of a "stable" ecosystem assumes that when the system is stressed, it either will return to its original state when the stress is removed or will change unalterably. However, other models of ecosystem stability include adaptation to the stress in a new stable state (Holling 1986; Watt and Craig 1986; Wiens et al. 1986), as well as predictable or unpredictable cycling between stable states. The impact of contaminant stress on ecosystem stability may be even more complex in successional ecosystems or ecosystems that are adapted to periodic disturbance (Rice and Jain 1985; Walter and Chapin 1987; George et al. 1992). In these cases, it is unclear how ecosystem response to the additional stress induced by contaminants should be measured. Depending upon one's view of ecosystem stability, the goal may be to protect (preserve) the ecosystem in its state as observed at time = 0 or to preserve its ability to move between states. It is by far more tractable to assume ecosystem stability when developing NSQGs. One can assume, for the purposes of screening, that if ecosystem structure and function are preserved at time = 0, the potential for natural dynamics is also preserved. However, as we move towards the use of population-level and community-level methods of determining ecosystem response to contaminants in latter tiers of an ERA, assumptions concerning ecosystem stability must be revisited.

Conceptual models on how ecosystems work require knowledge of how these systems are structured and how they function. "Structure of an ecosystem" refers to the types and numbers of organisms in the system (composition) and how they relate to one another (trophic level, connectivity). Structure of an ecosystem is often viewed in terms of a food web, as discussed in Chapter 4. Knowledge of ecosystem structure is clearly important in terms of identifying potential exposure pathways to contaminants. "Function of an ecosystem" refers to the natural processes conducted by the resident organisms that are normally observed in unperturbed systems, such as nutrient cycling or primary production. The services provided by soil, as identified in "Soils and Society" (p 76), can be considered in terms of whether they support ecosystem structure or ecosystem function. The goal of NSQGs is to protect these soil services, irrespective of intended land use. Thus, there is a need to identify the receptors that are associated with the defined soil service and to identify potential exposure pathways to these organisms. By focusing on developing exposure pathways and NSQGs for the receptor most important to a defined soil service, the number of receptors or pathways that must be handled can be reduced but still be protective. It is clearly unreasonable to develop an exposure pathway and resulting NSQG for every organism within a system.

Receptor arrays and routes of exposure

Table 3-4 lists a set of potential receptors associated with the identified soil services. The receptors shown represent a generic set of receptor groups that can reasonably be expected to be seen in an ecosystem. It is important to have all major receptor groups represented, although not all groups may be observed in a given ecosystem. Again, a great deal of combining of receptors is required. In the scheme shown in Table 3-4, the starting point is the physical or chemical structure of the soil itself, and movement is from lower organisms to higher vertebrates. Aquatic organisms are included in recognition that transport of contaminants from contaminated soil through the subsurface to groundwater or surfacewater runoff could result in contaminated surface water. Lower organisms tend to be grouped together (i.e., microorganisms, soil invertebrates), whereas higher organisms are further subdivided. There are several reasons for this. First is the recognition that society tends to value individuals and populations of higher organisms to a greater degree than lower organisms. In other words, the structure of higher organisms is an important assessment endpoint with respect to public perception. Because of this, lower organisms are often considered to be more important with respect to ecosystem function versus ecosystem structure, as shown in Table 3-4. For these organisms, species substitution may play a major role in maintaining the functioning of an ecosystem. Species substitution has been recognized in certain cases where species have become extirpated from a system, but the function of the system is maintained through adaptation of new or remaining species (Moore and DeRuiter 1993). Certain soil invertebrates or microorganisms may be susceptible to soil contamination and may be reduced in total number or biomass, but the ecosystem process of concern may not be impacted due to the ability for other soil species to compensate and perform the function of the reduced or eliminated species (i.e., functional redundancy). However, there is a danger to this view. Soil microorganisms and soil invertebrates represent a substantial portion of the world's biodiversity, but due to the focus on function instead of community or ecosystem structure for these organisms, biodiversity for these receptors is ranked only somewhat important in Table 3-4. There is always the potential that the loss of species, even if function is momentarily retained, may have future implications for ecosystem stability, as well as unknown human uses in areas such as pharmaceuticals (Tilman 1997).

Higher vertebrates in Table 3-4 are broken into 5 groupings. These groupings are an attempt to reflect major differences in potential exposure pathways. Small resident land vertebrates tend to be fossorial, living directly in or on the soil, and to have very small home ranges. Larger vertebrates may also be resident in an area but have larger home ranges. These vertebrates may also den within the soil (foxes,

Table 3-4 Receptors associated with soil services

<————————————> <————————————>

– Not important, + somewhat important, ++ important, +++ very important

	<——Ecosystem structure——>			<——Ecosystem function——>				
Receptor	Biodiversity	Support wildlife	Support plant communities	Regulation of element or nutrient cycle	Buffering of hydrologic cycle	Protection of water resources	Decomposition	Physical support
Soil physical or chemical structure	+[a]	+	++[b]	++	+++[c]	+++	+	+++
Soil micro-organisms	+	+	++	+++	++	+++	+++	+
Soil invertebrates	+	++	+	++	++	++	++	++
Terrestrial plants	++	++	+++	++	++	++	+	+
Small resident land vertebrates	++	+++	+	+	–	–	+	–
Large resident land vertebrates	+++	+++	+		–	–	–	–
Migratory land vertebrates	+++	+++	+		–	–	–	–
Resident avian species	++	+++	+		–	–	–	–
Migratory avian species	+++	+++	+		–	–	–	–
Aquatic species	+++	+++	+	+	–	+++	+	–

[a]+ Some link between land use and service is expected.

[b]++ Strong link between land use and service is expected.

[c]+++ Very strong link between land use and service is expected.

coyotes) or live primarily upon the soil (grazing ungulates). Migratory land vertebrates may have only seasonal contact with the contaminated site, although this may be during critical life stages (raising young). Resident avian species are likely to have fairly small home ranges and may be either ground or tree nesters and predators or granivores. Migratory avian species, like migratory vertebrates, may have only seasonal contact with the area, but life stage consideration is necessary. This is only 1 example of a way to group vertebrate receptors; many others are possible. The important consideration is to ensure the grouping has some meaning with respect to exposure pathways, and use of the grouping in NSQGs will ensure sufficient protection to all organisms within the group.

Obviously, even within these vertebrate groupings, a decision concerning what type of vertebrate species and the pathways to be used will need to be made. Figure 3-3 shows a subset of the many species and pathways that could be considered for potential inclusion into the development of NSQGs. In many screening-level risk assessments and various efforts at developing NSQGs, a single to a few vertebrate species are selected as the most potentially exposed organisms or most toxicologically sensitive (CalEPA 1994; Carlsen 1996; CCME 1996a; USEPA 1998). Table 3-5 is an example of the types of exposure scenarios most typically assumed in the development of NSQGs. As discussed in Chapter 9, the consideration of a particular receptor and/or pathway should be dependent on the soil contaminant in-

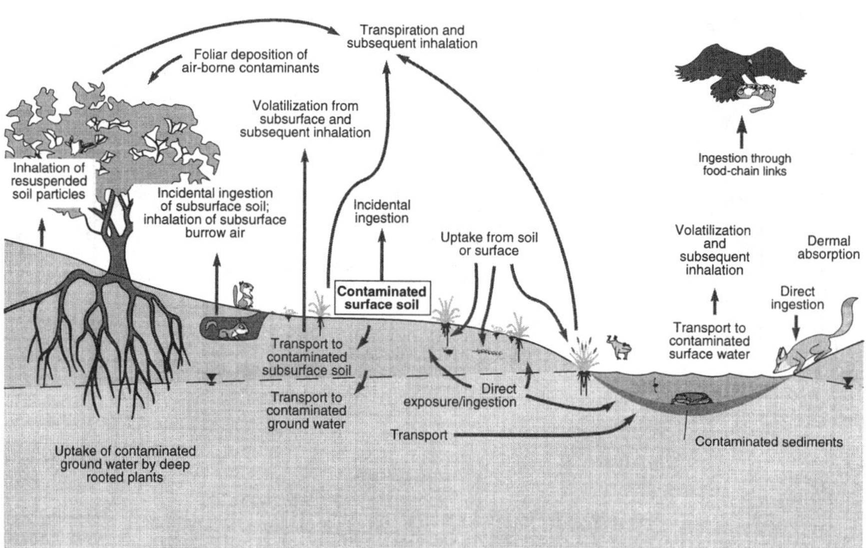

Figure 3-3 Potential ecological exposure pathways with contaminated soil (adapted from Carlsen 1997)

Table 3-5 Potential exposure pathways for consideration in the development of NSQGs

Receptors	Typical exposure pathways assumed in NSQGs
Soil microorganisms and invertebrates	Direct contact, direct ingestion
Terrestrial plants	Direct uptake, bioaccumulation
Vertebrate herbivore	Incidental ingestion of contaminated soil, direct ingestion of contaminated forage, bioaccumulation
Vertebrate carnivore	Incidental ingestion of contaminated soil, direct ingestion of contaminated prey, bioaccumulation

volved. For example, inhalation pathways of subsurface burrow air may be important for fossorial vertebrates in areas contaminated with volatile compounds (Carlsen 1996) but would not be important for soils contaminated with heavy metals. The vertebrate carnivore receptor and associated bioaccumulation pathways are important for known bioaccumulative compounds, such as certain highly chlorinated organic compounds (Moriarty 1999).

For many of the exposure pathways listed in Table 3-5, exposure models must be used to estimate final expected tissue concentrations for comparison to the toxicological literature. As discussed in Chapter 2, these models can range from the use of bioaccumulation factors estimated from known chemical properties (Travis and Arms 1988) to more sophisticated models, such as physiologically based pharmacokinetic models that account for bioavailability and depuration rates (Gerlowski and Jain 1983). Considerations for improving the estimates of final tissue concentration or body burden, and thus exposure, are discussed in Chapters 5 and 8. As new and better methods and models become available, their selection for use will depend on the quality of the data available to parameterize the model. If the data are available, the model providing the best estimate of true exposure should obviously be utilized. However, in current practice, the paucity of data requires the use of simplistic models. When this is the case, these models should be chosen to ensure the desired level of protection. For now, this means the resulting estimates of uptake and exposure will necessarily be conservative.

Spatial scales and contexts

Scale is an important consideration in the context of the spatial distribution of the contaminants in the soil, both vertically and horizontally, on macro- and microscales; the spatial distributions of home ranges of ecological receptors; and the spatial scale of the population of the ecological receptor. Because of their intended use, exposure scenarios used in the development of NSQGs are necessarily conservative. The ecological receptor of interest is often assumed to reside in the area of contamination for 100% of the time. This may be unrealistic with respect to sites with a small areal extent of contamination compared to the size of home ranges of many vertebrate species. In addition, with the exception of endangered species, it is largely recognized that the unit of protection for most plant and vertebrate species is the population (USEPA 1989; Suter 1993; Moriarty 1999). Thus, at least in wild populations, the loss of 1 animal is typically not significant, provided the overall population is not impacted. One simplistic way to address this is to eliminate from consideration sites that are small compared to a given species population boundary. Figure 3-4A shows an example of such a site. Here, simple comparison of site size to overall population suggests impact from the site would be negligible. However, the danger in setting a minimum site size is illustrated in Figure 3-4B. In this case, several small sites, perhaps owned by separate parties, occur within the population boundary for the given species. In this case, there is a high potential that the cumulative effect of these sites may have a significant impact on the species.

For this reason, the elimination of sites based solely on size is not necessarily protective. Should the ability to eliminate such small sites be desired, an infrastructure would need to be developed and put into place to identify and track such sites to ensure cumulative population impacts are not significant. This would include the determination of important population parameters (e.g., population age structure, population rate of increase) for use in models developed by the conservation biology community (Burgman et al. 1993) to ensure the loss of X number of animals at Y life stage does not significantly impact the population. For the generation of generic NSQGs, the conservative assumption of 100% residence of the species on the site is necessary. However, should a site such as that illustrated in Figure 3-4A exceed a generic NSQG, several options are available. The section "Adjusting Benchmark Values Based of Site-Specific Data" (p 99) presents suggestions for modifying the assumptions on which the NSQGs are based. Chapter 2 discusses concepts involving local populations that can be used in Tier 2 or 3 assessments that could provide further spatial context with respect to the wildlife population of interest.

Temporal scales and contexts

Temporal scales are important in ERAs because time plays a role in the rate of release of a contaminant into the environment, transformation, sequestration, and biodegradation of the contaminant within the environment, as well as uptake and metabolism of the contaminant by ecological receptors. In general, contaminant concentration, particularly the bioavailable fraction, will be reduced over time (Chapters 5 through 9). Such considerations, although seldom made in the past, are important to include in higher-tier risk assessments. However, because of their intended use, NSQGs must be designed to provide for a certain level of protection to the ecosystem for the contaminant at time = 0.

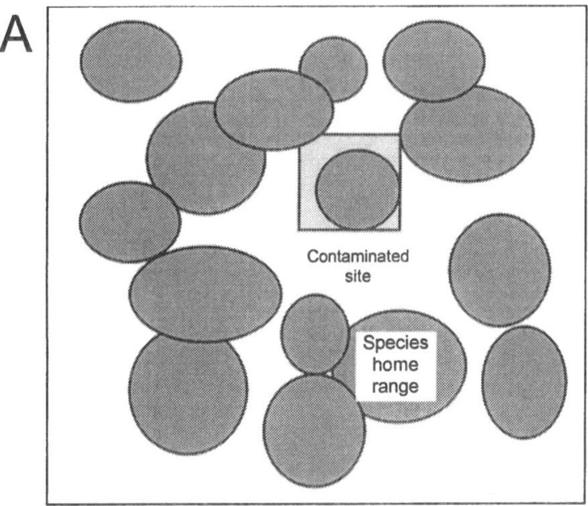

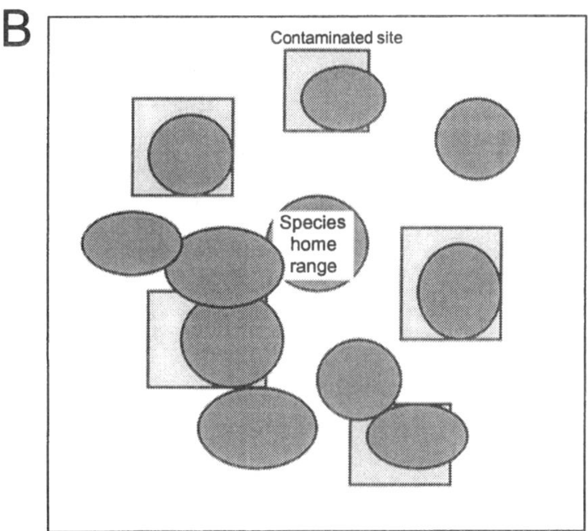

Figure 3-4 Relationship between species' home range and areal extent of contamination as a determinant of exposure

Temporal scales can also influence exposure pathways. Diet can change seasonally, numbers of individuals within a population can change over time, home ranges can fluctuate. Again, such considerations would be incorporated into higher-tier risk assessments, while the conservative, protective assumption of static pathways will need to be made for development of NSQGs.

Development of Numerical Soil-Quality Guidelines

Terrestrial ecotoxicology has seen rapid development in the last decade. Ten years ago, earthworm and plants tests were more or less the only terrestrial ecotoxicological tests available for developing NSQGs for the soil environment. However, today a number of tests exploring effects on soil invertebrate and soil function processes are available (Chapter 4). Relevant tests are hence available for deriving NSQGs for several soil services and assessment endpoints (Table 3-6).

The methods make use of toxicity and uptake data for terrestrial species, that is, soil microorganisms or functions, soil invertebrates, flora, and wildlife. In some countries (e.g., the Netherlands), NSQGs have occasionally been developed on the basis of background heavy metal concentrations or on the basis of aquatic toxicity data extrapolated to soil conditions using equilibrium partitioning theory.

Numerical soil-quality guidelines development methods typically specify a preference for information from relatively well-defined measurement endpoints, preferably no-observed-effect concentration (NOEC), lowest-observed-effect

Table 3-6 Examples of measurement endpoints used for developing NSQGs and their association with soil services

Soil services	Examples of relevant biological tests
Physical support	
Support biodiversity	Acute and sublethal invertebrate tests, e.g., earthworms, springtails, enchytraeids
Support wildlife	Toxicity tests with rodents and other mammals, avians, phytotoxicity tests, bioaccumulation studies
Support plant communities	Emergence and growth tests with plants
Regulation of element and nutrient cycles	Functional tests, e.g., nitrification
Decomposition	Earthworm tests, cotton strip, litter bag, carbon mineralization
Buffering of hydrologic cycle	Earthworm tests
Protection of water resources	Biodegradation tests

concentration (LOEC), ECx, or LC50 values. The accuracy of NOEC and LOEC values depends on the number of concentrations used and the variability of the data investigated. Thus, laboratory-derived NOEC and LOEC values do not necessarily represent "true" no-effect concentrations and lowest-effect concentration values. An alternative approach is the use of ECx point estimation or the "bounded effect concentration" (Hoekstra and Van Ewijk 1993). Some of the advantages of the ECx estimations compared to NOEC values are that

- the values are interpolated and less sensitive to the choice of test concentrations,
- the precision can be quantified with confidence intervals,
- the ECx values are statistically comparable,
- the NOEC values may correspond to large effects due to large variations in a data set,
- the toxic responses of organisms may be characterized if a model can be fitted to the data.

It is essential to consider the assessment endpoint or soil services prior to using the toxicity data. Although alternate endpoints are being developed, mortality is still the most common measurement endpoint among the existing toxicity data for soil invertebrates and terrestrial wildlife. This is important because it appears inappropriate to use EC50 or LC50 values as measurement endpoints if the intended aim was to protect the multifunctional use of soils. At present, when median effects data are used, it is necessary to apply safety factors to reach the desired protection level for sensitive land uses.

Numerical soil-quality guidelines have been derived and published by a number of western countries or institutions (VROM 1994; CCME 1997; Crommentuijn et al. 1997; Jensen et al. 1997; see also Table 3-1). Ideally, NSQGs should be developed for all relevant assessment endpoints connected to the soil environment, but this is seldom done. Most countries do, however, consider several receptors (e.g., soil functions, soil invertebrates, plants, and wildlife) and then develop an NSQG on the basis of the governing receptor–pathway combination. This collective value is typically the one that is found in look-up tables used as decision tools by risk assessors. A single generic guideline value based on the most sensitive receptor within a range of assessment endpoints indicates a need for a diligent assessment of site factors to determine the relevance of the limiting receptor–pathway. Some systems provide summaries of individual or grouped responses in order to assist the site assessor in this regard.

For soil-dwelling organisms or soil processes, different approaches have been used. These generally consist of 1 or more of the following methods: The NSQGs may be developed using statistical extrapolation methods based on a sensitivity distribution assumed to be log-logistic (e.g., the Netherlands, U.S.) or lognormal (e.g., Denmark, Germany) (Wagner and Løkke 1991; Aldenberg and Slob 1993). In

Canada, when sufficient data exist, SQGs for soil-dwelling organisms are derived by a weight-of-evidence approach. This method uses a percentile (25%) of the effect data set, or a combined effect and no-effect data set, to estimate a concentration in the soil expected to cause, respectively, a tolerable risk or no adverse ecological effects to soil-dwelling organisms such as plants, invertebrates, and microbes. The SQGs for agricultural and residential land uses may be derived on the basis of the 25th percentile of all available data (NOEC, LOEC, and EC50 or LC50 values), whereas the guideline value for commercial and industrial areas may be derived on the basis of the 25th percentile of all effect values.

In cases where data are insufficient for a statistical extrapolation or weight-of-evidence method, NSQGs may be developed by applying safety factors and/or assessment factors to the data set. For a discussion of the use of safety factors or assessment factors, the reader is referred to Chapman et al. (1998). In other cases, the NSQGs are developed in relation to background values or by a critical expert judgment of available data. Depending on the required protection level, different values may be derived using the same data set, for example, the Dutch target values and intervention values are set on a level aimed at protecting 95% and 50% of the species in a soil ecosystem, respectively.

In most countries, wildlife is considered when deriving benchmarks. The development of an NSQG for wildlife is typically based on simple assumptions regarding exposure to soil contaminants via soil and/or food uptake. Two broad areas of uncertainty are associated with the development of NSQGs for wildlife: one dealing with threshold effect doses (e.g., species sensitivity, long-term effects, population and community effects) and another one associated with a number of processes controlling exposure (e.g., bioavailability, bioconcentration factors, food preferences, rate of food intake).

Numerical soil-quality guidelines may typically be developed for several receptors, such as soil-dwelling organisms, soil functions, plants, and wildlife. Generic values may be set for each of these, or as observed in Canada, Denmark, and the Netherlands, as 1 single value based on the lowest-observed value. The assessment endpoint may not necessarily be the same for all land uses. In Canada, for example, the NSQG for wildlife is developed only for agricultural and residential or parkland uses, whereas toxicity to plants and soil-dwelling species is considered for all land uses. Because NSQGs may be used for different purposes, it is essential that the database, the level of protection, and the general rationale behind the development of the NSQGs is explicitly stated and made available. This will provide the assessor with the necessary information for judging whether the NSQGs fulfill the requirements as to protection level and assessment endpoints for the current purpose.

Improving numerical soil-quality guidelines

Numerical soil-quality guidelines are derived by extrapolating results from single-species toxicity tests in the laboratory to soil concentrations in the field that, with a certain level of protection, aim to prevent adverse effects in soil ecosystems. However, in many cases, there are insufficient data available to evaluate the toxicity of a chemical and develop an NSQG. Even when sufficient data exist, they may not be of a form that allows a refined assessment (e.g., missing data on soil type and soil chemistry such as organic matter, clay content, and pH). In other cases, insufficient information is available regarding relevant bioaccumulation factors (BAFs), depreciating the assessment of risk to wildlife as a result of bioaccumulation or biomagnification. Bioaccumulation factors may be very site specific, changing according to chemical concentrations and soil physicochemical characteristics. Typically the highest BAFs are found at concentrations close to background levels. Therefore, site-specific measurement of BAFs may lead to an adjustment of the NSQGs.

Several problems, uncertainties, and limitations are associated with deriving NSQGs. However, not all of them have the same importance. First, as presented in Chapter 4, the significant differences between bioavailability of chemicals in laboratory tests and field conditions make a reliable comparison of numerical soil guideline values with specific field situations difficult. Most of the toxicity data available do not include information on exposure concentrations other than total chemical levels. Exposure estimates incorporating some measure of chemical bioavailability, such as porewater chemical levels or critical body residues, are lacking. In many cases, the data do not even contain information about soil type or soil conditions controlling bioavailability. Because most toxicity tests are conducted using soluble salts, bioavailability is assumed to be high, at least initially. However, rapid sorption decreases bioavailability quickly, and in chronic tests an assumption of high bioavailability throughout the test is probably erroneous, with actual bioavailable exposure concentrations decreasing during the course of the test.

Without a doubt, establishing linkages between bioavailability, toxicant uptake, and toxicity to soil biota, soil processes, or wildlife will significantly improve the predictive value of laboratory toxicity tests and hence, the assessment of soil quality. Therefore, the expression of NSQGs as a bioavailable chemical level (e.g., $CaCl_2$-exctractable metal concentrations, porewater concentrations) should be prioritized. In some cases, normalizing NSQGs to soil organic matter or clay content and/or pH may also improve guideline interpretation. A change in the units of expression for NSQGs will require that future toxicity tests include measurements of exposure expressed as some measure of bioavailability in addition to total chemical concentrations or that it will be possible to identify and quantify the influence of the most critical parameters in soil controlling bioavailability. This information will also be needed to allow site and bioassay conditions to be compared on a common basis.

Bioavailability is not the only issue that may greatly vary among contaminated sites. Other factors influencing the reliability of extrapolating data from laboratory to field includes additive, antagonistic, or synergistic effects of toxicants in mixtures, multiple stressors, interspecies differences, adaptation, and ecological recovery. Some of these issues are briefly addressed below.

Mixtures of chemicals

The presence of more than 1 pollutant is the common situation at most contaminated sites. However, toxicity tests are usually conducted with single chemicals. The toxic response to a mixture of chemicals may be antagonistic, synergistic, or additive depending on the concentration and mode of action of the different chemicals. For soil-dwelling organisms, there is no evidence for synergistic effects of chemicals being a common situation in lab or field (Posthuma et al. 1996, 1997; Van Gestel and Hensbergen 1997). Unless available information suggests the presence of chemical properties likely to cause synergistic or antagonistic effects in chemical mixtures, an assumption of additivity of toxicity is assumed to offer sufficient protection.

Interspecies differences

How representative are the available test organisms or processes? Heimbach (1985) concluded, after comparing the toxicity of 23 pesticides with *Lumbricus terrestris,* that *Eisenia fetida* could be used as a representative species for testing the toxicity of chemicals to earthworms. Edwards and Coulson (1992) also concluded, on the basis of a large review of literature data, that the compost worm was suitable for an initial screening of chemicals. According to the review by Edwards and Coulson (1992), an application of a factor 10 would bring *E. fetida* in line with the most sensitive species, normally *Apporectodea caliginosa*. Chapman et al. (1998) reviewed a number of reports investigating the difference in wildlife sensitivity to chemicals. In general, the sensitivity of birds and mammals to pesticides was within a range of 10 and 20, respectively. However, larger differences have been observed in plants.

Combinations of stresses

Laboratory experiments usually are conducted under standardized conditions to ensure optimal survival, growth, and reproduction of the test animals. In nature, however, organisms may be confronted with large fluctuations in environmental conditions. Large variations in temperature, humidity, food supply, or predation may exist throughout the year. It has been observed that chemical toxicity to soil organisms increases under suboptimal conditions, such as food shortage (Bengtsson et al. 1985; van Beelen and Doelman 1997) or drought (Demon and Eijsackers 1985). Observations have also been made indicating that chemical stresses influence the drought tolerance or cold hardiness of organisms (Zacharariassen and Lundheim 1995; Holmstrup 1997).

Although not insignificant, the uncertainties discussed above generally are considered to be less important than changes in chemical bioavailability due to contaminant aging and weathering in soil and the shortage of relevant toxicity data or measurement endpoints for several receptors. In the same manner as bioavailability, these other factors are often site specific and not easily incorporated into the derivation of generic NSQGs. However, in most cases, they can be taken into consideration and incorporated in a site-specific assessment or modification of the guideline value. This may be done by using native species or mimicking local conditions in site-specific bioassays.

Taking into consideration all these and other uncertainties in the extrapolation from laboratory to field, it is clear that NSQGs are useful yardsticks in a first-tier approach indicating whether further assessment or action is needed. However, first-tier comparisons should be viewed with caution—it should be possible to increase the accuracy of assessment through the application of appropriate site information in a second tier.

Adjusting Benchmark Values on the Basis of Site-Specific Data

Numerical soil-quality guideline values for soils are developed generically with intended use in the broadest applications and are conservative wherever the guideline development goal has been to be protective of most soils in most applications. Conservatism results from the development of a sensitive exposure scenario, as discussed in "Development of Numerical Soil-Quality Guidelines" (p.94). Depending upon the specific calculation procedures and checks used, benchmarks may be below soil background concentrations for some geographic regions. Other challenges may arise in the application of NSQGs to specific sites. For example,

- certain receptors or pathways considered in the development of NSQGs may not be present or operative at a particular site;
- stratigraphy, soil properties, or groundwater regime may vary significantly from the generic case in ways that attenuate exposure;
- contaminant-soil interactions may reduce bioavailability below levels influencing development of the NSQG; and
- sensitive ecological conditions such as rare or endangered species, critical range, or protected land status may not have been considered in the development of the NSQG.

Nevertheless, if NSQGs are developed with a sound basis in risk analysis and if assumptions, methods, and calculations are transparent and organized, many opportunities will exist for site-specific adjustments that increase realism and utility.

Two broad types of adjustments appear possible. First, site-specific information can be collected for purposes of calibrating the exposure assessment. Parameters and assumptions used in fate and transport analyses and exposure equations are most often set as conservative defaults in the development of NSQGs; site-specific information can be used to improve assumptions and set parameters more accurately. Such adjustment procedures do not alter assessment endpoints and thus do not depend on or present an alternative view of soil services or beneficial uses underpinning the NSQG. Second, on the basis of site-specific information or knowledge, conceptual model adjustments might be performed that define different management objectives keyed to different assessment endpoints. Typically, conceptual model adjustments pose a greater challenge to risk managers and stakeholders than do calibration adjustments. Regardless of the type of calibration adjustment, it is necessary to involve all managers and stakeholders in the discussion to avoid slanting the calibration adjustment towards vested interests, whether they are perceived or real.

Before discussion of the use of these 2 adjustment mechanisms, it is appropriate to step back and consider the use of NSQGs within a general assessment framework. As discussed in the introduction, NSQGs are most useful within an overall site-assessment framework that guides a user to appropriate options and actions as information and understanding accrues at a site. The tiered frameworks presented by CCME (1996a, 1996b) and the American Society for Testing and Materials (ASTM) Risk-Based Corrective Action (RBCA) (1995) are prominent North American examples. In the first tier of each, initial site characterization data are compared to generic NSQGs to identify compliant areas and any areas or sites requiring further study or action (Figure 3-2) (CCME: comparison to soil-quality guidelines; ASTM: comparison to risk-based screening levels). Where exceedances occur, a proponent could choose to remediate to the Tier 1 value but most likely would choose such action only after a careful examination of higher-tier options to see what NSQG adjustments are possible and whether site-specific risk assessment or management is appropriate.

Several jurisdictions and systems facilitate Tier 2 procedures through use of "look-up tables" that summarize information on exposure pathways. Table 3-7 presents a generic look-up table for 3 hypothetical contaminants representing differing combinations of mobility and toxicity. Table 3-7 includes information on human health, which, while not the focus of this volume, must be considered in contaminated site management decision-making and may be relevant to establishing the scope of NSQG adjustment procedures. Although ASTM RBCA has been based solely on human health risks, ecological risk protocols have recently been developed (ASTM 1999). Table 3-7 shows that, for chemical 1, human health considerations govern the value of the NSQG. There is, therefore, no point in exploring Tier 2 adjustments for ecological receptors unless an acceptable 50-fold reduction in

Table 3-7 Generic "look-up table" for 3 hypothetical substances of differing mobility and toxicity to ecological receptors and humans

Exposure pathway	Chemical 1 (limited mobility, e.g., metal)	Chemical 2 (mobile, e.g., polar organic chemical)	Chemical 3 (immobile, e.g., nonpolar organic chemical)
		mg substance/kg soil	
Human health			
Ingestion	X	4Y	10000Z
Inhalation	100X	90Y	–
Dermal contact	5X	100Y	50000Z
Ecological			
Direct contact	40X	10Y	Z
Soil ingestion and grazing	85X	50Y	50Z
Transport to ground water or surface water	12000X	Y	–

human exposure can be shown for the site. In the case of chemical 2, transport to ground or surface water is the limiting ecological pathway, and there would be reason to believe that collection and application of site-specific information on factors affecting leaching and attenuation in ground water would be helpful in adjusting the NSQG to produce a site-specific benchmark (CCME: site-specific objective; RBCA: site-specific target level). Note, however, that in this illustration only a 4-fold increase in the NSQG in relation to fate and transport to water is possible before a secondary, governing pathway is encountered on the human health side. Chemical 3 has low toxicity to humans; the principal concern is direct contact exposure for plants and/or soil fauna. Additionally, the next most limiting pathway provides 50-fold "headroom" in the direct contact pathway. This provides an opportunity to explore direct exposure in more detail. Relevant species and bioavailability are 2 potential factors for Tier 2 investigation. It may well be cost-effective and environmentally acceptable to adopt unaltered benchmarks for all other pathways addressed in the NSQG. Some Tier 2 adjustment procedures and further considerations are discussed below.

Calibration adjustments

The adjustment by calibration can be specific to transport and fate parameters and to background conditions, bioavailability, and other factors governing exposure. Parameters that represent good candidates for adjustment procedures influence the risk determination and can be altered easily and verified through site studies. As discussed in "Development of Numerical Soil-Quality Guidelines" (p 94), NSQGs are generic and conservative; site-specific concerns including fate cannot easily be accommodated within the NSQG. These values can serve an important role in the soil-screening process. Often, however, these values are taken as default cleanup goals with no consideration of the inherent conservatism built into their derivation.

In weight-of-evidence approaches to NSQG development, values are often derived from many species using different measurement endpoints. Using NOEL (units as mg/kg soil) and LOEL in the same NSQG can result in confused interpretation and weighting of the value. Similarly, using ECx values that describe different "x" levels of effect (e.g., 10%, 20%, 25%, or 50%) is inherently protective to a certain (x) level. Some protocols (CCME 1996a) also provide for use of LCx (lethal concentration) data. Combinations of data from different assessment endpoints using different measurement endpoints may be regarded as building inconsistency into the NSQG, and depending on the balance between no effects and definitive effects information, with unknown but variably conservative results. The final statement or derivation of the NSQG often has a safety or assessment factor built in and is typically from a factor of 1 to 1000, as related to data quantity and quality, scaling effects, and laboratory-to-field extrapolation considerations. These factors are representative of possible problems with singular generic NSQGs and provide opportunities for an adjustment for site-specific applications. In particular, recalculation procedures based on deletion of irrelevant species (CCME 1996b) and bioassays using relevant site species are attractive options.

Bioavailability is affected by numerous physical, chemical, and biological parameters (Chapters 7, 8, 9). Accordingly, laboratory-derived toxicity results may differ substantially from results obtained at a particular site. This laboratory-to-field variability is characteristic of toxicity tests. Aquatic toxicity tests typically show a 2- to 3-fold variation from laboratory to field when single chemicals are tested. Repeatability is fairly high with coefficients of variation within 100% (DeGraeve et al. 1991). The sediment toxicity test system containing aquatic receptor, water phase, and sediment phase becomes predictable for nonpolar organic chemicals within a factor of 10-fold variation (Meyer et al. 1993). This is also reflected in work by Di Toro et al. (1991) leading to the development of guidance on adjusting the environmental availability of nonpolar organic compounds in sediment for organic C content according to equilibrium partitioning theory.

Relative to aquatic systems, soils are extremely heterogeneous, with physicochemical characteristics and toxicants not as homogeneously distributed, resulting in increased variability in toxicity test results. At best, we may expect a coefficient of variation ranging from 8% to 103% in invertebrate tests with a complex mixture such as fresh oil amended into different soils of variable organic C content (Dorn et al. 1998). However, we can make some assumptions that a nontoxic, toxic, and "gray" zone may be defined for field soils. For example, a recent study of 8 field soils contaminated primarily with hydrocarbons showed that for total petroleum hydrocarbons measured using infrared analysis, a concentration of about 9000 mg/kg or above was toxic to earthworms, and a concentration below approximately 1000 mg/kg was not toxic to earthworms. Earthworm response to concentrations in between was variable and did not conform to any commonly used dose–response models (Saterbak et al. 1999). This example highlights the probability that field tests, and especially laboratory tests adjusted for field conditions, may produce variable and uncertain results. However, test data may produce a demarcation of expected toxicity that can be used in a screening assessment. Highly significant correlations between laboratory and field results are not likely to emerge in the short term. Nevertheless, recommendations must be made for developing site risk decisions. This discussion offers some approaches to making field decisions before, or as an alternative to, entering a full site-specific risk assessment.

Transport and fate parameters

The calibration adjustment can be made for factors such as 1) transport and fate parameters (ASTM 1995), 2) background condition and concentrations in the soil (CCME 1996b), and potentially, 3) bioavailability. Collection of site information on basic soil properties such as total organic C, porosity, and pH enables certain exposure or bioavailability adjustments to be made via use of techniques such as partitioning theory. For example, if soil organic C is considered along with a contaminant that partitions to organic C, the available fraction toxicity may be reduced by a factor directly related to C content and partitioning behavior of the chemical. An organic chemical that follows nonpolar equilibrium partitioning will partition proportional to the fraction of organic soil C as described by a linear partitioning coefficient (K_d) up to perhaps 80% of the solubility of the chemical in water (Hassett and Banwart 1989). An example of bioavailability being modified by soil organic C content after application of hydrocarbon (applied as oil) is shown in Figure 3-5; Norwood and Norwood/Baccto soils differ in their organic C fractions. In order to make use of this approach, one must know the fraction of organic C in the original tests underpinning the NSQG and that the toxicity of the contaminant of concern does substantially follow the pattern predicted from partitioning theory. Other sorption characteristics might be used to assess or adjust mobility and bioavailability. For example, cation exchange capacity and pH are important factors in the sorption of metals (Evans 1989). Again, in order to take advantage of this knowledge on a site-specific basis, we need a reliable functional relationship and information on governing parameters in both the base case (used for development of the NSQG) and at the site in question.

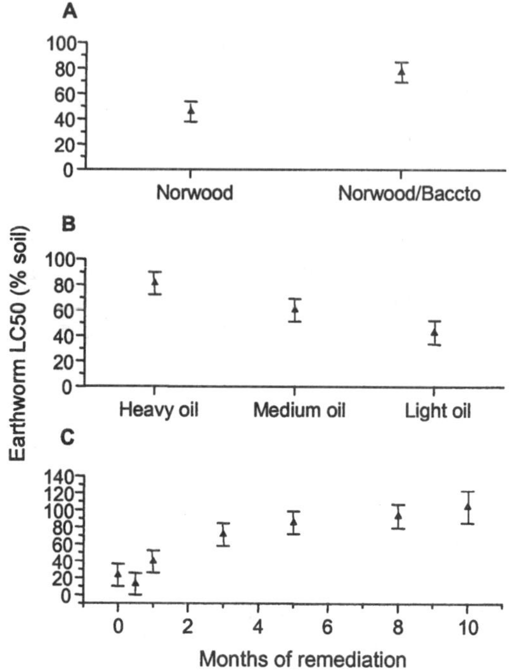

Figure 3-5 Effects of soil type (A), oil type (B), and bioremediation (C) on toxicity of petroleum hydrocarbons to earthworms. Panel (C) represents the combined response for both soils and 3 oil types (modified from Dorn et al. 1998).

Background

Background concentrations of naturally occurring inorganic substances in soils vary primarily in relation to geological and pedogenic factors (McKeague and Wolynetz 1980; Shacklette and Boerngen 1984; Chesworth 1991). Local-to-regional background may also be affected by diffuse contamination sources such as atmospheric deposition. Background concentrations are handled in various ways in the development of NSQGs, and sites are occasionally found where background concentrations exceed the NSQG. Because of this, many jurisdictions using NSQGs provide for site-specific background adjustments. The most common approach of incorporating background is to adjust the assessment benchmark to background if the soil is similar to other "natural" soils in the region of the site. Note that this procedure does not actually make any clear statement regarding ecological risk. Depending on bioavailability, it is possible that an anomalously high background does affect the local ecosystem. Adjusting for background considerations related to essential elements or unusually high geological concentrations can reduce the inappropriate application of generic NSQGs.

Bioavailability

It is clearly evident that the bioavailability of contaminants introduced to soil declines with time as a result of "aging" and "weathering" (Chapters 7, 8, and 9). Aging and weathering are largely a function of various sorption or stabilization processes that may be described mathematically through kinetic modeling (Chapter 7). Stabilization and other fate processes such as biodegradation are sometimes lumped under the general rubric of "dissipation" and kinetic parameters can be estimated for the combined processes. First (k_1), second (k_2), and zero (k_0) order rate constants may be introduced for the chemicals of concern.

Where kinetics of dissipation are characterized, potential exists for demonstrating the decrease in concentration of the chemicals in soil at times removed from the spill event, or from time = 0, or from the time the contamination is assessed. Kinetic rates can be coupled to decreases in soil toxicity if data are available or can be estimated.

An example of declining toxicity resulting from aging under laboratory conditions is shown in Figure 3-5. In Figure 3-5C, the combined toxicity response data for 3 oils in 2 soils over a 10-month period is shown. An abundance of evidence, reviewed in Chapters 8 and 9, shows a decrease in toxicity with weathering and time. Fate processes underlying such decreases in toxicity vary among contaminants and soils (Chapters 5, 6, 7) but in the case of organic contaminants such as petroleum hydrocarbons, are understood to be primarily biodegradation and stabilization. In order to make explicit use of such information in an NSQG adjustment process, certain conditions should be satisfied. First, a functional relationship between toxicity (the assessment endpoint) and bioavailability must be identified. Second, estimates of bioavailability are needed for both the bioassays supporting the NSQG and the site soils under consideration. With these in hand, a site manager could estimate expected effects without detailed site-specific bioassay information.

Another possibility for adjustment of NSQGs is to develop relationships between toxicity and time under field conditions and use, where available, knowledge of time since release as a modifying factor. Considerable challenges lie in identifying appropriate boundary conditions (e.g., initial extent and intensity of contamination) and identification and documentation of factors influencing rates of fate processes (e.g., temperature, moisture and nutrient regimes, soil texture, organic matter content, pH, presence of nonaqueous-phase liquid [NAPL]).

Conceptual model adjustments

Eliminating irrelevant pathways

Conceptual model adjustments may be made to the NSQGs to produce a tailored site-specific assessment. In this procedure, the NSQGs are adjusted to respond only to relevant assessment endpoints where pathways are identified as present or likely to be present in future. The specific land use and beneficial uses for a site represent only a subset of those used in development of the generic NSQG. For example, at an industrial site in a large industrial area with no beneficial use for wildlife, a recalculation procedure could be performed to delete any wildlife exposure pathways from the NSQG. Similarly, while most NSQGs do attempt to protect against cross-contamination of ground water, there may be instances where this consideration can be diminished or eliminated. Examples include sufficient depth to ground water that natural dissipation removes the chemical of concern, insufficient yield or quality of ground water for any identifiable use, and land-use zoning and infrastructure that militate against groundwater use (e.g., existence of municipal mains water system obviates use of potable wells).

It is important to note that many conceptual model adjustments involve a change in assessment endpoints, and as noted in "Soils and Society" (p 76) and "Developing Exposure Scenarios for Numerical Soil-Quality Guidelines" (p 86), these are intimately linked to societal values and expectations for the soil resource. These philosophical linkages trigger very practical issues when the revamped risk analysis either calls for or permits movement and management of contamination beyond the site boundaries under the direct care and control of the proponent. For example, in urban environments isolated from surface water and serviced by municipal mains, transport of contamination from soil to ground water may be identified as an "irrelevant" pathway. However, if this means a contaminant plume originating on the proponent's property will extend downgradient to other properties, there likely will be practical risk management and perhaps compensation issues related to the downgradient parties' expectations and values for their land. Nevertheless, this is not a limitation peculiar to application of NSQG; it is a general issue in risk management.

Spatial issues

The conceptual model can be modified by bringing into play spatial information on contaminant distribution and receptor range. The NSQG approach assumes soil concentrations on the site are uniform—the site should not present unacceptable risks even if contaminated throughout at the NSQG value—and constant over time. If one or more high concentrations drives the assessment, an averaged approach using statistical terms can be employed. Different scenarios can be developed to determine what type of site remediation would be required to reduce chemical levels below NSQG concentrations. A more thorough assessment using spatially weighted averaging techniques may be more appropriate when uncertainty over exposure is high. A spatial view of the contamination can be used to average or weight the site contamination if data are available to adjust concentrations in the NSQG (Clifford et al. 1995). As an example of the use of this technique, at the Rocky Mountain Arsenal, a comprehensive soil contaminant sampling was statistically evaluated to develop spatially weighted averages covering the entire post and was overlaid with wildlife and beneficial use criteria in areas. Resulting from that evaluation, "hot spot" areas were identified for sensitive populations within the pathway of exposure, and spatially weighted contaminant concentrations were calculated. Carlsen (1994) showed that examining the overlap between the proportion of a site that is contaminated and the home range of animal populations at the site can be an important factor in assessing risks and appropriate actions. For that study, a spatially derived concentration was determined and used in a risk assessment. Certainly, few sites are of the magnitude of complexity or contamination that these examples are based upon, but these examples highlight the use of averaging techniques for a site of concern. The effect of averaging essentially lowers the exposure concentration to the receptor of concern, as shown in Figure 3-6.

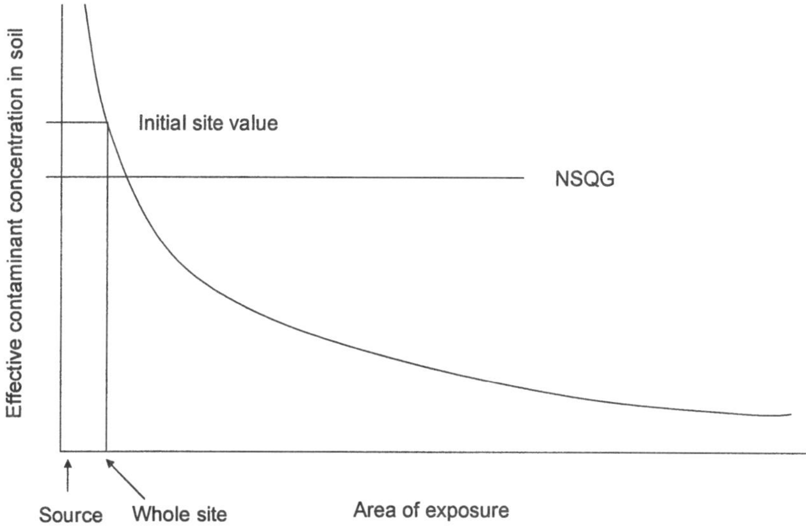

Figure 3-6 Interpretation of NSQGs using site and spatial averaging

A hypothetical example of this procedure is illustrated in Figure 3-7, where an industrial site is located in sensitive habitat populated by wildlife. The site is accessible to the wildlife as indicated by home ranges (shaded areas) overlapping the site. The spill area (indicated by the darkened area) is within the home range accessible to the wildlife. Due to wildlife access, the land use considered cannot be industrial but natural (see Table 3-2). The approach for assessing this site includes taking soil samples to determine contaminant levels that could be compared against an NSQG. It is likely that the site would fail the screening test of comparison to an NSQG. Adjustments could be made to the NSQG that would include averaging for the population range with the contaminant, which may result in adjustment to a de minimus risk to wildlife.

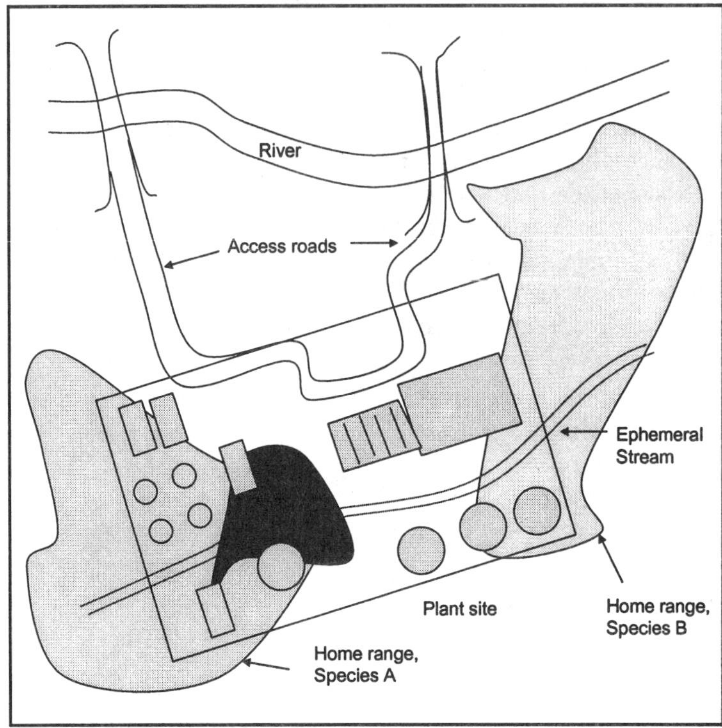

Figure 3-7 Spatial consideration of land use and beneficial use for contaminated industrial site. Shaded areas represent range of wildlife both on and off site. Dark area is contaminated soil in the assessment.

It should be noted that this adjustment relies on infrastructure remaining "as is" at the site. Removal of buildings, pads, etc. is likely to increase wildlife access and may require a reevaluation of model and exposure assumptions. This emphasizes the crucial importance of the conceptual model underlying the NSQG or site-specific RA–RM approach. Explicit articulation of future site conditions is prerequisite to development of a successful conceptual model and any subsequent risk management plan.

Additional Approaches to Numerical Soil-Quality Guidelines

As discussed earlier, NSQGs are of greatest use when they are a component of an overall assessment and management framework that provides a full suite of options. Two additional opportunities for a preliminary assessment are the use of ecological site screening checklists and the use of simple site survey or laboratory or in situ bioassays. Both these approaches allow for the inclusion of some biological measures in preliminary site assessments, rather than relying strictly on the chemical analysis of soils.

Checklists

The checklist approach couples an initial site walk and a determination of sensitive or high-quality habitat with basic questions leading to conclusions regarding exposure pathways and sensitive ecological receptors. A checklist can include assessment of spatial dimensions of the affected area, depth, general type of habitat, and whether ground water or surface water is impacted. This information, coupled with limited samples from the site, can be combined with benchmark screening levels to determine further activity. This process may be particularly useful for small sites in areas of industrial activity where beneficial use and land use options are limited. The use of checklists is employed by several jurisdictions' ERA procedures. The checklist enables a site manager to reach a decision and certify if the contamination or spill is of sufficient severity to warrant further action. Regulatory authorities may also use such checklists or classification tools to determine the priority of such sites for further consideration. Ultimately, the information collected is used to make a "go" or "no-go" or other rank ordinal priority decision (CCME 1992; TNRCC 1996).

Information required for using existing checklists includes landscape setting, site description, photographs of the site, satellite or aerial reconnaissance for past activities, description of local ecological resources, and interpretation of actual and potential exposure pathways. A tier is usually incorporated to determine obvious ecological risk through apparent pathways and receptors. A professional assessment would be required for the ecological information. If the site is concluded to be of de minimus or no ecological risk, the form can be signed, certified, and filed as evidence of activity and action concerning the site. This can then be incorporated in regional regulatory files in a system such as geographic information system (GIS) to allow a more regional tracking and assessment of active or inactive contaminated soil sites. This overview would in essence allow regulatory bodies to track sites. An additional alternative assessment may be made of the site using limited bioassays or biosurveys to make a "pass/fail" assessment of the site for further consideration.

Bioassays

Results from bioassays are not usually intended to provide a comprehensive basis for a risk management decision. However, bioassays can provide "acceptable" or "unacceptable" soil information that could drive discussions to proceed to a formal ERA of the site, exit the site with no action, or use the preliminary data for site prioritization. Also, there may be situations where the "risk driver" is direct soil contact by ecological receptors (e.g., chemical 3 in Table 3-7). In either situation, bioassay information would be attractive to regulatory risk managers who have limited resources and wish to set priorities for a potentially vast number of contaminated sites requiring rapid assessment and remediation decisions. This process may be also applicable to focus on sites where there are limited or no chemical data, or the site is complex with multiple chemicals or unknown soil properties. It is probable that either laboratory or in situ toxicity tests using site soil would demonstrate considerably lower toxicity than predicted using the screening NSQG value. Limited data on some assessment endpoints could enable a rapid decision on the site, such as exclusion from further consideration, minor remediation, or proceeding into an ERA.

Exit to a full risk assessment

Figure 3-2 (p 77) provides a general model for phased acquisition and evaluation of site information leading to various risk management options. Under this model, a full site-specific risk assessment is triggered only after a series of screening and calibration steps have confirmed that no low-cost, environmentally acceptable solution is available using "blunt instruments." Once a decision is reached to proceed with a full risk assessment, sufficient information will have been collected to allow focused and efficient stakeholder discussions and planning that is critical to the initial problem formulation phase (Chapter 2). However, if the site is exempted or relegated to low priority, or contaminant concentrations do not exceed raw or adjusted NSQGs, or only minor remedial work is needed to bring the site into compliance with raw or adjusted NSQGs (including those adjusted on bioassay results), a site-specific risk assessment can be avoided altogether.

This framework for application of NSQGs allows some potential "off ramps" from the site and the ERA process, as well as the ability to prioritize the site for regulatory activity. The use of NSQGs has a role in the site assessment process but should not be considered a definitive step to proceeding into a risk assessment. The procedures for adjustment either through recalibration or alteration of the conceptual model supporting the NSQGs offer powerful opportunities for early site decisions.

Summary and Conclusions

Numerical soil-quality guidelines have become an important tool in the contaminated soil assessment toolkit. Early NSQGs were developed informally using professional judgment, conservative principles, and unknown but sparse public input. They were originally intended as a tool to prioritize investigative action but soon were applied and misapplied in various jurisdictions as remedial targets. Because their scientific and policy basis was usually poorly documented, they could not be adapted to site-specific circumstances.

Throughout the 1990s, significant activity and progress has been made in Europe, North America, Australia, and New Zealand on development and application of NSQGs. NSQGs are now developed transparently using risk analysis techniques. This approach entails elaboration of a defined exposure scenario that reflects regulatory goals for a particular set of beneficial uses of the soil and/or site. Usually, the NSQGs are presented along with results for specific, common exposure pathways that are summarized in a look-up table. NSQGs constructed in this way provide important advantages over earlier efforts based on professional judgment. First, because exposure and risk goals of the NSQGs are conservative and known, they can be used for screening sites from further evaluation when contaminant concentrations are lower. Second, where sites fail a screening test, the look-up table provides a quick indication of what exposure or receptor pathway is likely to require additional study and/or management. Third, and this is where great potential lies, NSQGs can be made more relevant to a particular site through adjustment procedures driven by site data.

In practice, users of NSQGs are likely to focus on acquisition of site data that are easily obtained and influential to exposure via the governing pathway identified from the look-up table during the initial screening. Data may be collected to calibrate exposure models, provide additional or alternative response information, or as a basis for adjustment of the site conceptual model. Once adjustments are made, site conditions may be compared again to the modified NSQGs. This procedure may screen out the site from further action or provide a choice between remediation to the modified NSQGs or further evaluation through a site-specific ERA. NSQGs can thus make an important contribution to an overall assessment and management framework that provides several options for investigation and numerous “off ramps” that can be accessed under different risk management arrangements. When used with classification tools such as checklists, sites may be deferred or excluded from further assessment.

Numerical soil-quality guidelines may be developed for a single, multifunctional land use or be based on land uses or sensitivity. What route is chosen depends on land use management goals, which vary with jurisdiction. Small countries with high population densities favor multifunctional land use scenarios, whereas larger, less populous countries, such as Canada, have developed NSQGs for various land uses.

A significant and seriously underestimated challenge in the development of NSQGs and other risk-based tools lies in collecting and interpreting intelligence on societal values for and expectations of the soil resource. Whereas soil is a crucial component of the environment, it is also a commodity that is bought and sold as "land." Landowners feel entitled to great latitude in its uses and care, including applications in waste treatment and disposal. Yet, we generally expect to acquire land (i.e., soil) in excellent condition, with little or no encumbrance from toxic substances. Diverse or polarized views exist for several other aspects of the soil quality question as well.

We have provided a framework for discussion of the values and expectations societies might attach to soils under some common land uses. We anticipate lively stakeholder discussions will be occurring in many jurisdictions struggling with contaminated site issues over the next few years. Focusing those discussions so as to deliver clear assessment endpoints for soil will be a challenge, but one that must be met if we are to see further improvement in the efficiency and effectiveness of contaminated sites decision-making.

References

Aldenberg T, Slob W. 1993. Confidence limits for hazardous concentrations based on logistically distributed NOEC toxicity data. *Ecotox Environ Saf* 25:48-63.

[ASTM] American Society for Testing and Materials. 1999. Standard guide for risk-based corrective action for the protection of ecological resources. In: Annual book of ASTM standards. Volume 11.04. Philadelphia PA, USA: ASTM. E2081-00. p 1238-1240.

[ASTM] American Society for Testing and Materials. 1995. Standard guide for risk-based corrective action applied at petroleum release sites. In: Annual book of ASTM standards. Volume 11.04. Philadelphia PA, USA: ASTM. E 1739-95e1. p 610-660.

[BCE] British Columbia Environment. 1996. Matrix values numerical soil standards. Victoria BC, Canada: British Columbia Ministry of the Environment, Lands and Forests. BC Reg 375/96. 41 p.

Bengtsson G, Ohlsson L, Rundgren S. 1985. Influence of fungi on growth and survival of *Onychiuruis armatus (Collembola)* in a metal polluted soil. *Oecologia* 68:63-68.

Burgman MA, Ferson S, Akcakaya HR. 1993. Risk assessment in conservation biology. London, England: Chapman and Hall. 328 p.

[CalEPA] California Environmental Protection Agency. 1994. California Department of Toxic Substances Control guidance for ecological risk assessment at hazardous waste sites and permitted facilities. Sacramento CA, USA: Department of Toxic Substances Control Office of Scientific Affairs Human and Ecological Risk Section. 47 p.

Carlsen TM. 1994. Ecological assessment. In: Webster-Scholten CP, editor. Final site-wide remedial investigation report. Livermore CA, USA: Lawrence Livermore National Laboratory. UCAR-AR-108131. p 6.1-6.276.

Carlsen TM. 1996. Ecological risks to fossorial vertebrates from volatile organic compounds in soil. *Risk Anal* 16:211-219.

[CCME] Canadian Council of Ministers of the Environment. 1992. National classification system for contaminated sites. Winnipeg MB, Canada: CCME. EPC-CS39E. Publ nr 1005. 54 p.

[CCME] Canadian Council of Ministers of the Environment. 1996a. A protocol for the derivation of environmental and human health soil quality guidelines. Winnipeg MB, Canada: CCME. EPC-101E. Publ nr 1207. 69 p.

[CCME] Canadian Council of Ministers of the Environment. 1996b. Guidance manual for developing site-specific soil quality remediation objectives for contaminated sites in Canada. Winnipeg MB, Canada: CCME. Publ nr 1197. 50 p.

[CCME] Canadian Council of Ministers of the Environment. 1997. Recommended Canadian soil quality guidelines. Winnipeg MB, Canada: CCME. Publ nr 1268. 185 p.

Chapman PM, Fairbrother A, Brown D. 1998. A critical evaluation of safety (uncertainty) factors for ecological risk assessment. *Environ Toxicol Chem* 17:99-108.

Chesworth W. 1991. Geochemistry of micronutrients. In: Mortvedt JJ, Cox FR, Shuman LM, Welch RM, editors. Micronutrients in agriculture. Madison WI, USA: SSSA. Book Series nr 4. p 1-29.

Clifford PA, Barchers DE, Ludwig DF, Sielken RL, Klingensmith JS, Graham RV, Banton MI. 1995. An approach to quantifying spatial components of exposure for ecological risk assessment. *Environ Toxicol Chem* 14:895-906.

Crommentuijn T, Polder MD, van de Plassche EJ. 1997. Maximum permissible concentrations and negligible concentrations for metals, taking background concentrations into account. Bilthoven, NL: RIVM. Report nr 601-501-001.

Daily GC, Matson PA, Vitousek PM. 1997. Ecosystem services supplied by soil. In: Daily GC, editor. Nature's services: Societal dependence on natural ecosystems. Washington DC, USA: Island Press. p 113-132.

DeGraeve GM, Cooney JD, McIntyre DO, Pollock TM, Reichenbach NG, Dean JH, Marcus MD. 1991. Variability in the performance of the seven-day fathead minnow (*Pimephales promelas*) larval survival and growth test: An intra, interlaboratory study. *Environ Toxicol Chem* 10:1189-1203.

Demon A, Eijsackers H. 1985. The effects of lindane and zainphosmethyl on survival time of soil animals, under extreme or fluctuating temperature and moisture conditions. *Z Angew Entomol* 100:540-510.

Di Toro DM, Zarba CS, Hansen DJ, Berry WJ, Swartz RC, Cowan CE, Pavlou SP, Allen HE, Thomas NA, Paquin PR. 1991. Technical basis for establishing sediment quality criteria for nonionic organic chemicals using equilibrium partitioning. *Environ Toxicol Chem* 10:1541-1583.

Dorn PB, Vipond TE, Salanitro JP, Wiesniewski HL. 1998. Assessment of the acute toxicity of crude oils using earthworms, microtox, and plants. *Chemosphere* 37:845-860.

Edwards PJ, Coulson JM. 1992. Choice of earthworm species for laboratory tests. In: Greig-Smith PW, Becker H, Edwards PJ, Heimbach F, editors. Ecotoxicology of earthworms. Andover, UK: Intercept Ltd. p 36-43.

Evans LJ. 1989. Chemistry of metal retention in soils. *Environ Sci Technol* 23:1947-1056.

Ferguson C, Darmendrail D, Freier K, Jensen B, Jensen J, Kasamas H, Urezelai A, Vegter J. 1998. Risk assessment for contaminated sites in Europe. Volume 1. Scientific basis. Nottingham, UK: LQM Press. p 1-165.

George MR, Brown JR, Clawson WJ. 1992. Application of nonequilibrium ecology to management of Mediterranean grasslands. *J Range Manag* 45:436-440.

Gerlowski LE, Jain RK. 1983. Physiologically based pharmacokinetic modeling: Principles and applications. *J Pharm Sci* 72:1103-1127.

Hassett JJ, Banwart WL. 1989. The sorption of nonpolar organics by soils and sediments. In: Sawney BL, Brown K, editors. Reactions and movement of organic chemicals in soils. Madison WI, USA: SSSA. Special Publication 22. p 31-44.

Heimbach F. 1985. Comparison of laboratory methods, using *Eisenia fetida* and *Lumbricus terrestris*, for the assessment of the hazard of chemicals to earthworms. *Z Pflanzenkr Pflanzenschutz* 92:186-193.

Hoekstra JA, Van Ewijk PH. 1993. Alternatives for the no-observed-effect level. *Environ Toxicol Chem* 12:187-194.

Holling CS. 1986. Resilience of ecosystems; local surprise, and global change. In: Clark WC, Munn RE, editors. Sustainable development of the biosphere. Cambridge, UK: Cambridge University Press. p 292-317.

Holmstrup M. 1997. Drought tolerance in *Folsomia candida* Willem (Collembola) after exposure to sublethal concentrations of three soil-polluting chemicals. *Pedobiologia* 41:354-361.

Jensen J, Kristensen HL, Scott-Fordsman JJ, Pedersen MB. 1997. Soil quality criteria for selected compounds. Copenhagen, DK: Ministry of Environment and Energy, Danish EPA. Working Report nr 83/97.

McKeague JA, Wolynetz MS. 1980. Background levels of minor elements in some Canadian soils. *Geoderma* 24:299.

Meyer CL, Suedel BR, Rodgers Jr JH, Dorn PB. 1993. Bioavailability of sediment-sorbed chlorinated ethers. *Environ Toxicol Chem* 12:493-506.

Milloy SJ. 1995. Science-based risk assessment: A piece of the Superfund puzzle. Washington DC, USA: National Environmental Policy Institute.

Moen JET. 1988. Soil protection in the Netherlands. In: Wolf K, van den Brink WJ, Colon FJ, editors. Contaminated soil '88. Dordrecht, NL: Kluwer Academic. p 1495-1504.

Moore JC, DeRuiter PC. 1993. Assessment of disturbance in soil ecosystems. *Vet Parasitol* 48:75-85.

Moriarty F. 1999. Ecotoxicology: The study of pollutants in ecosystems. 3rd ed. London, England: Academic Press. 347 p.

[OMEE] Ontario Ministry of Environment and Energy. 1996. Guideline for use at contaminated sites. Toronto ON, Canada: Queen's Printer for Ontario, Ontario Ministry of Environment and Energy. Publ nr 3161E01. 159 p.

Posthuma L, Baerselman R, van Veen RPM, Dirven-Van Breemen EM. 1997. Single and joint toxic effects of copper and zinc on reproduction of *Enchytraeus crypticus* in relation to sorption of metals in soils. *Ecotox Environ Saf* 38:108-121.

Posthuma L, Weltje L, Anton-Sanchez FA. 1996. Joint toxic effects of cadmium and pyrene on reproduction and growth of the earthworm *Eisenia andrei*. Amsterdam, NL: RIVM. Report nr 607506001. p 1-38.

Rice K, Jain S. 1985. Plant population genetics and evolution in disturbed environments. In: Pickett STA, White PS, editors. The ecology of natural disturbance and patch dynamics. Orlando FL, USA: Academic Press. 472 p.

Saterbak A, Wong DCL, McMain BJ, Williams MP, Dorn PB, Brzuzy LP, Chai EY, Salanitro JP. 1999. Ecotoxicological and analytical assessment of hydrocarbon-contaminated soils and application to ecological risk assessment. *Environ Toxicol Chem* 18:1591-1607.

Shacklette HT, Boerngen JG. 1984. Element concentrations in soils and other surficial materials of the coterminus. Alexandria VA, USA: USGS. Prof Pap 1270.

Siegrist RL. 1989. International review of approaches for establishing cleanup goals for hazardous waste contaminated land. Aas, Norway: Institute for Georesources and Pollution Research. 81 p.

Suter GW. 1993. Ecological risk assessment. Ann Arbor MI, USA: Lewis Publisher. 538 p.

[TNRCC] Texas Natural Resource Conservation Commission. 1996. Guidance for conducting ecological risk assessments under the Texas risk reduction program. Austin TX, USA: Office of Waste Management. Draft. November 15, 1996.

Tilman D. 1997. Biodiversity and ecosystem function. In: Daily GC, editor. Nature's services, societal dependence on natural ecosystems. Washington DC, USA: Island Press. p 93-112.

Travis CC, Arms AD. 1988. Bioconcentration of organics in beef, milk and vegetation. *Environ Sci Technol* 22:271-274.

[USEPA] U.S. Environmental Protection Agency. 1989. Ecological assessment of hazardous waste sites: A field and laboratory reference. Corvallis OR, USA: USEPA, Environmental Research Laboratory. EPA-600-3-89-013. 259 p.

[USEPA] U.S. Environmental Protection Agency. 1994. Managing ecological risks at EPA: Issues and recommendations for progress. Washington DC, USA: USEPA, Office of Policy, Planning and Evaluation. EPA-600-R-94-183. 125 p.

[USEPA] U.S. Environmental Protection Agency. 1997. Ecological risk assessment guidance for Superfund: Process for designing and conducting ecological risk assessments. Washington DC, USA: USEPA, Office of Solid Waste and Emergency Response. June, 1997. Interim Final. EPA-540-R-97-006. 230 p.

[USEPA] U.S. Environmental Protection Agency. 1998. Guidelines for ecological risk assessment. Washington DC, USA: USEPA, Office of Research and Development. EPA-630-R-95-002F PB98-117849.

Van Beelen P, Doelman P. 1997. Significance and application of microbial toxicity tests in assessing ecotoxicological risk of contaminants in soil and sediment. *Chemosphere* 34:455-499.

Van Gestel CAM, Hensbergen PJ. 1997. Interaction of Cd and Zn toxicity for *Folsomia candida* in relation to bioavailability. *Environ Toxicol Chem* 16:1177-1186.

Visser WJF. 1993. Contaminated land policies in some industrialized countries. Hague, NL: Technical Soil Protection Committee. TCB Report nr 2. 186 p.

[VROM] Ministerie van Volkshuisvesting, Ruimtelijke Ordering en Milieubeheer. 1994. Environmental objectives in the Netherlands. A review of environmental quality objectives and their policy framework in the Netherlands. Hague, NL: VROM. Samsom-Sijthoff, Alphen a/d Rijn. 465 p.

Wagner C, Løkke H. 1991. Estimation of ecotoxicological protection levels from NOEC toxicity data. *Water Res* 25:1237-1242.

Walter LR, Chapin FS. 1987. Interactions among processes controlling successional change. *Oikos* 50:131-135.

Watt KEF, Craig PP. 1986. System stability principles. *Syst Res* 3:191-201.

Wiens JA, Addicott JF, Case TJ, Diamond J. 1986. Overiew: The importance of spatial and temporal scale in ecological investigations. In: Diamond J, Case T, editors. Community ecology. New York NY, USA: Harper and Row. 665 p.

Zachariassen KE, Lundheim R. 1995. Effects of environmental pollutants on the cold-hardiness of arctic and boreal ectothermic animals. In: Munawar M, Luotola M, editors. The contaminants in the Nordic ecosystem: Dynamics, processes and fate. Amsterdam, NL: SPB Academic Publishing. p 71-83.

CHAPTER 4

Effects of Contaminants on Soil Ecosystem Structure and Function

Randall S. Wentsel, W. Nelson Beyer, Clive A. Edwards, Lawrence A. Kapustka, Roman G. Kuperman

Introduction

Soil communities contain populations of organisms that differ greatly in size, numbers, habits, life-cycles, food sources, distribution, and interactions. Different ecosystems have greatly differing soil communities containing diverse populations of decomposers, predators, microbial symbionts, pathogens, and parasites. These organisms can facilitate soil formation, organic matter breakdown, nutrient cycling, and thus, overall soil fertility.

Organisms that inhabit soil are very interdependent, and their populations form complex interactive systems referred to as the "soil community." Populations of soil-inhabiting organisms are characterized by their diversity and abundance, and these characteristics differ greatly among communities. Their collective actions drive the dynamic soil processes that influence soil fertility. Contaminants such as pesticides, metals, and acid precipitation can affect both the structure and function of natural and managed ecosystems. Contaminants can have drastic and markedly differential effects on the populations of the various taxa of soil fauna and microflora that often lead to dramatic changes in the relative abundance of soil organisms. Contaminants can decrease the numbers of some groups of organisms due to direct toxicity and greatly increase the numbers of others due to release from pressure by predators and other antagonists. Such changes affect the function of soil communities, particularly in terms of changing the rates of organic matter breakdown and nutrient cycling.

Mycorrhizae fungi and bacteria are important components of the rhizosphere. Except in high-intensity agriculture where fertilizers limit the degree of colonization, mycorrhizae strongly influence the uptake of water and nutrients into plants. The extent of mycorrhizal colonization in nonagricultural species, as well as the extensive colonization of roots by rhizosphere bacteria, may be important in the uptake of contaminants into plants from soil. Because many plant uptake studies have been done in hydroponic media, which precludes formation or maintenance

Contaminated Soils: From Soil-Chemical Interactions to Ecosystem Management. Roman P. Lanno, editor.
 ISBN 1-880611-31-7

of mycorrhizae, our knowledge of mycorrhizal- or associated bacteria-mediated uptake of contaminants is limited.

What happens below ground largely determines what occurs above ground. A few generalities about plants and associated microbes illustrate the importance of the belowground environment. The net photosynthate produced by the aboveground portion of plants is distributed to various plant parts (stems, reproductive tissues, and belowground parts) according to developmental stage and genetic makeup. The amount distributed below ground varies from 40% to 85% of net photosynthate (Fogel 1985). However, the pattern of allocation is highly dependent on communities of microorganisms inhabiting the rhizosphere and penetrating root tissues. Under gnotobiotic conditions, the addition of nonpathogenic bacteria to grass seedlings can result in overall changes in net primary production ranging from 40% to 370% of controls with no apparent alteration of the shoot but virtually all the growth response in the roots (Kapustka et al. 1985). Kapustka (1987) illustrated the dynamic properties of the soil environment of an ordinary herbaceous young plant with 10 g aerial mass as follows:

> We can expect, for simplicity, a root mass of 20 g distributed in a soil volume of 1000 cc. This soil volume harbors some 10 to 2000 billion microorganisms, one million nematodes, thousands of insects in various stages of development, a few hundred seeds of potentially interfering plants, and roots of a few neighboring plants. If this plant is to grow at a moderately high relative growth rate of 8% per day for 30 days, the aerial portion of the plant will increase 10-fold. Most likely, so will the roots, extending into a proportionately new soil volume with its attendant populations. During (and in response to) this growth, the microbial population will multiply 3- to 25-fold per unit volume of soil.

Estimates of bacterial biomass to a soil depth of 30 cm range from 32 to 76 g/m^2 and of fungal biomass from 84 to 117 g/m^2 (Dommergues and Krupa 1978). Though significant microbial biomass occurs in the litter layer and on organic matter incorporated into soil horizons, a large microbial biomass also occurs in the rhizosphere where it performs functions critical to soil system nutrient cycling.

The numbers and biomass of microorganisms and invertebrates in agroecosystems are enormous. Indeed, they total more in biomass than the vertebrates that live above ground (Table 4-1). Clearly, with the exception of earthworms, the total biomass of the microorganisms greatly exceeds that of the invertebrates. Nevertheless, there is increasing evidence that the importance of the role of invertebrates in controlling organic matter decomposition processes greatly exceeds their relative numbers and biomass in soils and litter (Moore and Hunt 1988).

Table 4-1 Relative numbers and biomass of soil invertebrates and microorganisms

	Nr. m^{-2}	Nr. gm^{-1}	kg/ha
Microflora			
Bacteria	10^{13}–10^{14}	10^{8}–10^{9}	400–5000
Actinomycetes	10^{12}–10^{13}	10^{7}–10^{8}	400–5000
Fungi	10^{10}–10^{11}	10^{5}–10^{6}	1000–15,000
Algae	10^{9}–10^{10}	10^{4}–10^{6}	10–500
Invertebrates			
Protozoa	10^{9}–10^{10}	10^{4}–10^{5}	20–200
Nematoda	10^{6}–10^{7}	10–10^{2}	10–150
Acarina	10^{3}–10^{6}	1–10	5–150
Collembola	10^{3}–10^{6}	1–10	5–150
Earthworms	10–10^{3}		100–15,000
Others	10^{2}–10^{4}		10–10^{2}

The interactions between microorganisms and different functional groups of invertebrates are extremely complex. Over the last 25 years, there have been many attempts at linking and better understanding these interactions through construction of complex food web models, which allow energy and material flows to be estimated, but we are a long way from a full trophic-level analysis of any soil ecosystem (Edwards 1999). Moreover, identifying interactions between different groups of soil organisms is made more difficult because of our lack of knowledge of the food and feeding habits of many soil-inhabiting invertebrates. Additionally, although invertebrates can be separated conveniently into the functional groups of plant feeders, predators, detritivores (saprophages, fungivores, bacterivores) and carnivores, this classification can be confusing because particular groups of invertebrates may fall into several of these categories or may even be omnivores. Even a single species can change its food and feeding habits in response to food availability and environmental stress.

The maintenance of soil quality, fertility, and structure is essential for sustainable agricultural production and for the protection and maintenance of the biodiversity and dynamics of terrestrial ecosystems. Central to achieving this aim is the need for a vastly improved understanding of the potential effects of chemical contaminants on the structure and function of ecosystems. Chemical contaminants of soils can exert their effects directly through toxicity to soil organisms or indirectly by altering specific interactions and by disrupting soil food webs (Kuperman et al. 1998). Ultimately, these effects can interfere with soil processes that may be important in the regulation, flow, and cycling of C and nutrients in ecosystems (Parmelee et al. 1993; Edwards and Bohlen 1995). The effects of chemicals on particular species or groups of soil organisms, dynamic soil processes, and more rarely on whole-soil systems, have all been used to evaluate the potential overall environmental impact of chemicals reaching soil using both laboratory and field experiments.

Soil organisms and C and nutrient cycling processes are intimately linked. Effects of chemicals on one aspect are likely to impact many of the others. Furthermore, the intensity and duration of the environmental effects of chemicals may often depend strongly upon processes that influence activity, persistence, and movement of chemical contaminants through soil and into microbes, soil animals, and plants. Soil organisms, including plants, interact with chemicals in soil and influence their degradation, bioaccumulation, and persistence (Edwards 1994). Because of the complexity of soil ecosystems, and the tight linkages between soil organisms and soil processes, it is difficult to assess the influence of chemical contaminants on soil ecosystems without considering their potential interactive effects.

Assessing baseline characteristics of soil invertebrate and microbial communities or measuring changes resulting from exposure to biological, chemical, or physical stressors has largely been ignored in U.S. regulatory circles. This is despite the fact that recent U.S. Environmental Protection Agency (USEPA) guidelines for ecological risk assessment (ERA) list ecological relevance and susceptibility to known or potential stressors as the 2 essential criteria for selecting scientifically defensible assessment endpoints (USEPA 1998). Techniques to monitor changes in soil organism communities are used extensively in the applied fields of soil ecology, agriculture, forestry, horticulture, and range management, where the research focus is well defined (e.g., crop yield). There are few examples of their use in risk assessment or statutory environmental management programs administered by the USEPA or state environmental agencies. This is more of a reflection of the actual or presumed valued resources and practices of risk assessors than a limitation of the field. Indeed most ERAs are, in practice, assessments of potential exposure of wildlife to toxic substances. If the practice changes so that ERAs focus on ecosystem functions, then the tools of plant, microbial, and invertebrate ecology, as well as techniques from various applied agronomic fields, could be adapted with relative ease provided there is agreement on what assessment endpoints are important or relevant.

Contaminants that impact soil communities include pesticides, metals, metalloids, petroleum products, various industrial mixed wastes, and acid precipitation. Pesticides reach soils through extensive application not only on agroecosystems but also on many other managed ecosystems. They may also reach natural communities and ecosystems through spray drift from treated areas and in precipitation after long-range atmospheric transport. Insecticides, acaricides, fungicides, nematicides, and herbicides differ greatly in their persistence and mobility in the environment (Edwards 1973; Edwards and Thompson 1973; Edwards and Bohlen 1992). The organochlorine insecticides, though banned in most developed countries, are particularly persistent chemicals with potentially long-term toxic effects on soil organisms. Certain organophosphate and carbamate insecticides and triazine herbicides can also persist in soils for several years and have diverse influences on soil communities. Most soil contamination by pesticides has resulted from the deliberate application of such chemicals to agricultural and horticultural land with the aim of controlling pests, although there is some evidence of global transport (Edwards 1973).

Elevated concentrations of various metals and metalloids occur in the vicinity of mining complexes. Naturally high concentrations due to mineralized parent material, plus emissions from smelters and debris piles, have long-term effects on soil systems. Elsewhere, naturally occurring soils with high concentrations of metals (e.g., serpentine soils) have evolved unique, tolerant, biotic assemblages.

Soil acidification from compounds in precipitation has been implicated in the recent decline of many forest ecosystems in both Europe and North America (Kuperman 1996, 1999; Kuperman and Edwards 1997). The main components of acidic deposition include oxides of S and N, weak organic acids, ammonia, nitric acid, and hydrochloric acid. These compounds are carried mainly on particulate matter and arise from a wide range of industrial emissions. Other chemicals can also be washed out of the atmosphere and can contaminate nearby land. These chemicals include halogenated phenols and other halogenated organic compounds generated by combustion processes in automobiles and industry (Paasivirta 1991).

Petroleum products and mixed industrial wastes have localized effects on soil communities. Depending upon soil concentrations, the effects of these contaminants may be physical or chemical. At higher levels of bioavailable chemicals, direct toxic effects on organisms may occur. At low bioavailable levels, soil structure may be altered, preventing proper water and nutrient uptake by plants. In addition to detrimental effects on soil biota from soil contamination with petroleum products, soil processes may be accelerated as biodegradation converts some toxic substances to sources of organic C.

Linkage of Soil Processes and Tests to Assessment Endpoints

Effective environmental management of valued resources requires clear understanding of linkages among system components. The general public readily comprehends linkage of aquatic insect populations as food for valued fish populations, or even as indicators of clean water. Outside agricultural-based communities, public awareness of linkage of soil organisms and processes to crops, vegetation, or wildlife populations is generally limited. To foster better understanding of the relationships, we offer a few illustrative examples in the framework of ERA procedures. In particular, we show the linkage between the ultimate point of evaluation, the assessment endpoint, and the tools used to characterize the resource or process, the measurement endpoint.

To communicate this information more clearly, we propose a 3-step process:

Measurement Endpoint → Linkage Communication → Assessment Endpoint.

Assessment endpoints are defined in the problem formulation stage of ERAs (USEPA 1998). Articulation of assessment endpoints should include ecological relevance, sensitivity to the stressor, and social concerns. Before a risk assessment can have a chance of succeeding, it is incumbent on the scientists to communicate in unambiguous terms precisely why such relationships are important. Earthworms, nematodes, mycorrhizae, and other soil biota are critical to sustaining terrestrial ecosystems, which support populations of warblers, beech trees, finches, tortoises, robins, deer, deer mice, and owls. Despite the abundance of technical knowledge supporting various tests to date, soil ecologists have been less successful at illustrating the critical connections between what can be measured and what is valued by society.

A linkage communication step may clarify the value of soil biota information as measurement endpoints. The examples in Table 4-2 present how various terrestrial test methods can be linked to support broad assessment endpoints. Assessment endpoints, such as song birds, medium-sized mammals (e.g., fox), large mammals (e.g., bear, deer, elk) threatened showy plant species (e.g., sego lily) or rare insects (Fender's butterfly), will be the primary concern. Community or ecosystem properties such as "suitable habitat for wildlife," "species diversity," or "forest productivity" are also likely assessment endpoints. Functional guilds of soil invertebrates, litter decomposition, and nutrient cycling are measurement endpoints that can be linked through better communication to soil fertility, forest productivity, or plant community structure.

A few examples of successfully linking assessment endpoints and measurement endpoints are presented to foster a better understanding of the use of terrestrial ecology in risk assessment.

Table 4-2 Linkage communication of soil measurement endpoints to assessment endpoints

Assessment endpoint	Linkage communication	Method-measurement endpoint
Sustained agricultural productivity	Plant growth potential	Bioassays to measure plant growth
	Soil fertility	Soil processes: organic matter breakdown, respiration, nutrient cycling, soil biota toxicity tests
Sustained forest production	Plant establishment and growth	Soil processes: organic matter breakdown, respiration, nutrient cycling, soil biota toxicity tests
		Mycorrhizae production and growth
Sustained wildlife populations	Habitat and diet	Vegetation composition
		Community structure (layers, etc.)
		Species diversity
		Tissue concentration of contaminants in food items

Example 1

The first example is an ERA of an area with soils influenced by mining activities, with established wildlife populations, including elk, deer, insectivores, and carnivores, as the principal valued ecological resources. Two broadly structured questions were posed:

1) Were soil conditions at the site adversely affecting plant community structure and function (i.e., structural habitats)?
2) Were the animals being exposed to excessive levels of metals and metalloids in their diets?

Though the ultimate focus of this risk assessment was on wildlife species, the first question was addressed through a classical vegetation sampling approach that determined percentage cover and species composition across the project area. The contemporary vegetation composition, as well as successional trends described from several earlier measures of plant cover, led to the conclusion that metals and metalloids in the soils were not impeding expected successional processes and that the contemporary vegetation represented a diverse and dynamic situation supportive of wildlife populations. The second question (i.e., dietary exposure) was addressed through the analysis of metal and metalloid concentrations in various food items and soils. Species that were collected and analyzed for metal and metalloid levels (wheatgrass, dropseed, curly cup, crickets, spiders, Lepidoptera, mice) were food items for the wildlife species of interest, and by themselves, their relevance to the ecosystem might not be clear. However, by communicating their linkage as food items for elk, kestrels, fox, and mountain bluebirds, the importance of measuring the metal content of these species was understood. The successful linkage effectively communicated the scientific information to the risk manager.

Example 2

Another large-scale assessment of soil contamination at an abandoned mining site also focused on wildlife. In this case, elk, pine marten, and grouse were selected as surrogates for a broader suite of birds and mammals of interest. The assessment incorporated field sampling of vegetation cover, species composition, and vertical structure of the vegetation. Laboratory phytotoxicity tests were performed on composite soil samples collected from locations that had been identified using a stratified random process. An early seedling growth test with 1 species of grass and 2 species of forbs was used to quantify plant shoot and root growth. A soil triad approach (soil-chemical concentration–phytotoxicity results–vegetation cover) was used. Plant growth measured in laboratory tests was correlated with measures of soil metal and metalloid content and with field measures of plant cover. These correlations were communicated in a weight-of-evidence approach to demonstrate linkage between metals, toxicity, and vegetative effects that were responsible for poor wildlife habitat condition. The resource manager was able to develop mitigation steps to enhance wildlife populations.

Example 3

An assessment of bioremediation efficacy was conducted at a crude oil production site in the Rocky Mountain area. The petroleum content of oil-contaminated soil was reduced approximately 80% by composting for 4 months. To evaluate the productivity of the biotreated soil, earthworm survival and plant germination tests were conducted. The plant tested was little bluestem (*Schizachyrium scoparium*) because this species is a major component of the seed mix recommended by the Bureau of Land Management for the region. Earthworm survival was 100%; however, seed germination was impaired in the remediated soil. Further analysis of the soil indicated that an elevated salt content was responsible for inhibition of germination rather than the residual hydrocarbon content. Remediation methodologies were then focused to treat the soil with calcium sulfate and water washing to reduce the salt content. By using a suite of biological tests, the proper mitigation technique was identified.

Use of Soil Processes and Community Structure in Contaminant Assessment

Environmental toxicology has expanded in the past decade, with most progress achieved in studies of aquatic and aboveground terrestrial ecosystems. With few exceptions, soil ecosystems have received little attention in ecotoxicological studies, and in many conceptual models, they are presented as a "black box." This is largely explained by the immense complexity of soil ecosystems and the interdependence of the diverse soil community and the processes it regulates in soil.

The main aim of studies into the effects of chemicals on soil ecosystems is to identify those chemicals that are most hazardous to important groups of key soil organisms and their activity, particularly those that may influence aspects of nutrient cycling such as the breakdown of organic matter (mineralization), respiration, and other metabolic processes. Laboratory tests can provide quantitative information to determine impacts on measurement endpoints. Methods of studying soil biota, community structure, and soil ecosystem function have been developed (Crossley et al. 1991) and used by soil ecologists. However, many environmental toxicologists are not aware of the variety of test methods available. The purpose of this section is to present common soil testing methods and their utilization. Soil is a net sink for many pollutants, and the functioning of the overall terrestrial ecosystem is dependent on soil processes (organic matter decomposition and release of nutrients). It is clear that better application of existing tools, as well as refinement of methods, would improve the characterization of risk for terrestrial systems and provide useful information to decision-makers in terms of focusing site cleanup and remedial efforts.

Ecological risk assessments at most contaminated sites in the U.S. rely extensively on chemical measurements and ratios of the measured chemical concentration and literature-based toxicity thresholds (i.e., risk quotients). If biological testing is done at all, it is often limited to a few methods such as an earthworm toxicity test or phytotoxicity tests (e.g., lettuce seed germination and root elongation tests) (Greene et al. 1988; Edwards and Bohlen 1992). Several other terrestrial tests have been standardized in the past 10 years and are available for use. Compilations of test methods have greatly expanded the suite of tests readily available and increased flexibility with respect to test species selection, such that greater ecological relevance can be achieved during biological site assessment (Linder et al. 1992; Kapustka 1997; Løkke and Van Gestel 1998; ASTM 1999a, 1999b, 1999c, 1999d). Standardization of invertebrate sampling and toxicity tests, as well as microbial process measurements, would improve the utility of many more tools germane to assessing soil ecology parameters.

Ecological risk assessments are focused on key questions intended to provide a technically sound basis for deciding among management options. As such, investigations should rely on a balance of field and laboratory work. Seldom do these investigations entail experimental research or fine-scale characterization of spatial or temporal patterns. Rather than be overwhelmed by the great magnitude of spatial and temporal variation in soils, assessments of contaminant effects on soil systems must be tailored to answer specific questions. Disciplines of basic and applied sciences that deal with soil have produced libraries of methods to evaluate physical, chemical, and biological processes in soil. Some of the biological procedures deal primarily with characterization of structural features of populations and communities. Other methods are aimed at characterizing functional attributes. In the following sections, we present a limited set of techniques that can be used effectively in ERAs targeting contaminant effects in terrestrial systems.

Methods and procedures for invertebrates, plants, and microorganisms

Field and laboratory methods

Field methods

Due to the great biodiversity among soil-dwelling invertebrates and their wide range in size, numbers, and habits, different sizes of samples and methods of separation of invertebrates from soil are necessary for assessing various populations (Edwards 1991). Soil samples can be taken with augers that range in diameter from 2.5 to 20 cm. Most soil invertebrates spend most of their lives in the top 10 cm of soil, so samples taken to a vertical depth of 10 to 15 cm are commonly used and these probably recover 80% of the total populations.

Most species of arthropods can be extracted from soil using modified high-gradient Tullgren samplers or by flotation methods based on methods developed by Raw (1959). Nematodes can be extracted from soil using Baermann funnels or by elutriation techniques with high efficiency. These techniques can recover up to 80% of the organisms in the soil sample. Enchytraeids can be sampled using soil samples in water suspensions in funnels or by flotation methods (Phillips et al. 1999). Earthworm populations can be sampled in the field by pouring a dilute formaldehyde solution (50 ml 100% formalin in 9 L water) onto the soil surface in defined quadrants (Raw 1959). This stimulates the earthworms to emerge to the soil surface, where they can be collected and counted. The efficiency of this technique approaches 100% for deep-burrowing species in spring and autumn and about 70% for other species (Raw 1959).

Laboratory methods

The suite of soil toxicity test methods available covers a broad range of effects measurements for different components of soil systems. Just as analytical chemists select certain extraction protocols and particular instruments and make specific adjustments to the instrument settings to measure different analytes in different media, the soil ecotoxicologist chooses among test methods, test organisms, and test conditions, depending on the chemicals of concern and the physicochemical makeup of the test soil. Considerations for the suite of tests available include

- species from different taxonomic groups to represent a range of behavior, sensitivity, and physiological mechanisms;
- spectrum of diverse ecological functions in the soil community (plants, herbivores, predators, saprovores, bacterivores, and fungivores);
- species that are common in a broad range of soils and geographical areas;
- species moderately sensitive to a wide range of chemicals;
- different exposure routes to chemicals (epidermal contact, inhalation, or soil ingestion); and
- life-cycle tests to identify different sensitivities related to the developmental stages of the test organisms (juvenile and adult survival, reproduction).

In addition to ecotoxicological criteria, soil toxicity test methods should accommodate a number of practical considerations. These may include

- easy to maintain in laboratory culture,
- high reproductive rate,
- short generation time,
- low cost of conducting the test (staff time and equipment),
- easy to standardize,
- noncontroversial (animal rights group concerns),
- reproducible results,
- clearly defined test validity criteria,
- unambiguous measurement endpoints, and
- sufficient literature information on the use and interpretation of the test.

Soil invertebrates

Soil-inhabiting invertebrates are extremely diverse in form, size, numbers, and function. They play an important role in the degradation of organic matter in soil and the recycling of nutrients or may be pests of plants or predators and/or parasites of plant pests and other soil-inhabiting invertebrates. They differ greatly in their size, life history, ecological niche, spatial distribution in soil, seasonal abundance, and sensitivity to chemicals. The types of invertebrates involved in key soil processes include microarthropods, particularly Collembola (springtails) and

Acarina (mites), nematodes, symphylids, Diplopoda (millipedes), Chilopoda (centipedes), Isopoda (sowbugs), Insecta (insects), earthworms, and enchytraeids.

Some of the key features that relate to interactions between populations of soil-inhabiting invertebrates in communities are

- the availability of appropriate food for various taxa of soil-inhabiting invertebrates (including plant material, decaying organic matter, other invertebrates, and microorganisms),
- the species diversity of microorganisms and invertebrates,
- the predator–prey interactions between soil-inhabiting invertebrates, and
- the uptake of contaminants into food chains.

It is difficult to demonstrate experimentally how contaminants affect the availability of soil food resources because chemicals not only affect the amount of food available but also its palatability. However, it is possible to demonstrate experimentally how particular contaminants, such as pesticides or heavy metals, affect species diversity and predator–prey interactions and are taken up into food chains of soil-inhabiting invertebrates (Parmelee et al. 1993, 1997).

Chemicals may influence populations of soil-inhabiting invertebrates in different ways. These include direct acute toxicity, chronic toxicity such as effects on reproduction, indirectly by changing their food supply, or indirectly by killing their predators and parasites. Few of the studies reported in the literature attempt to separate these 4 types of effects.

It is extremely difficult to separate the direct toxicity of contaminants to soil-inhabiting invertebrates from the indirect effects of the toxicant. Indirect effects can occur because numbers of predators or prey of a specific invertebrate are drastically changed because its food supply is affected or because its habitat is changed. Such interactions have shown that the overall impact of a contaminant can sometimes lead not only to decreases in populations of invertebrates but sometimes to quite large increases. For instance, Collembola populations often increase after exposure to organochlorine or organophosphate insecticides because the main effect of these pesticides is the elimination of mesostigmatid mites, the main predators of Collembolans (Edwards and Thompson 1973).

Summary of invertebrate toxicity test methods used in practice

Earthworm survival toxicity test (acute)

The standard 14-day acute survival assay utilizes the earthworm *Eisenia fetida* (Edwards 1983, 1984; Greene et al. 1988; ASTM 1999a). This species was chosen as a test species for toxicity assessment due primarily to the relative ease with which it can be cultured in the laboratory. Uncertainty remains, however, regarding the scientific justification for using this earthworm species as a typical representative of the soil community.

Earthworm reproduction toxicity test (chronic)
A 21-day chronic reproduction assay has also been proposed as a routine test (Van Gestel et al. 1989; ISO 1997). This protocol is similar to the standard 14-day acute survival assay, except earthworms are fed during the test. After 28 days, the number of surviving earthworms is recorded, and cocoons are recovered and counted. In contrast to the earthworm acute test, the chronic reproduction test has greater ecological relevance and justifies additional effort.

Enchytraeid survival and reproduction test
A potential alternative to the standard earthworm test is the enchytraeid reproduction test (ERT). Based upon the methods of Kasprzak (1982), it was developed to determine the effects of test chemicals on the reproductive potential of the enchytraeid *Enchytraeus albidus* as representative of the soil fauna. This species has several advantages over *E. fetida*:

1) It is more ecologically relevant because it actually lives in soil, whereas *E. fetida* inhabits compost or manure piles and is not a "true" soil species.
2) It is easier to culture.
3) It has a shorter generation time.
4) It has intimate contact with pore water.
5) It requires much less (1/10) soil per toxicity test, thereby reducing the quantity of contaminated material generated and the expenses incurred during disposal.

Enchytraeid worms, including *E. albidus*, have been used in a variety of ecotoxicological studies for more than 3 decades (Weuffen 1968; Kaufman 1975; Huhta 1984; Römbke 1989; Römbke and Federschmidt 1995), with recent efforts to standardize enchytraeid test protocols for routine chemical toxicity testing (Römbke 1996). The results of an International Ring test (Rombke and Moser 2002), including more than 100 tests by participants from 14 countries, have been reviewed, and a draft guideline was prepared for distribution within the Organization for Economic Cooperation and Development (OECD) in the spring of 1998.

Nematode life-cycle toxicity test
The soil-dwelling, bacterivous, rhabditid nematode *Caenorhabditis elegans* is a toxicological test organism that can be utilized as a surrogate for bacterivore components of soil food web (Donkin and Dusenbery 1993). It provides a genetically homogeneous test population that minimizes random variations in response due to differences among individuals. It is easy to culture and has a short life-cycle (Wood 1988). Another potentially useful species of free-living, bacterivorous nematode is *Plectus acuminatus*, which has the advantage of being cultured in standard artificial soil, enabling comparison with other standardized test systems using earthworms and springtails (Kammenga et al. 1996). Acute and life-cycle toxicity tests can be conducted using either species.

Collembolan life-cycle toxicity test

Collembola (springtails) are the most studied group of soil microarthropods, with several species used for soil toxicity testing. The species used most often in ecotoxicological investigations is *Folsomia candida*. The test with *F. candida* was designed to investigate the effects of soil contamination on adult and juvenile survival and reproduction (Thompson and Gore 1972). This parthenogenetic species is both sensitive to a wide range of disturbances and signals changes in the soil community, due in part to the fact that *F. candida* belongs to the fungal energy channel (all food chains originating from fungi), which is more sensitive to disturbance than the bacterial or root energy channels (Moore and DeRuiter 1991). The *F. candida* test has the potential to adequately determine the effects of soil contamination on the structure of the soil microarthropod community and also is well standardized for artificial soil (ISO 1998).

Oribatid mites

Oribatid mites are one of the most abundant groups of soil microarthropods and comprise hundreds of species. Denneman and Van Straalen (1991) described a reproduction toxicity test using the parthenogenetic species *Platynothrus peltifer*. This test was designed to assess the effects of chemicals in food on egg production and is not a true soil toxicity test because test individuals are exposed to a contaminated paste of algae on a filter paper. The low rate of egg production in this species (only 2 eggs per week) and its long life-cycle (1 year) make it difficult for use in risk assessment, although it was the most sensitive soil invertebrate tested for Cd and more sensitive to Cu and Pb than Collembola (Denneman and Van Straalen 1991).

Isopods

Isopods are an important group of arthropods in heavy metal research because of their unique ability to concentrate high amounts of metals in their bodies (Hopkin 1989). Most isopods are easy to culture and do not require special conditions. Disadvantages of isopod toxicity testing include a relatively long (1 year) life-cycle and the fact that they live in the surface litter, not the soil itself. The most suitable test species *Trichoniscus pusillus* is parthenogenetic (Van Straalen and van Gestel 1993). The use of isopods as a test group for soil toxicity testing is restricted to a few studies, and there were no attempts made to standardize this test (Hopkin 1989; Donker and Bogert 1991; Van Straalen and Verweij 1991).

Multispecies test systems: Soil microcosms and terrestrial model ecosystems

The single-species tests described above can provide important ecotoxicological information for chemicals and contaminated soils. However, they have 1 major limitation: they do not assess population interactions or community-level and system-level effects of soil contamination on soil ecosystems. A simple and inexpensive multispecies microcosm technique can be used to supplement a battery of single-species tests (Kuperman et al. 2002). Multispecies microcosms were shown to have the potential to detect more subtle indirect effects of chemicals on soil food web structure (Parmelee et al. 1993, 1997). The most standardized method is the terrestrial model ecosystem (TME), an open, undisturbed soil core used primarily for examining contaminant transport (Checkai et al. 1993; UBA 1994). Recently, TMEs have been tested at 2 levels of complexity: a small, integrated soil microcosm, with sieved soil, selected introduced and indigenous invertebrates, single plant species, with large numbers of replicates under controlled laboratory conditions, and a larger TME with intact soil cores, indigenous invertebrates, greater biodiversity, mixed plant flora, and less replication under semi-field laboratory conditions (Bogomolov et al. 1996; Edwards et al. 1996, 1998). Such a multispecies test system can offer high resolution of the ecotoxicological testing of chemicals in complex soil ecosystems.

Plant toxicity tests

The use of plant tests to assess soil conditions dates back to at least 1840, when Leibig investigated the effects of metals and other plant nutrients on plant growth. Rice (1984) quotes from Theophrastus (300 BC) and Pliney (1 AD) on the use of plants to detect "soil sickness." Through this long, albeit patchy, history of phytotoxicity, tests have become standardized and used for many purposes. Though regulatory standardized tests list a limited suite of agronomic species, nearly a hundred species have been tested routinely to study phytotoxicity (Kapustka 1997). Fletcher et al. (1988) reported phytotoxicity test results for 1569 plant species from 682 genera and 147 families in the PHYTOTOX database. Measured endpoints in standardized tests included percentage germination, percentage emergence, shoot mass, shoot height, root mass, root length, total plant mass, survival, fruit and seed set, and general observations of plant condition. The general literature on plant toxicity expands the measurement endpoint list to include chlorophyll content, chlorophyll fluorescence, chlorosis, peroxidase activity, and chemical residues.

In addition to the regulatory testing guidelines of the USEPA, U.S. Food and Drug Administration (USFDA), OECD, and others, the American Society for Testing and Materials (ASTM) has published standard test procedures that are intended to have broader use than those devised for regulatory purposes (ASTM 1999b, 1999c). The latest ASTM standards present a framework for conducting phytotoxicity tests using either contaminated soils or clean soils to which a test substance is added. It includes annexes on seedling emergence, root elongation, Brassica life-cycle, and woody plant tests. The depth and breadth of test procedures in the ASTM standards should help expand the range of phytotoxicity testing beyond the nearly exclusive reliance on lettuce germination and radish root elongation tests in the evaluation of contaminated soils. Each method can be described with numerical and narrative data quality objectives and quality assurance–quality control (QA–QC) features that meet the standards for regulatory or litigation use.

The mass and distribution of roots in soil are influenced by physical, chemical, and biological conditions of the soil. Contaminants can inhibit root growth and greatly reduce the fitness of plants, even though the aerial portion of the plants shows little or no overt signs. Direct observation of root condition can yield valuable information for ERA. Quantitative measures of root mass or length per unit volume of soil can range from relatively simple to exceedingly complex. Methods ranging from simple excavation to obtain static measurements to sophisticated viewing scopes that provide growth rate information can be found in Fitter et al. (1985).

Mycorrhizal colonization and spore counts can provide useful indications of the interactive functions of symbiotic fungi and vascular plants. More than 95% of terrestrial plant species and nearly 99% of individual terrestrial plants have mycorrhizal associations. Uptake of soil nutrients, water, and contaminants is typically modulated by mycorrhizal fungi. Therefore, assessing the status of mycorrhizae in terms of contaminant effects has great ecological relevance. Methods to quantify mycorrhizal colonization of plant roots and to quantify mycorrhizal spore populations are presented in several publications (Schenck 1982; Harley and Smith 1983; Fitter et al. 1985; Allen 1991; Linder et al. 1992)

Vegetation surveys

Plants and soils interact dynamically with the physical and chemical characteristics of soils, greatly influencing the types of plants and level of primary productivity within a given climatic setting. Simultaneously, contributions of litter and the sloughing of roots, as well as secretions from plants, contribute to soil organic matter characteristics. The suite of microbes and soil invertebrates that colonize living and dead plant matter catabolize these organic products, resulting in marked differences in soil organic content and chemical character among deciduous forests, coniferous forests, and grasslands. Over decades, or perhaps centuries, the soils diverge sufficiently to the point of being identifiable according to forest type or grassland, each supporting different suites of aboveground and belowground communities.

Vegetative cover is an important indicator of soil quality. A robust, diverse plant community indirectly signals that general soil services such as decomposition, mineralization, and secondary productivity of soil invertebrates are functioning normally. Soil contamination that interferes with basic soil community processes or has direct toxic effects to plants is likely to be manifested in plant community structure or function. Therefore, one approach to characterize chronic effects of soil contamination is to assess plant community parameters such as cover, biomass, species diversity, functional diversity, and structural features such as life forms or layers.

The basic tools of plant ecology sampling have been described in many books and articles (Mueller-Dombois and Ellenberg 1974; Green 1979; Meyers and Shelton 1980; Avery and Burkhart 1983; Greig-Smith 1983; Moore and Chapman 1986;

Barbour et al. 1987; Bonham 1989; Kapustka and Reporter 1993). Basic vegetation sampling methods have been abstracted into an ASTM standard that includes annexes on specific sampling techniques (ASTM 1999d). The various techniques provide a range of rigor and sophistication in sampling that can be selected to meet data requirements for risk assessments. Relatively simple reconnaissance methods can yield estimates of vegetative cover and species richness (e.g., relevé method; see ASTM 1999d). More involved defined-area sampling methods may be selected to provide quantitative estimates of plant cover, biomass, density, and frequency for each taxon.

Alternative measurement techniques to assess "functional quality" of plant communities have been employed to describe suitability of vegetation cover for wildlife. Various habitat evaluation procedure (HEP) models quantify vegetation structure in terms of layers and species composition (Schroeder and Haire 1993; Rand and Newman 1998). Differences in vegetation structure and the functional importance to wildlife have been linked to soil contaminants (Gailbraith et al. 1995).

Soil microorganisms

Microbial populations in soil are extremely large. Bacteria are the most numerous soil microorganisms, ranging from 10^6 to 10^9 bacteria per gram of soil. The numbers of actinomycetes range from 10^5 to 10^8 per gram of soil, and fungi range from 10^4 to 10^7 per gram of soil. This represents a biomass of 300 to 3000 kg/ha of bacteria, 700 kg/ha of actinomycetes, and 500 to 5000 kg/ha of fungi. They differ greatly in function, with various groups involved in organic matter breakdown, degradation of xenobiotics, nutrient cycling, and as pathogens of plants and invertebrates. The methods commonly used to assess the effects of chemicals on soil microorganisms were reviewed and summarized by Edwards (1989), and further details of the following recommendations can be found in that paper.

Chemicals can be applied to field plots or soil samples collected from the field, and the resulting effects can be assessed by counting populations of microorganisms in the plots or samples at intervals after treatment and comparing them with populations in untreated control samples. However, such an assessment of whole populations of microorganisms is both difficult and time-consuming. It is possible to count microorganisms directly under a microscope using soil suspensions from treated soils. Such tests have the advantage of being carried out under ecological conditions that resemble those in the field. However, when assessing effects, it is difficult to distinguish between living microorganisms and dead ones, and dead organisms are not usually included in the characterization of the active populations in a soil. The use of fluorescence and coloration techniques can reduce such errors to a certain extent (Edwards 1989). There are several methods for estimating microbial biomass measurements of adenosine triphosphate (ATP) (Edwards 1999), and muramic acid is most often used. The biomass may also be calculated from the biovolume by multiplying the number of organisms in a soil sample by the mean volume of an organism.

Microorganisms are usually counted after being grown on suitable media, which are usually synthetic organic media, either liquid or solid (with gelose) or mineral (silicagel). Microorganisms are introduced from inoculum suspensions, more or less diluted with water, into the medium or onto its surface. After a period of incubation, the numbers of microorganisms are estimated from the count of the colonies on the plates or from an estimation of the highest dilution that permits growth of the microorganisms. The number of microorganisms per gram of soil can be calculated on the basis of observed growth. This method of determination may be used on soil samples taken directly in the field from treated experimental plots, or in trials carried out in vitro in a laboratory or greenhouse. The total microflora is usually counted in an aqueous soil extract (Paul and Clark 1996). Some researchers add yeast and mineral salts to this medium. Other media used for the determination of the total microflora include Kaunat (mineral gelose medium with casein, peptone, and glucose) and Thoronton's medium. Fungal biomass can be estimated by determining the numbers, diameters, and lengths of fluorescein diacetate-stained fungal hyphae in soil-agar films using phase-contrast microscopy (Söderström 1977; Ingham and Klein 1984). Counts of bacteria are often made on a medium called "nutritive gelose" or on Czapek's gelose medium (Wainwright and Pugh 1974). Spore-forming bacteria can be separated from the others by heating the soil for 10 minutes at 90 °C or by heating a soil suspension for 15 minutes at 75 °C. Actinomycetes can be counted on media containing antibiotics such as nystatine and actidone or albamycine and streptomycin. Such plating methods can provide estimates of the total populations of the major groups of microorganisms in control plots and samples.

Fumigation-extraction procedures can also be used to estimate soil microbial biomass. In these procedures, microbial cells are killed (lysed) by fumigation, and the C and/or N thereby rendered extractable are calculated as the difference between the amount extracted from the treated and untreated soils (Ross 1989). Chloroform ($CHCl_3$) fumigation of soil that causes increases in total (inorganic and organic) K_2SO_4-extractable N (Brookes et al. 1985) is a commonly used method to measure soil microbial biomass.

The numbers of active and total soil bacteria and fungal hyphal lengths can be determined using direct counting methods. Bacteria are determined microscopically using a europium chelate with fluorescent brightener as a differential stain (Anderson and Westmoreland 1971). Fungal hyphal lengths are determined by counting water soluble aniline blue-stained hyphae collected on irgalan black-stained Gelman membrane filters (Newell and Hicks 1982).

Bioluminescence-based bacterial assays (e.g., Microtox) used for assessing the toxicity of many different types of environmental samples (Steinberg et al. 1995) have proved to be simple, reproducible, and cost-effective. An assay with the bacterium *Vibrio harveyi* (Lapota) showed that it was a sensitive and reproducible test that can be used for rapid screening of large numbers of samples, including field screening using portable equipment (Kuperman et al. 1997). The endpoint in this bacterial assay is the reduction in bioluminescence due to the inhibition of enzyme activity as an expression of changes in cell metabolism resulting from exposure to toxicants. Both acute (1 hour) and log-phase growth (6 hours) tests can be used in this assay. The acute test detects toxicity primarily due to inhibition of bioluminescence due to effects on general metabolism, while lack of luminescence in the growth assay may also be due to the inhibition of growth, inhibition of luciferase synthesis, or inhibition of the luminescence enzymes as they are synthesized. However, the interpretation of this type of bioluminescence assay is limited to comparing the relative toxicity of samples and should not be misconstrued as having significant relevance to soil ecosystem function. Limitations of the Microtox assay for soil systems include the use of a marine bacterium for testing soil toxicity, and the assays are conducted using soil extracts.

These limitations have resulted in the development of other bioluminescence assays based upon the insertion of *lux* genes into more ecologically relevant bacteria such as *Pseudomonas fluorescens* and *Rhizobium leguminosarum* biovar *trifolii* (Paton et al. 1995; Paton, Rattray et al. 1997; Paton, Palmer et al. 1997). Lux genes may either be linked to no specific reporter gene and provide a general toxicological response based upon the inhibition of general cellular metabolism or be linked to reporter genes involved in the metabolism of specific chemicals (e.g., Hg, naphthalene) (Heitzer et al. 1992; Selifonova et al. 1993). Linking lux genes with specific reporter genes offers great promise as a technology that may be used to assess contaminant-specific bioavailability and toxicity in contaminated soils.

Soil ecological functions

Information on the impacts of toxicants on component processes in nutrient cycling is important because it provides ecologically relevant data on a complex system. Use of these tests as an additional line of evidence can provide assessors linkage to more complex assessment endpoints. However, for utilization in ERAs, their usefulness may be limited unless issues such as data quality, sensitivity, precision, and accuracy are fully addressed. Additional research and validation are needed to further apply these methods in regulatory arenas.

Organic matter breakdown

In many natural and managed ecosystems, the bulk of the annual net primary productivity is transferred directly to the decomposer subsystem. Decomposition and mineralization of nutrients contained in this organic material are critical processes affecting soil fertility and primary productivity of ecosystems. The distribution and turnover of organic matter contributes to ecosystem functioning by forming soil organic matter pools and nutrient exchange sites for root uptake. The recycling of nutrients contained in litter is an important aspect of soil ecosystem dynamics, and the regulation of rates of nutrient release plays an integral role in the functioning of soil ecosystems (Blair and Crossley 1988). Organic matter decomposition is one of the most integrated processes within the soil ecosystem because it involves complex interactions of soil microbial and faunal activity with the soil-chemical environment. Species from different taxonomic groups contribute to the breakdown of plant and animal debris. Any disturbance that alters organic matter decomposition can result in nutrient losses and decreased soil fertility. Therefore, an assessment of how chemical contamination may alter rates of organic matter decomposition and the rates of nutrient retention and release is critical to understanding chemical impacts on overall ecosystem structure and function (Edwards 1979). Methods of determining the effects of contaminants on organic matter decomposition are summarized below.

Bait-lamina test

The bait-lamina test (Von Törne 1990) is the simplest and most practical test for assessing whether organic matter decomposition has been affected or not. The production of bait-lamina is well standardized, but the major drawback is that it does not allow one to determine the relative contribution of different groups of soil biota to the decomposition process, to quantify the decomposition rate, or assess nutrient mineralization or immobilization dynamics in decomposing material. However, the bait-lamina method does provide a measure of overall feeding activity of soil invertebrates such as mites, Collembola, Enchytraeids, and earthworms.

Cotton strip assay

The cotton strip assay (Latter and Howson 1977; Kratz 1996) uses the loss of tensile strength of a cotton strip exposed to soil as a surrogate for the decomposition process. An advantage of this method is the availability of a well-standardized cotton material that is very consistent in composition and tensile strength. Disadvantages include the need for special equipment that is limited to measuring tensile strength loss and the unclear relevance of the degradation of pure cotton to the degradation of natural litter.

Litter bag method

The litter bag method (Crossley and Hoglund 1962; Edwards and Heath 1963; Heath et al. 1964) allows the determination of chemical effects on both mass loss and patterns of nutrient dynamics in decomposing litter. Although it requires more effort than the 2 methods described above, this technique provides ecologically relevant information and is preferable. The disadvantages of this method include

- substrate packed in a litter bag may create a microclimate condition different from bulk soil,
- substrate in a litter bag does not come into contact with contaminated soil (this problem may be overcome by spiking litter with test chemicals at the same rate as the bulk soil), and
- litter in the bag may attract soil organisms, which in turn may overestimate the actual rate of decomposition.

Soil respiration

Soil respiration, as represented by oxygen uptake and carbon dioxide evolution, is an excellent index of overall biological activity in soil. Respiration measurements are probably the most common methods for assessing microbial activity in soil because respiration usually correlates well with other soil activities, such as transformations of C, P, and N. The use of radiolabeled isotopes has alleviated many of the problems of nonspecificity involved in traditional methods of measuring respiration.

Carbon mineralization potential can be determined using the substrate-induced respiration method. The addition of substrate (glucose) to a soil sample induces a maximal respiratory response from the soil microbial biomass, measured as CO_2 evolution. Microbial biomass is the main acting agent for most soil biogeochemical processes in terrestrial ecosystems, which interacts with the primary productivity of ecosystems by regulating nutrient availability and degradation pathways of soil contaminants.

Substrate-induced respiration is determined using a soil respiration measuring system with continuous gas flow involving an incubation chamber, an air flow controlling unit, an air flow measuring unit, and a CO_2 analyzer (Cheng and Coleman 1989). An additional advantage of this method is that selective biotic inhibitors, such as streptomycin and cyclohexamide, can be added to the glucose solution to determine the contribution of fungal and bacterial groups to overall soil respiration (Anderson and Domsch 1973).

Oxygen uptake can be measured using electrolytic respirometers. The general principle of these is the incubation of soil samples in closed cells with periodic introduction of oxygen by electrolysis to compensate for the soil uptake. In order to work under more natural ecological conditions, some workers have made their activity measurements in situ directly, in contaminated soil, for instance, by Dommergues' technique (Dommergues 1960).

One of the most serious criticisms of using respiration measurements to assess the effects of chemicals on soil respiration, whether by CO_2 production or O_2 uptake, is that a measured nil or small effect may hide large, counterbalancing stimulation and inhibition of different major parts of the soil microflora. This criticism (e.g., a toxic effect on respiration that is balanced by extra respiration resulting from utilization of the chemical as a substrate) certainly leads to the conclusion that, in the cases with no significant reductions in CO_2 evolution, respiration measurements should be used as a supplementary index in conjunction with other methods of assessing more specific side effects on microbial function. Despite these drawbacks, a soil respiration test is currently being considered by the OECD as a standard test (OECD 1999) for investigating the effects of chemicals on C transformations in soil.

Nitrogen transformations

Many free-living microorganisms can fix atmospheric N. In some systems, such as rice paddy soils, the most important dinitrogen-fixing organisms are cyanobacteria. Cyanobacteria also contribute significant portions of the annual N input in grasslands and desert environments (Kapustka and Rice 1978; Kapustka 1990). Several genera of heterotrophic bacteria associated with roots are important contributors of N in grasslands. N fixation by symbiotic organisms associated with legumes is of great importance in tilled pasturelands. Bacteria in the genus *Rhizobium*, which form symbiotic relationships with legumes, are highly susceptible to chemicals and have been investigated by a variety of methods, ranging from plate counts of *Rhizobium* cells to examination of nodulation of legumes (Bergersen 1980). Similarly, actinorrhizal associations with actinomycetes (*Frankia* spp.) in a diverse range of plant families (e.g., alder) are important sources of new N in forests (Gordon et al. 1979). Relatively less research has been done on actinorrhizal groups, but they appear to be nearly as sensitive to chemical stressors as rhizobia.

Nitrogen utilization requires distinct enzymatic steps for extracellular hydrolysis, uptake, deamination, and intracellular catabolism, each of which could be regulated differently (Smith et al. 1989). The dynamics of soil N transformations are determined by measuring changes in the sizes of soil N pools. Rates of net N mineralization are calculated as the difference in soil inorganic N at the beginning and end of a particular time period, plus the amount of N taken up in plants and lost through leaching during that period. Leaching losses of inorganic N are measured in soil leachate samples using a N analyzer.

Because mineral N is a major plant nutrient, there have been many studies on the effects of chemicals on the mineralization of soil N. Another reason for the abundance of such studies is that chemical analytical methods for following N transformations, especially for measuring nitrate-, nitrite-, and ammonium-N, are sensitive and reproducible. They are also readily automated and thus allow optimum replication within experiments.

Ammonification and nitrification, as parts of the N-mineralization process, are the most useful systems in studying the side effects of chemicals on N cycles. Ammonificiation is the indicator for the release of N bound in organic matter and its availability for plant nutrition. Nitrification is the oxidation of ammonium to nitrite and nitrate. Unlike C mineralization, nitrification is one of the most sensitive processes related to soil function because conversion of ammonia to nitrate is carried out only by a limited number of bacterial genera (mostly *Nitrosomonas* and *Nitrobacter*), and impairment of nitrification may be indicative of harmful influences to chemicals.

The effects of chemicals on the release of ammonium from organic matter can be studied in the same way as the mineralization of C, for example, by addition of 0.5% lucern-meal to soil. If ammonification is not affected, the 2 steps of nitrification can be studied in the same experiment. If ammonification is inhibited, a separate experiment is required to study nitrification independently. This is achieved by adding ammonium sulfate (100 mg N/kg) to the soil and following the disappearance of NH_4 and the appearance of NO_2 and NO_3.

Tests aimed at assessing the effects of chemicals on N fixation depend upon the unique symbiotic relationship between the host plant and the dinitrogen-fixing microorganisms (e.g., *Rhizobium* or actinomyete) and can determine, in a single test, both plant and bacterial responses to a chemical. The endpoints that are measured include plant growth, plant yield, and estimates of healthy nodulation over time. These experiments can be conducted in pots containing a soil type suitable for growth of the plant and treated with chemicals. The plant seeds need to be inoculated with *Rhizobium* only if there are no suitable bacteria present naturally in the soil. If appropriate, the plants receive only a mineral fertilizer that excludes N. If the chemical causes a critical effect in these tests, further data can be obtained on the direct effects of the chemical on the *Rhizobium* organism, the nodulation process, and the N fixation (e.g., acetylene reduction) ability of the whole plant or intact root system. Other laboratory methods are available for assessing the effects of chemicals on N fixation using the acetylene assay and for nitrification by assessing NO_3 production by the phenoldisulfonic acid method or an ion probe. Denitrification can be assessed using disappearance of NO_3 ions. With appropriate culturing modifications, similar tests on actinorrhizal species can be conducted with woody plants.

Soil enzymatic activity

Enzymes present in soil can be of extracellular and intracellular origin and can be evaluated after chemical treatment of a soil to find correlations between enzymatic activity and chemical concentration. Enzymatic test methods are attractive because of their simplicity and reproducibility compared to classical techniques used in soil microbiology; yet interpretation of the data requires considerable care because we have limited knowledge of soil enzymes, their origins, and their relationships to soil fertility. USEPA guidelines state specifically the need for the measurement of activities of enzymes such as phosphatase. Kuperman and Carreiro (1997) reported a strong correlation between soil heavy metal concentration and activities of soil extracellular enzymes involved in C, N, and P cycling.

The potential activities of the extracellular enzymes -1,4,-glucosidase (C-acquiring enzyme), n-acetylglucosaminidase (N-acquiring enzyme), and acid phosphatase (P-acquiring enzyme) can be easily quantified. Assays are conducted using soil slurries suspended in acetate buffer, and enzyme activities can be measured spectrophotometrically using substrates bound to the chromogen, *p*-nitrophenol (Sinsabaugh and Linkins 1990). Urease is also an important enzyme in biochemical transformations of N in the soil. N limitation to microbial and plant growth may be important when soil water and organic C sources are abundant. Urease activity can be assayed by determining the release of ammonium-N following urea additions to 1 gram soil subsamples.

The intracellular dehydrogenase enzyme plays an important role in cellular respiration, and therefore can provide a good index of the respiratory status and biomass of soil microorganisms. Casida (1977) described a modified method to determine soil dehydrogenase activity.

Integration

Interpretation of soil biota methods

The field of soil ecology has developed methods to sample and identify soil organisms in most taxonomic groups (Dindal 1990; Edwards 1991). Studies in polluted soil systems have begun to show the diverse responses of soil communities to contaminants, and some ecologists have advocated using soil community structure to evaluate ecosystem condition. Some groups of organisms, such as earthworms, have been widely studied, whereas others have been studied by relatively few specialists. One of the major problems is that the extreme complexity and variability in soil community structure precludes comparison of communities from different sites. Population estimates are also highly dependent on sampling procedures and local environmental conditions. Moreover, soil ecologists rarely study all taxa present, but select a subset that differs from study to study. Beyond these practical difficulties, soil ecologists have the greater problem of interpreting data on populations of soil organisms once they are measured and deciding whether observed deviations from the presumed norm are meaningful ecologically.

Soil ecologists bring different assumptions and values to their studies, influencing the way that they interpret their data and complicating the selection of appropriate assessment endpoints. For some ecologists, maintaining the structure of the ecosystem is an end in itself, and any reduction in population size, species diversity, or even genetic diversity is detrimental (Büchs 1994; Pankhurst and Lynch 1994). Other ecologists may consider a change in ecosystem structure detrimental only when it affects function. Structure and function are of course related, and various ecologists have argued that reductions in soil organism populations caused by contaminants have led to harmful consequences such as a reduction in plant productivity or a disruption of nutrient cycling (Watson et al. 1976; Witkamp and Ausmus 1976; Hoy 1980; Hutton 1984). However, the linkage may be weak and difficult to predict. Often soil ecologists frame their arguments in terms of sustaining ecosystems, while other soil scientists prefer to use the term of "soil quality." Although the concepts ought to overlap broadly, the discussions and the published literatures of the 2 views have remained largely separate. Soil scientists acknowledge that soil fauna have a role in soil quality, but in practice, they rely mainly on traditional chemical and physical measures of the soil (Doran and Safley 1997).

Earthworms, of all soil invertebrates, are most likely to be useful as indicators of soil health, albeit limited to those soils in regions that normally harbor earthworms. Extensive research has demonstrated that, although there are some exceptions to the rule, earthworms are in general beneficial to soil (Guild 1955; Van de Westeringh 1972; Van Rhee 1977; Edwards and Lofty 1980; Edwards 1981; Lee 1985; Logsdon and Linden 1992; Blair et al. 1995; Edwards and Bohlen 1996). Earthworms tend to promote soil health by forming soil aggregates, aerating soil, improving drainage, and increasing fertility. A review of studies from 6 countries, including 14 plant species, 6 groups of soils, and 13 earthworm species, demonstrated significant increases in plant growth in the presence of earthworms (Brown et al. 1999). The mean increases in shoot and grain biomass due to earthworms were 56% and 36%, respectively. Benefits of earthworm introduction were greatest in perennial crops. However, in reviewing 10 agronomic case studies, Doube and Schmidt (1997) concluded that earthworm abundance was not a good predictor of plant productivity. Although the earthworms present may have been beneficial to soil, these authors concluded that the many interacting environmental and anthropogenic factors affecting both plant productivity and earthworm abundance in agroecoystems precluded a simple relation.

Soil ecologists, similarly to most ecological disciplines, lack a simple, accepted means of interpreting the significance of population estimates of soil organisms and of their community structure. As with most fields of ecology, debates occur between a recognition of the complexity of soil ecosystems and the goal to develop simple measures of soil ecosystem condition. Diversity indices, such as the Shannon-Weiner index, have been promoted in some discussions of soil ecotoxicology as a means of summarizing population data into a single number that represents the condition of the ecosystem (Gupta and Yeates 1997). In extremely contaminated soils, diversity indices certainly decrease (Hoy 1980; Bengtsson and Rundgren 1984). The use of diversity indices has been criticized, however, because they may be insensitive to pollutants and because there are no expected values to compare to observed values (Forbes and Forbes 1994). In practice, simple diversity measures have frequently failed to show expected decreases in response to contaminants (Watson et al. 1976; Dindal 1977; Myrold 1990; Beyer and Linder 1995; Paoletti et al. 1995; Spurgeon et al. 1996). Some have suggested that the more complex indices will eventually prove to be more useful than the simpler ones, and Sheehan (1984) provides a list of 18 indices that may be suitable under particular circumstances, as well as a few alternative statistical approaches to evaluating populations. From an agronomic point of view, soil fertility may be maintained as diversity is decreased. Pankhurst and Lynch (1994) state that some loss of species diversity will inevitably occur in managed agriculture. Some biologists have expressed concern over the usefulness of soil invertebrate bioindicators, stating that their use is appealing but unproven (Van Straalen 1997).

Soil communities change as soil becomes severely polluted, affecting ecosystem function. Decreases in populations of soil biota have been associated with an increase in the mass of partially decomposed litter (Watson et al. 1976; Hutton 1984). In studies on microorganisms, both the numbers of species present and their activity was reduced, although some resistant species survived even in the most severely polluted sites (Jordan and Lechevalier 1975; Kelley and Tate 1998). Populations of earthworms and arthropods were reduced in metal-contaminated sites, decreasing both the rate of litter decomposition and its incorporation into the mineral soil (Bengtsson and Rundgren 1984; Tyler 1984; Bengtsson et al. 1988). Studies on soils contaminated with pesticides have shown similar effects (Hirst et al. 1961; Cook and Swait 1975). Reductions in earthworm populations have caused an increase of organic matter in turf and buildup of litter in intensively managed orchards (Turgeon et al. 1975; Potter et al. 1990). At several contaminated sites, there has been a shift from organisms associated with a mull litter to those associated with a mor litter (Van de Westeringh 1972; Beyer and Linder 1995). Species composition, trophic structure, and abundance of nematodes also change predictably in response to pollution, at least on the gross level (Yardim and Edwards 1998). Nematodes have been classified in terms of their tendency to be colonizers or persisters on a scale of 1 to 5. The fraction of the population in each category multiplied by the value for each category was suggested as an index of chemical or physical disturbance of the soil (Bongers 1990; Korthals et al. 1996). From these studies, it was concluded that soil ecologists may be able to predict qualitative population changes caused by contaminants and to relate these changes to meaningful functional changes in soil ecosystems. However, Sparling (1997) and Doube and Schmidt (1997) suggest that because soil organism populations are naturally affected by so many soil variables, attributing changes in populations to contaminants may be difficult.

Soil biota testing for measurement endpoints

This section provides additional information on the use of soil biota methods to supply information on the soil ecosystem. The selection of a suitable laboratory toxicity test method is determined by the proposed future site use and the goals of ERA formulated early in the process. In cases where the contaminated area will be used for industrial development with minimal areas of land supporting natural habitats, the battery of toxicity tests may be limited to just a few assays such as earthworm toxicity, organic matter breakdown, and/or plant toxicity tests. In contrast, when a future site use entails managing a diverse habitat, the list of test methods should be extended to provide information on how chemical contamination affects both structure and function of soil ecosystems (Table 4-3).

Table 4-3 Summary of toxicity test methods available for assessing structural and functional endpoints in soil systems

Measurement endpoint	Method
Structural endpoints	
Earthworm survival and reproduction	Standard 14-day toxicity test; Cocoon production test
Enchytraeid survival and reproduction	Enchytraeid reproduction test
Collembolan survival and reproduction	Collembolan toxicity test
Nematode survival and reproduction	Nematode life-cycle test
Plant growth	Seed germination, root elongation, biomass
Microbial biomass	Direct counts; Fumigation-direct extraction; SIR[a]
Functional endpoints	
Organic matter breakdown	Litter bag method; Bait-lamina test; Cotton strip assay
Carbon transformations	SIR; Basal respiration
Nitrogen cycling	Mineral N concentrations
Soil enzymes	C-, N-, P-acquiring enzyme activities; Dehydrogenase activity

[a] SIR = Substrate-induced respiration.

Guidance for Use of Soil Testing Methods

Table 4-4 presents qualitative evaluations of the stage of development, the relative cost, the technical skill required, and the utility of the soil testing methods described in this chapter. This information can guide risk assessors in the selection of appropriate tests or sampling methods.

Table 4-4 Comparison of the stage of development, relative cost, technical skill required, and utility of toxicity methods for assessing the toxicity of contaminated soils

Category Method	Stage of development, acceptance[a]	Relative cost[b]	Technical skill required[c]	Utility[d]
Vegetation surveys				
Relevé	H	L	M	M
Point frame	H	L	M	M
Line intercept	H	L	M	L
Point quarters	H	M	M	H
Variable angle	H	L	M	L
Defined area	H	M	M	H
Root distribution	L	L	L	M
Root observation	L	M	M	M
Phytotoxicity tests				
Early seedling growth	H	M	M	H
Seedling emergence	H	L	L	L
Root elongation	H	L	L	M
Brassica life cycle	M	H	M	H
Woody plant assay	M	M	M	M
Plant community test	L	H	H	M
Invertebrate surveys				
Pitfall traps	M	L	M	L
Sticky traps	M	L	M	M
High gradient extractor	M	L	M	M
Soil sorting	M	M	H	M
Wet funnel	M	M	M	M
Invertebrate toxicity test				
Earthworm survival	H	L	L	M
Earthworm reproduction	H	M	M	M
Enchytraeid reproduction	M	L	L	M
Nematode life cycle	L	M	M	L
Collembolan life cycle	M	L	M	M
Orbatid mites	L	M	M	L
Isopods	M	M	L	M
Terrestrial model ecosystem	L	H	H	M

-continued-

Table 4-4 continued

Category Method	Stage of development, acceptance[a]	Relative cost[b]	Technical skill required[c]	Utility[d]
Microbial surveys				
Most probable number (MPN)	H	L	L	L
Selective media: bacteria	H	L	M	M
Selective media: fungi	H	L	M	M
Direct counts: bacteria cells	H	L	M	L
Direct counts: fungal spores	H	L	M	L
Bioluminescence	M	L	M	M
Hyphal length	M	L	M	M
Mycorrhizal spore count	M	L	M	L
Mycorrhizal colonization	M	M	M	M
Microbial processes				
Bait-lamina	M	L	L	M
Cotton strip	M	L	L	M
Litter bag	M	L	L	L
Carbon mineralization	M	L	L	M
Nitrification	M	L	M	M
Dinitrogen fixation	M	L	M	M
Denitrification	M	L	M	M
Cellulase	M	L	M	M
Amylase	M	L	M	M
ß-1,4-glucosidase	M	L	M	M
N-acetylglucosaminidase	M	L	M	M
Phosphatase	M	L	M	M
Urease	M	L	M	L
Lipase	M	L	M	L
Dehydrogenase	M	L	M	L
Microbial toxicity tests				
Microtox	H	L	L	L
Lux-based bioassays	M	L	L	M-H

[a] State of development: H = Published standardized method either in ASTM or USEPA format; M = widely used procedure though not available in "Standard" Format; L = limited use, procedure described in primary research papers only.

[b] Relative cost: L = $10 to $500 per unit on tests or <$2000 for survey; M = $500 to $1000 per test or $2000 to $20,000 for survey; H = >$1000 per test or >$20,000 for survey.

[c] Technical skill required: L = entry level technician; M = highly trained technical staff required; H = highly trained and highly specialized staff required.

[d] Utility: L = "noisy" results that may pose some difficulty linking to contaminant effects; M = relatively consistent results and relatively easy to link to contaminant; H = consistent results and readily linked to contaminant effects.

Relating Concentrations of Contaminants in Soil to Wildlife

Wildlife may ingest soil deliberately, such as when they visit salt licks (Kreulen and Jager 1984), or inadvertently when they eat vegetation with adhering soil or prey containing soil. Soil ingestion has been shown to be the most important route of exposure for some contaminants in domestic animals feeding in contaminated fields (Fries et al. 1982). Soil ingestion was shown to be important in estimating exposure of pronghorn and black-tailed jackrabbits to elements such as V, Na, Fe, and F (Arthur and Gates 1988). This route of exposure is generally most important for those contaminants, like Pb, that are not bioaccumulated. The soil ingestion rate of an animal may vary substantially as it selects different foods or when its food becomes contaminated with soil, such as after a heavy rain.

Average soil ingestion rates showing general tendencies have been measured in a few species of wildlife (Beyer et al. 1994). In estimating the importance of this exposure route at a contaminated site, a risk assessor often selects a species with a high rate of soil ingestion and then decides to model exposure to either an individual with an average ingestion rate, or an individual that has greater than average exposure, at the 90th percentile, for example. Both ingestion of soil and exposure to poorly absorbed contaminants may be estimated for wildlife from digesta taken from carcasses or from scat (Beyer et al. 1998). The assessor should be aware that ingestion rates may be highly seasonal and that collections made at one time may not be representative of the highest exposures, or they may substantially overestimate exposure. Concentrations detected in the scat may be converted to estimates of concentrations in the diet (Beyer et al. 1998), which in turn may be compared to concentrations reported in the literature as having caused toxicity. Possible differences in bioavailability between a contaminant in soil at a site and contaminant administered in a published toxicity test should also be considered.

In the absence of collections from the site, soil ingestion estimates may be taken from the literature for an animal selected as an appropriate model. As an example, consider armadillos who ingest an average of 17% soil in their diet as they feed (Beyer et al. 1994). An armadillo feeding in a soil containing 100 mg/kg Pb might be expected to ingest 17 mg/kg Pb in its diet (from the soil), plus whatever Pb was in its prey. It is assumed that the Pb concentration of the ingested soil matches that of the soil as it might be sampled. This assumption is not valid in studies of humans exposed to Pb, in which ingested particles have been shown to be much finer than those found in the ambient soil as a whole (Sheppard 1995), but it is probably reasonable for animals ingesting a lot of soil.

Wildlife may also be poisoned by soil-borne contaminants through their food chains or from prey or vegetation. The vegetation route may be relevant for some elements, such as Se, and the evaluation should be straightforward. The transfer of contaminants from soil to earthworms and to wildlife may be a valuable model useful in risk assessments of predators exposed to contaminated soils, especially for contaminants that do bioaccumulate. The importance of this route of exposure was demonstrated years ago by Barker (1958), who showed that DDT remaining in soil from the previous year's application bioaccumulated in earthworms and was biomagnified during trophic transfer, poisoning songbirds feeding on the earthworms.

Invertebrates besides earthworms take up contaminants, but in general, the earthworm is probably the best choice for a model in risk assessments: Earthworms are eaten at least occasionally by animals belonging to many species, they are large enough to yield enough tissue for analysis, and the literature contains bioaccumulation data on many contaminants. When conducting a site assessment, it is best to collect earthworms at that site, but rough estimates may be made from published studies relating soil concentrations to earthworm concentrations. However, there may be considerable error associated with these estimates from the literature because uptake will depend greatly on the soil conditions at the site, and the bioaccumulation factor is dependent upon the bioavailable fraction of chemical in the soil.

Contaminants such as dieldrin and some other organochlorine pesticides accumulate to higher concentrations in earthworm tissue than in soil. In an experimental study, dieldrin was applied to soil and measured in soils and earthworms for 11 years. After 11 years, the soil dieldrin concentration was about 2 mg/kg and the dieldrin concentration in earthworms was about 8 mg/kg, which in laboratory studies was shown to be toxic to birds (Beyer and Gish 1980). This simple approach associates a soil concentration with a presumed hazard to birds. It was assumed that the bird ate only worms, or that the bird ate other invertebrates as well, but on average, these other invertebrates contained concentrations that were similar to those in earthworms.

Heavy metals, especially Cd and Zn, can bioaccumulate in earthworms (Edwards and Bohlen 1992, 1996), however, Pb does not usually bioaccumulate. Most of the Pb in an earthworm is found in the soil in its gut, rather than in its tissue. Consequently, the exposure of a bird to Pb from eating earthworms depends upon the mass of soil in the earthworm gut, the concentration of Pb in the soil ingested by the earthworm, and the bioavailability of the Pb to the bird (Beyer and Stafford 1993). This simplifies the risk assessor's job because soil contamination may be directly linked to exposure in wildlife. However, estimating the toxicity to wildlife through this route of exposure should take into account the bioavailability of the contaminant, information that may not be available for some metals. For Pb, the best studied metal in this respect, bioavailability compared to Pb acetate was estimated to be 50% in one study, before concluding that 670 mg/kg in soil is associated with 300 mg/kg in earthworms, which theoretically would cause Pb toxicity in songbirds feeding at the site (Beyer and Stafford 1993).

However, there are a number of problems associated with both estimating soil ingestion and using the earthworm as a model soil organism. Soils contain high concentrations of elements such as Al and Fe, which are rarely, if ever, hazardous to terrestrial wildlife but if considered in a risk assessment context, might appear to be toxic based on laboratory tests using more soluble forms of the elements. Presumably, these elements have a low bioavailability in natural soils, and it is unlikely that animals ingesting a lot of soil suffer from Al poisoning. Finding high concentrations of metals in earthworms is not always evidence of serious contamination or that wildlife predators are at risk. Earthworms of the genus *Eisenoides* naturally accumulate concentrations of Pb that arguably could be called toxic (Beyer and Cromartie 1987). Some earthworms accumulate concentrations of Se, Ba, and Fe that are above the mineral tolerances for domestic animals recommended by the National Research Council (NRC 1980). However, determining concentrations of these elements in earthworms from closely matched reference areas will generally help reduce the possibility of confusing these peculiarities of invertebrate metal bioaccumulation with serious contamination.

Future Directions

Most ERAs of contaminated sites rely extensively on chemical measurements and ratios of the measured environmental concentration and literature-based toxicity thresholds, also known as risk or hazard quotients. Biological testing, if conducted at all, is usually limited to a few standardized methods such as earthworm toxicity tests or lettuce seed germination and root elongation tests. Over the last decade, European initiatives have led to the standardization of several other terrestrial tests, and North American agencies have followed suit. Compilations of these test methods are available (Linder et al. 1992; Løkke and Van Gestel 1998), providing increased flexibility with respect to test species selection (Kapustka 1997; ASTM 1999b, 1999c), and greatly expanding the suite of tests so that greater ecological relevance can be achieved. Further standardization of invertebrate sampling techniques, invertebrate toxicity testing, and microbial process measurements provides a research initiative that will increase the number of tools available for assessing soil ecology and reduce reliance on chemical analysis alone for the characterization of contaminated sites.

During the assessment and cleanup of contaminated sites, management decisions are ultimately made as to whether or not active remediation is warranted, and if so, what remediation technology best achieves project goals. Most often, such decisions are based on the measurement of contaminants in soil and whether or not the contaminant level exceeds some sort of predetermined value (e.g., hazard quotient). However, the information generated from ecological effects measurements can also be used effectively to both select the appropriate technology and evaluate performance of the technology, rather than just measuring contaminant levels in soil. In a simplified sense, the choice of effects measurements should reflect as closely as possible the potential or actual harm contaminants have on valued resources. As remediation technologies are considered, each should be examined (at least on a theoretical basis) in terms of how the technology would influence those parameters used to measure ecological effects. For example, if soil microbial processes were measured to understand effects on soil fertility and ultimately forest growth potential, these questions could be framed:

- What are the consequences to microbial processes of excavating the top 50 cm of soil?
- What are the consequences of incinerating the soil?
- What are the consequences of soil amendments that promote the biodegradation of specific contaminants by microbes?
- What are the predictions for the future if a "no remediation" or "natural attenuation" option is selected?

This approach recognizes that there are potential ecological risks associated with all remedial options. Although digging and hauling contaminated soil may be a simple approach to reducing the concentration of a contaminant in a confined area, the actual removal of the upper horizons of soil and the compaction of remaining soil by heavy equipment used in soil removal may have a greater impact on soil processes over the long term than leaving contaminants in place. Posing questions from this perspective would shift the focus of remediation to preserving the ecological resource to be protected, rather than holding to a narrower objective of lowering the concentration of contaminants. Maintaining the ecological integrity of soil would certainly be beneficial to the overall goal of preserving soils and soil services for future generations. With the advent of new methods for assessing soil as a valued biological resource, rather than a matrix for contaminant deposition, it is possible to provide quantitative biological data alongside the chemical analysis of soil for the management of contaminated sites.

References

Allen MF. 1991. The ecology of mycorrhizae. Cambridge, UK: Cambridge University Press. 184 p.

Anderson J, Domsch K. 1973. Quantification of bacterial and fungal contributions to soil respiration. *Arch Microbiol* 93:113-127.

Anderson JR, Westmoreland D. 1971. Direct counts of soil organisms using a fluorescent brightener and a europium chelate. *Soil Biol Biochem* 3:85-87.

Arthur III WJ, Gates RJ. 1988. Trace element intake via soil ingestion in pronghorns and in black-tailed jackrabbits. *J Range Manag* 41:162-166.

[ASTM] American Society for Testing and Matcrials. 1999a. Standard guide for conducting a laboratory soil toxicity test with the Lumbricid earthworm *Eisenia foetida*. In: Annual book of ASTM standards. Volume 11.05. E47, Committee on Biological Effects and Environmental Fate. Philadelphia PA, USA: ASTM. E1676-97. p 1062-1079.

[ASTM] American Society for Testing and Materials. 1999b. Standard guide for conducting terrestrial plant toxicity tests. ASTM E-1963-98. In: Annual book of ASTM standards. Volume 11.05. E47, Committee on Biological Effects and Environmental Fate. West Conshohocken PA, USA: ASTM. E1598-94. p 1481-1500.

[ASTM] American Society for Testing and Materials. 1999c. Standard practice for conducting early seedling growth tests. In: Annual book of ASTM standards. Volume 11.05. E47, Committee on Biological Effects and Environmental Fate. West Conshohocken PA, USA: ASTM. E1598-94. p 1000-1006.

[ASTM] American Society for Testing and Materials. 1999d. Standard guide for sampling terrestrial and wetland vegetation. In: Annual book of ASTM standards. Volume 11.05. E47, Committee on Biological Effects and Environmental Fate. Philadelphia PA, USA: ASTM. E-1923-97. p 1449-1466.

Avery TE, Burkhart HE. 1983. Forest measurements. 3rd ed. New York NY, USA: McGraw-Hill Book Company. 331 p.

Barbour MG, Burk JH, Pitts WD. 1987. Terrestrial plant ecology. 2nd ed. Menlo Park CA, USA: Benjamin/Cummings Publishing Company. 634 p.

Barker RJ. 1958. Notes on some ecological effects of DDT sprayed on elms. *J Wildl Manag* 22:269-274.

Bengtsson G, Berden M, Rundgren S. 1988. Influence of soil animals and metals on decomposition processes: A microcosm experiment. *J Environ Qual* 17:113-119.

Bengtsson G, Rundgren S. 1984. Ground-living invertebrates in metal-polluted forest soils. *Ambio* 13:29-33.

Bergersen F. 1980. Methods for evaluating biological nitrogen fixation. New York NY, USA: John Wiley & Sons. 702 p.

Beyer WN, Audet DJ, Morton A, Campbell JK, LeCaptain L. 1998. Lead exposure of waterfowl ingesting Coeur d'Alene River Basin sediments. *J Environ Qual* 27:1533-1538.

Beyer WN, Connor EE, Gerould S. 1994. Estimates of soil ingestion by wildlife. *J Wildl Manag* 58:375-382.

Beyer WN, Cromartie EJ. 1987. A survey of Pb, Cu, Zn, Cd, Cr, As, and Se in earthworms and soil from diverse sites. *Environ Monit Assess* 8:27-36.

Beyer WN, Gish CD. 1980. Persistence in earthworms and potential hazards to birds of soil applied DDT, dieldrin, and heptachlor. *J Appl Toxicol* 17:295-307.

Beyer WN, Linder G. 1995. Making sense of soil ecotoxicology. In: Hoffman DJ, Rattner BA, Burton Jr GA, Cairns Jr J, editors. Handbook of ecotoxicology. Boca Raton FL, USA: Lewis Publishers. p 104-116.

Beyer WN, Stafford C. 1993. Survey and evaluation of contaminants in earthworms and in soils derived from dredged material at confined disposal facilities in the Great Lakes Region. *Environ Monit Assess* 24:151-165.

Blair JM, Crossley DA. 1988. Litter decomposition, nitrogen dynamics, and litter microarthropods in a southern Appalachian hardwood forest 8 years following clearcutting. *J Appl Ecol* 25:683-698.

Blair JM, Bohlen PJ, Edwards CA, Stinner BR, McCartney DA, Allen MF. 1995. Manipulation of earthworm populations in field experiments in agroecosystems. *Acta Zool Fenn* 196:48-51.

Bogomolov DM, Chen S-K, Parmelee RW, Subler S, Edwards CA. 1996. An ecosystem approach to soil toxicity testing: A study of copper contamination in laboratory soil microcosms. *Appl Soil Ecol* 4:95-105.

Bongers T. 1990. The maturity index: An ecological measure of environmental disturbance based on nematode species composition. *Oecologia* 83:14-19.

Bonham CD. 1989. Measurements for terrestrial vegetation. New York NY, USA: John Wiley & Sons, Inc. 338 p.

Brookes PC, Kragt JF, Powlson DS, Jenkinson DS. 1985. Chloroform fumigation and the release of soil nitrogen: The effects of fumigation time and temperature. *Soil Biol Biochem* 17:831-835.

Brown G, Pashanasi B, Gilot-Vilenove C, Patron J, Senapati B, Giri S, Borois I, Lavells P, Blakemore R, Spain A, Boyer J. 1999. Effects of earthworms on plant production in the tropics. 6th International Symposium on Earthworm Ecology; 1998 Aug 31-Sep 4; Vigo, Spain. Santiago, Spain: Graficolor Minerva. 239 p.

Büchs W. 1994. Effects of different input of pesticides and fertilizers on the abundance of arthropods in a sugar beet crop: an example for a long-term risk assessment in the field. In: Donker MH, Eijsackers H, Heimbach F, editors. Ecotoxicology of soil organisms. Boca Raton FL, USA: Lewis Publishers. p 303-321.

Casida Jr VLE. 1977. Microbial metabolic activity in soil as measured by dehydrogenase determinations. *Appl Environ Microbiol* 34:630-636.

Checkai R, Wentsel R, Phillips T. 1993. Controlled environment soil-core microcosm unit for investigating fate, migration, and transformation of chemicals in soils. *J Soil Contam* 2:229-243.

Cheng W, Coleman DC. 1989. A simple method for measuring CO_2 in a continuous air-flow system: Modifications to the substrate-induced respiration technique. *Soil Biol Biochem* 21:385-388.

Cook ME, Swait AA. 1975. Effects of some fungicide treatments on earthworm populations and leaf removal in apple orchards. *J Hortic Sci* 50:495-499.

Crossley DA, Coleman DC, Hendrix PF, Cheng W, Wright DH, Beare MH, Edwards CA, editors. 1991. Modern techniques in soil ecology. New York NY, USA: Elsevier. 510 p.

Crossley DA, Hoglund MP. 1962. A litterbag method for the study of microarthropods inhabiting leaf litter. *Ecology* 43:571-574.

Denneman CAJ, Van Straalen NM. 1991. The toxicity of lead and copper in reproduction tests using the oribatid mite *Platynorhus peltifer. Pedobiologia* 35:305-311.

Dindal DL. 1977. Influence of human activities on oribatid mite communities. In: Dindal DL, editor. Biology of Oribatid mites. Eastern Branch Meeting of the Entomological Society of America; 1975 Oct 2; Philadelphia, PA, USA. Syracuse NY, USA: State University of New York. p 105-122.

Dindal DL, editor. 1990. Soil biology guide. New York NY, USA: Wiley. 1349 p.

Dommergues Y. 1960. Influence du rayonnemen infra-rouge et du rayonnement-solaire sur la teneur en azote mineral et su quelques characteristiques biologiques des sols. *Agron Trop* 15:61.

Donker MH, Bogert CG. 1991. Adaptation to cadmium in three populations of the isopod *Porcellio scaber. Comp Biochem Physiol* 100C:143-146.

Donkin SG, Dusenbery DB. 1993. A soil toxicity test using the nematode *Caenorhabditis elegans* and an effective method of recovery. *Arch Environ Contam Toxicol* 25:145-151.

Dommergues YR, Krupa SV. 1978. Interactions between non-pathogenic soil microorganisms and plants. New York NY, USA: Elsevier Scientific. 475 p.

Doran JW, Safley M. 1997. Defining and assessing soil health and sustainable productivity. In: Pankhurst C, Doube BM, Gupta VVSR, editors. Biological indicators of soil health. Wallingford, UK: CAB International. p 1-28.

Doube BM, Schmidt O. 1997. Can the abundance or activity of soil macrofauna be used to indicate the biological health of soils? In: Pankhurst C, Doube BM, Gupta Sr VV, editors. Biological indicators of soil health. Wallingford, UK: CAB International. p 265-295.

Edwards CA. 1973. Persistent pesticides in the environment. Cleveland OH, USA: CRC Press. 161 p.

Edwards CA. 1979. Tests to assess the effects of pesticides on beneficial soil organisms. (Proceedings) Umweltbundesamt Seminar on Ecological Tests Relevant to Environmental Chemicals; Evaluation and Research Needs; 1977 Dec; Berlin, Germany. Berlin, Germany: Umweltbundesamt. p 240-253.

Edwards CA. 1981. Earthworms, soil fertility and plant growth. In: Appelhof M, editor. (Procceedings) Workshop on the Role of Earthworms in the Stabilization of Organic Residues, Volume 1; 1980 Apr 9-12; Kalamazoo, MI, USA. Kalamazoo MI, USA: Beech Leaf Press. p 61-85.

Edwards CA. 1983. Development of a standardized laboratory method for assessing the toxicity of chemical substances to earthworms. Environment and quality of life. Brussels, Belgium: Commission of the European Communities. Report EUR 8714 EN. 141 p.

Edwards CA. 1984. Report of the second stage in development of a standardized laboratory method for assessing the toxicity of chemical substances to earthworms. Environment and quality of life. Brussels, Belgium: Commission of the European Communities. Report EUR 9360 EN. 99 p.

Edwards CA. 1989. The impact of herbicides on soil ecosystems. *Crit Rev Plant Sci* 8:221-257.

Edwards CA. 1991. The assessment of populations of soil-inhabiting invertebrates. In: Crossley DA, Coleman DC, Hendrix PF, Cheng W, Wright DH, Beare MH, Edwards CA, editors. Modern techniques in soil ecology. New York NY, USA: Elsevier. p 145-177.

Edwards CA. 1994. Pesticides as environmental pollutants. In: Ekstrom G, editor. World directory of pesticide control organizations. 2nd ed. London, UK: Crop Protection Publications, Royal Society of Chemistry. p 1-24.

Edwards CA. 1998. Principles for the design of flexible earthworm field toxicity experiments and interpretation of results. In: Sheppard S, Bembridge J, Holmstrup M, Posthuma L, editors. Advances in earthworm ecotoxicology. (Proceedings) 2nd international workshop on earthworm ecotoxicology; 1997 Apr; Amsterdam, NL. Pensacola FL, USA: SETAC. p 313-326.

Edwards CA. 1999. Soil invertebrate controls and microbial interactions in nutrient and organic matter dynamics. In: Coleman DC, Hendrix P, editors. Webmaster functions in agroecosystems. Invertebrates, climate and substrate quality. Wallingford, UK: Commonwealth Agricultural Bureau. p 141-159.

Edwards CA, Bohlen PJ. 1992. The effects of toxic chemicals on earthworms. *Rev Environ Contam Toxicol* 125:23-99.

Edwards CA, Bohlen PJ. 1995. The effects of contaminants on the structure and function of soil communities. *Acta Zoologica Fennica* 196:48-49.

Edwards CA, Bohlen PJ. 1996. The biology and ecology of earthworms. 3rd ed. London, England: Chapman and Hall. 426 p.

Edwards CA, Heath GW. 1963. The role of soil animals in breakdown of leaf material. In: Doeksen J, Van der Drift J, editors. Soil organisms. Amsterdam, NL: North Holland Publishers. p 76-84.

Edwards CA, Lofty JR. 1980. Effects of earthworm inoculation upon root growth of direct drilled cereals. *J Appl Ecol* 17:533-543.

Edwards CA, Thompson AR. 1973. Pesticides and the soil fauna. *Residue Rev* 45:1-79.

Edwards CA, Knacker TT, Pokarzhevskii AA. 1998. The prediction of the fate and effects of pesticides in the environment using tiered terrestrial model ecosystems. (Proceedings) Brighton Pest Control Conference: Pests and Diseases. 4G1:267-272.

Edwards CA, Knacker TT, Pokarzhevskii AA, Subler S, Parmelee R. 1996. The use of soil microcosms in assessing the effects of pesticides on soil ecosystems. (Proceedings) International Symposium on Environmental Behavior of Crop Protection Chemicals. Vienna, Austria: International Atomic Energy Agency. p 435-352.

Fitter AH, Atkinson D, Read DJ, Usher MB. 1985. Ecological interactions in soil: Plants, microbes, and animals. London, UK: Blackwell Scientific Publications. Special Publication Series of the British Ecological Society Nr 4. 451 p.

Fletcher JS, Johnson FL, Mc Farlane JC. 1988. Database assessment of phytotoxicity data published on terrestrial vascular plants. *Environ Toxicol Chem* 7:615-622.

Fogel R. 1985. Roots as primary producers in below-ground ecosystems. In: Fitter AH, Atkinson D, Read DJ, Usher MB, editors. Ecological interactions in soils: Plants, microbes, and animals. London, England: Blackwell Scientific Publications. British Ecological Society. Special Publication Nr 4. p 23-36.

Forbes VE, Forbes TL. 1994. Ecotoxicology in theory and practice. London, UK: Chapman & Hall. 247 p.

Fries GF, Marrow GS, Snow PA. 1982. Soil ingestion by swine as a route of contaminant exposure. *Environ Chem* 1:201-204.

Galbraith H, LeJune K, Lipton J 1995. Metal and arsenic impacts on soils, vegetation communities, and wildlife habitats in southwest Montana uplands contaminated by smelter emissions: I. Field Evaluation. *Environ Toxicol Chem*. 14: 1985-1903.

Gordon JC, Wheeler CT, Perry DA. 1979. Symbiotic nitrogen fixation in the management of temperate forests. Proceedings of a workshop. Corvallis OR, USA: Forest Research Laboratory, Oregon State University. 501 p.

Green RH. 1979. Sampling design and statistical methods for environmental biologists. New York NY, USA: Wiley Interscience. 257 p.

Greene JC, Bartels CL, Warren-Hicks WJ, Parkhurst BR, Linder GL, Peterson SE, Miller WE. 1988. Protocols for short-term toxicity screening of hazardous waste sites. Corvallis OR, USA: USEPA. EPA-600-3-88-029.

Greig-Smith P. 1983. Quantitative plant ecology. 3rd ed. Berkeley CA, USA: University of California Press. 359 p.

Guild WJM. 1955. Earthworms and soil structure. In: Kevan DKM, editor. Soil zoology. (Proceedings) University of Nottingham Second Easter School in Agricultural Science; 1955; Nottingham, UK. London, England: Butterworths. p 83-98.

Gupta VVSR, Yeates GW. 1997. Soil microfauna as bioindicators of soil health. In: Pankhurst C, Doube BM, Gupta VVSR, editors. Biological indicators of soil health. Wallingford, UK: CAB International. p 201-233.

Harley JL, Smith SE. 1983. Mycorrhizal symbiosis. London, UK: Academic Press. 483 p.

Heath GW, Edwards CA, Arnold MK. 1964. Some methods of assessing the activity of soil animals in the breakdown of leaves. *Pedobiologia* 4:80-87.

Heitzer A, Webb OF, Thonnard JE, Sayler GS. 1992. Specific and quantitative assessment of naphthalene and salicylate bioavailability using a bioluminescent catabolic reporter bacterium. *Appl Environ Microbiol* 58:1839-1846.

Hirst JM, Le Riche HH, Bascomb CL. 1961. Copper accumulation in the soils of apple orchards near Wisbech. *Plant Pathol* 10:105-108.

Hopkin SP. 1989. Ecophysiology of metals in terrestrial invertebrates. London, UK: Elsevier Applied Science. 366 p.

Hoy JB. 1980. Effects of lindane, carbaryl, and chlorpyrifos on non-target soil arthropod communities. In: Dindal DL, editor. Soil biology as related to land use practices. (Proceedings) 7th International Colloquium of Soil Zoology; 1979 Jul 29-Aug 3; Syracuse NY, USA. Washington DC, USA: USEPA, Office of Pesticide and Toxic Substances. 880 p.

Huhta V. 1984. Response of *Cognettia sphagnetorum* (Enchytraeidae) to manipulation of pH and nutrient status in coniferous forest soil. *Pedobiologia* 27:245-260.

Hutton M. 1984. Impact of airborne metal contamination on a deciduous woodland system. In: Sheehan PJ, Miller DR, Butler GC, Bourdeau P, editors. Effects of pollutants at the ecosystem level. Chichester, UK: John Wiley and Sons. p 365-375.

Ingham E, Klein DA. 1984. Soil fungi: Relationships between hyphal activity and staining with fluorescein diacetate. *Soil Biol Biochem* 16:273-278.

[ISO] International Organization for Standardization. 1997. Soil quality-effects of pollutants on earthworms (*Eisenia fetida*): Part 2-Determination of effects on reproduction. Geneva, Switzerland: ISO. ISO/DIS 11268-2.2.

[ISO] International Organization for Standardization. 1998. Soil quality-effects of soil pollutants on Collembola (*Folsomia candida*): Method for the determination of effects on reproduction. Geneva, Switzerland: ISO. ISO/DIS 11267.

Jordan MJ, Lechevalier MP. 1975. Effects of zinc-smelter emissions on forest soil microflora. *Can J Microbiol* 21:1855-65.

Kammenga JE, Van Koert PHG, Riksen JAG, Korthals GW, Bakker J. 1996. A toxicity test in artificial soil based on the life-history strategy of the nematode *Plectus acuminatus*. *Environ Toxicol Chem* 15:722-727.

Kapustka LA. 1987. Interactions of plants and non-pathogenic soil microorganisms. In: Newman DW, Wilson KG, editors. Volume 3. Model building in plant physiology/biochemistry. Boca Raton FL, USA: CRC Press. p 49-56.

Kapustka LA. 1990. Dinitrogen fixation in central North American prairies. In: Yang H, editor. (Proceedings) International Symposium on Grassland Vegetation. 1987 Aug 15-20; Hohhot, The Peoples Republic of China. Beijing, CN: Science Press. p 481-486.

Kapustka LA. 1997. Selection of phytotoxicity tests for use in ecological risk assessments. In: Wang W, Gorsuch J, Hughes JS, editors. Plants for environmental studies. Boca Raton FL, USA: Lewis. p 515-548.

Kapustka LA, Arnold PT, Lattimore PT. 1985. Interactive responses of associative diazotrophs from a Nebraska sand hills grassland. In: Fitter AH, Atkinson D, Read DJ, Usher MB, editors. Ecological interactions in soils: Plants, microbes, and animals. London, England: Blackwell Scientific Publications. British Ecological Society. Special Publication Nr 4. p 149-158.

Kapustka LA, Reporter M. 1993. Terrestrial primary producers. In: Calow P, editor. Handbook of ecotoxicology. London, UK: Blackwell Scientific Publications. p 278-297.

Kapustka LA, Rice EL. 1978. Symbiotic and asymbiotic N_2-fixation in a tall grass prairie. *Soil Biol Biochem* 10:553-554.

Kasprzak K. 1982. Review of enchytraeid community structure and function in agricultural ecosystems. *Pedobiologia* 23:217-232.

Kaufman ES. 1975. Certain problems of phenol intoxication of *Enchytraeus albidus* from the view-point of stress [in Russian with English summary]. *J Hydrobiol* 11:44-46.

Kelley JJ, Tate III RL. 1998. Effects of heavy metal contamination and remediation on soil microcommunities in the vicinity of a zinc smelter. *J Environ Qual* 27:609-617.

Korthals GW, de Goede RGM, Kammenga JE, Bongers T. 1996. The maturity index as an instrument for risk assessment of soil pollution. In: Van Straalen NM, Krivolutsky DA, editors. Bioindicator systems for soil pollution. Dordrecht, NL: Kluwer Academic. p 85-93.

Kratz W. 1996. Die Cotton-strip Methode im Vergleich zu herkömmlichen Bodenbiologischen. *Untersuchungsmethoden im Freiland Mitt Deut Bodenkd Ges* 81:17-20.

Kreulen DA, Jager T. 1984. The significance of soil ingestion in the utilization of arid rangelands by large herbivores, with special reference to natural licks on the Kalahari pans. In: Gilchrist FMC, Mackie RI, editors. (Proceedings) International Symposium on Herbivore Nutrition in the Subtropics and Tropics; 1983; Pretoria, South Africa. Chraignall, South Africa: Science Press. p 204-221.

Kuperman R. 1996. Relationships between soil properties and community structure of soil macroinvertebrates in oak-hickory forests along an acidic deposition gradient. *Appl Soil Ecol* 4:125-137.

Kuperman R. 1999. Litter decomposition and nutrient dynamics in oak-hickory forests along a historic gradient of nitrogen and sulfur deposition. *Soil Biol Biochem* 31:237-244.

Kuperman R, Carreiro M. 1997. Soil heavy metal concentrations, microbial biomass and enzyme activities in a contaminated grassland ecosystem. *Soil Biol Biochem* 29:179-190.

Kuperman R, Checkai R, Thomulka K, Guelta M. 1997. Rapid assessment of ecotoxicological effects of soil contamination using luminescence-reduction assays with *Vibrio harveyi.* Society of Environmental Toxicology and Chemistry (SETAC). San Francisco CA, 16-20 November 1997 (published abstract).

Kuperman R, Edwards CA. 1997. Effects of acidic deposition on soil invertebrates and microorganisms. *Rev Environ Contam Toxicol* 48:35-138.

Kuperman R, Knacker T, Checkai R, Edwards C. 2002. Multispecies and multiprocess assays to assess the effects of chemicals on contaminated sites. In: Sunahara GI, Renoux AY, Gaudet CL, Thellon C, Pilon A, editors. Environmental analysis of contaminated sites: Tools to measure success or failure. UK: John Wiley and Sons. p 45-60.

Kuperman R, Williams G, Parmelee R. 1998. Spatial variability in the soil food webs in a contaminated grassland ecosystem. *Appl Soil Ecol* 9:509-514.

Latter PM, Howson G. 1977. The use of cotton strip to indicate cellulose decomposition in the field. *Pedobiologia* 17:145-155.

Lee KE. 1985. Earthworms, their ecology and relationships with soils and land use. Orlando FL, USA: Academic Press. 411 p.

Linder G, Ingham E, Brandt CJ, Henderson G. 1992. Evaluation of terrestrial indicators for use in ecological assessments at hazardous waste sites. Corvallis OR, USA: USEPA. EPA-600-R-92-183.

Logsdon SD, Linden DR. 1992. Interactions of earthworms with soil physical conditions influencing plant growth. *Soil Sci* 154:330-337.

Løkke H, Van Gestel CAM. 1998. Handbook of soil invertebrate toxicity tests. New York NY, USA: John Wiley & Sons, Inc. 281 p.

Meyers WL, Shelton RL. 1980. Survey methods for ecosystem management. New York NY, USA: John Wiley & Sons, Inc. 403 p.

Moore JC, DeRuiter PC. 1991. Temporal and spatial heterogeneity of trophic interactions within below ground food webs: An analytical approach to understanding multi-dimensional systems. *Agric Ecosys Environ* 34:371-397.

Moore JC, Hunt HW. 1988. Resource compartmentation and the stability of real food webs. *Nature* 333:261-263.

Moore PD, Chapman SB. 1986. Methods in plant ecology. 2nd ed. Oxford, UK: Blackwell Scientific Publications. 589 p.

Mueller-Dombois D, Ellenberg H. 1974. Aims and methods of vegetation ecology. New York NY, USA: John Wiley & Sons, Inc. 547 p.

Myrold DD. 1990. Effects of acidic deposition on soil organisms. In: Lucier AA, Haines SG, editors. Mechanisms of forest response to acidic deposition. New York NY, USA: Springer-Verlag. p 163-187.

Newell SY, Hicks RE. 1982. Direct count estimates of fungal and bacterial biovolume in dead leaves of smooth cordgrass (*Spartina altemiflora* Loisel). *Estuaries* 5:246-260.

[NRC] National Research Council. 1980. Mineral tolerances of domestic animals. Washington DC, USA: National Academy of Sciences. 577 p.

[OECD] Organization for Economic Cooperation Development. 1999. Draft guideline for testing of chemicals. Nr 117. Soil microorganisms: Carbon transformation test. Paris, France: OECD.

Paasivirta J. 1991. Chemical ecotoxicology. Chelsea MI, USA: Lewis Publishers. 210 p.

Pankhurst CE, Lynch JM. 1994. The role of the soil biota in sustainable agriculture. In: Pankhurst C, Doube BM, Gupta VVSR, editors. Biological indicators of soil health. Wallingford, UK: CAB International. p 3-9.

Paoletti MG, Schweigl U, Favretto MR. 1995. Soil macroinvertebrates, heavy metals and organochlorines in low and high input apple orchards and a coppiced woodland. *Pedobiologia* 39:20-33.

Parmelee RW, Phillips CT, Checkai RT, Bohlen PJ. 1997. Determining the effects of pollutants on soil faunal communities and trophic structure using a refined microcosm system. *Environ Toxicol Chem* 16:1212-1217.

Parmelee RW, Wentsel RS, Phillips CT. 1993. Soil microcosm for testing the effects of chemical pollutants on soil fauna communities and trophic structure. *Environ Toxicol Chem* 12:1477-1486.

Paton GI, Campbell CD, Glover LA, Killham K. 1995. Assessment of the bioavailability of heavy metals using lux modified constructs of *Pseudomonas fluorescens*. *Lett Appl Microbiol* 20:52-56.

Paton GI, Palmer G, Burton M, Rattray EAS, McGrath SP, Glover LA, Killham K. 1997. Development of an acute and chronic ecotoxicity assay using lux-marked *Rhizobium leguminosarum* biovar *trifolii*. *Lett Appl Microbiol* 24:296-300.

Paton GI, Rattray EAS, Campbell CD, Meussen H, Cresser MS, Glover LA, Killham K. 1997. Use of genetically modified microbial biosensors for soil ecotoxicity testing. In: Pankhurst CS, Doube B, Gupta V, editors. Bioindicators of soil health. Wallingford, UK: CAB International. p 397-418.

Paul EA, Clark FE. 1996. Soil microbiology and biochemistry. 2nd ed. San Diego CA, USA: Academic Press. 340 p.

Phillips C, Kuperman R, Checkai R. 1999. A rapid and highly-efficient method for extracting enchytraeids from soil. *Pedobiologia* 43:523-527.

Potter DA, Powell AJ, Smith MS. 1990. Degradation of turfgrass thatch by earthworms (Oligochaeta: Lumbricidae) and other soil invertebrates. *J Econ Entomol* 83:205-211.

Rand GM, Newman JR. 1998. The applicability of habitat evaluation methodologies in ecological risk assessment. *Human Ecol Risk Assess* 4:905-929.

Raw F. 1959. Estimating earthworm populations using formalin. *Nature* 184:161.

Rice EL. 1984. Allelopathy. 2nd ed. New York NY, USA: Academic Press. 422 p.

Römbke J. 1989. *Enchytraeus albidus* (Enchytraeidae, Oligochaeta) as a test organism in terrestrial laboratory systems. *Arch Toxicol* 13:402-405.

Römbke J. 1996. Recent advantages of the Enchytraeid Reproduction Test. *Newsletter on Enchytraeidae* 5:73-81.

Römbke J, Federschmidt A. 1995. Effects of the fungicide carbendazim on Enchytraeidae in laboratory and field tests. *Newsletter on Enchytraeidae* 4:79-96.

Römbke J, Moser T. 2002. Validating the enchytraeid reproduction test: Organization and results of an international ringtest. *Chemosphere* 46: 1117-1140.

Ross DJ. 1989. Estimation of soil microbial C by a fumigation-extraction procedure: Influence of soil moisture content. *Soil Biol Biochem* 21:767-772.

Schenck NC. 1982. Methods and principles of mycorrhizal research. St. Paul MN, USA: American Phytopathological Society. 244 p.

Schroeder RL, Haire SL. 1993. Guidelines for the development of community-level habitat evaluation models. Biological Report 8. Washington DC: US Department of Interior, Fish and Wildlife Service.

Selifonova O, Burlae R, Barkay T. 1993. Bioluminescent sensors for the detection of bioavailable Hg (II) in the environment. *Appl Environ Microbiol* 59:3083-3090.

Sheehan PJ. 1984. Effects on community and ecosystem structure and dynamics. In: Sheehan PJ, Miller DR, Butler GC, Bourdeau P, editors. Effects of pollutants at the ecosystem level. Chichester, UK: John Wiley & Sons, Inc. p 51-99.

Sheppard SC. 1995. Parameter values to model the soil ingestion pathway. *Environ Monit Assess* 43:27-44.

Sinsabaugh RL, Linkins AE. 1990. Enzymic and chemical analysis of particulate organic matter from a boreal river. *Freshwater Biol* 23:301-309.

Smith MS, Rice CW, Paul EA. 1989. Metabolism of labeled organic nitrogen in soil: Regulation by inorganic nitrogen. *Soil Sci Soc Am J* 53:768-773.

Söderström B. 1977. Vital staining of fungi in pure cultures and in soil with fluorescein diacetate. *Soil Biol Biochem* 9:59-63.

Sparling GP. 1997. Soil microbial biomass, activity and nutrient cycling as indicators of soil health. In: Pankhurst C, Doube BM, Gupta VVSR, editors. Biological indicators of soil health. Wallingford, UK: CAB International. p 97-119.

Spurgeon DJ, Sandifer RD, Hopkin SP. 1996. The use of macro-invertebrates for population and community monitoring of metal contamination-indicator taxa, effect parameters and the need for a soil invertebrate prediction and classification scheme (SIVPACS). In: Van Straalen NM, Krivolusky DA, editors. Bioindicator systems for soil pollution. Dordrecht, NL: Kluwer Academic Publishers. p 95-109.

Steinberg SM, Poziomek EJ, Englemann WH, Rogers KR. 1995. A review of environmental applications of bioluminescence measurements. *Chemosphere* 30:2155-2197.

Thompson AR, Gore FL. 1972. Toxicity of twenty-nine insecticides to *Folsomia candida*: Laboratory studies. *J Econ Entomol* 65:1255-1260.

Turgeon AJ, Freeborg RP, Bruce WN. 1975. Thatch development and other effects of preemergence herbicides in Kentucky bluegrass turf. *Agron J* 67:563-565.

Tyler G. 1984. The impact of heavy metal pollution on forests: A case study of Gusum, Sweden. *Ambio* 13:18-24.

[UBA] Umweltbundesamt. 1994. Workshop on Terrestrial Model Ecosystems. Berlin, DE: Umweltbundesamt. 89 p.

[USEPA] U.S. Environmental Protection Agency. 1998. Guidelines for ecological risk assessment. Federal Register 63(93):26846-26924.

Van de Westeringh W. 1972. Deterioration of soil structure in worm free orchard soils. *Pedobiologia* 12:6-15.

Van Gestel CAM, Van Dis WA, Van Breemen EM, Sparenburg PM. 1989. Development of a standardized reproduction toxicity test with the earthworm species *Eisenia fetida andrei* using copper, pentachlorophenol, and 2,4-dichloroaniline. *Ecotoxicol Environ Saf* 18:305-312.

Van Rhee JA. 1977. A study of the effect of earthworms on orchard productivity. *Pedobiologia* 17:107-114.

Van Straalen NM. 1997. Community structure of soil arthropods as a bioindicator of soil health. In: Pankhurst C, Doube BM, Gupta VVSR, editors. Biological indicators of soil health. Wallingford, UK: CAB International. p 235-264.

Van Straalen NM, Van Gestel CAM. 1993. Soil invertebrates and microorganisms. In: Calow P, editor. Handbook of ecotoxicology. London, UK: Blackwell. p. 251-277.

Van Straalen NM, Verweij RA. 1991. Effects of benzo(a)pyrene on food assimilation and growth efficiency in *Porcellio scaber* (Isopoda). *Bull Environ Contam Toxicol* 46:134-140.

Von Törne E. 1990. Assessing feeding activities of soil-living animals. I Bait-lamina tests. *Pedobiologia* 34:89-101.

Wainwright M, Pugh GJF. 1974. The effects of fungicides on certain chemical and microbiological properties. *Soil Biol Biochem* 6:263.

Watson AP, Van Hook RI, Jackson DR, Reichle DE. 1976. Impact of a lead mining-smelting complex on the forest floor litter arthropod fauna in the new lead belt region of southeast Missouri. Oak Ridge TN, USA: Oak Ridge National Laboratory, Environmental Sciences Division. Publication 881. 178 p.

Weuffen W. 1968. Zusammenhänge zwischen chemischer Konstitution und keimwidriger Wirkung. *Arch Exp Vet Med* 22:127-132.

Witkamp M, Ausmus BS. 1976. Processes in decomposition and nutrient transfer in forest systems. In: Anderson JM, Macfadyen A, editors. The role of terrestrial and aquatic organisms in decomposition processes. (Proceedings) 17th symposium of the British Ecological Society; 1975 Apr 15-18; Oxford UK. Oxford, UK: Blackwell Scientific. p 375-396.

Wood WB. 1988. Introduction to *C. elegans* biology. In: Wood WB, editor. The nematode *Caenorhabditis elegans*. Cold Spring Harbor NY, USA: Cold Spring Harbor Laboratory. p 1-16.

Yardim EN, Edwards CA. 1998. Effects of chemical, pest, disease, and weed management practices on the trophic structure of nematode populations in tomato agroecosystems. *Appl Soil Ecol* 7:137-147.

Section II

Chemical–Soil–Organism Interactions

Chapter 5

Fate of Soil Contaminants

Dennis E. Rolston, Allan S. Felsot, Kurt D. Pennell, Kate M. Scow, Hans F. Stroo

Importance of Contaminant Fate to Exposure Assessment

Meaningful risk assessment of contaminated soils strongly depends on understanding and predicting the fate of contaminants in soil and the availability of these chemicals for transport to ground water and/or the atmosphere. With increasing emphasis on risk-based evaluations of contaminants in soil, the ability to measure and model contaminant availability for transport (e.g., leaching and volatilization) and for uptake by microbial, animal, and plant receptors is strongly dependent upon our understanding of the processes controlling the fate of contaminants in soil. Because all biological components of terrestrial ecosystems are intimately related to the soil, exposure to soil contaminants may result in harmful effects on the ecosystem.

One facet of risk assessment relies upon information regarding the distribution and concentration of contaminants within locations in the environment where receptors reside or in environmental media that receptors come into contact with. Contaminant concentrations can be determined from direct measurements (monitoring) or estimated from mathematical models that integrate the various fate and transport processes, as well as contaminant properties, to determine concentration as a function of time and space. However, exposure concentrations used in risk assessment often are assumed to be constant in time and applicable to the entire exposure medium that the receptor is exposed to. This assumption is made because generally there are insufficient site-specific data to make reasonable estimates of the fate of particular chemicals or because risk managers do not trust the reliability of simulation models that often have little validation. Quantitative and reliable exposure assessment depends on accurate predictions of contaminant concentrations in the soil that will be in contact with biological receptors. Improved understanding and more realistic simulation models of fate and transport are needed to increase the reliability of exposure estimates in risk assessment. This chapter offers some insight and suggestions on how to increase the reliability of these estimates.

Contaminated Soils: From Soil-Chemical Interactions to Ecosystem Management. Roman P. Lanno, editor.
 ISBN 1-880611-31-7

Conceptual Model of Contaminant Fate

Contaminant fate deals with those processes that distribute the contaminant among various compartments in the soil and transform the original contaminant to other chemical forms. Evaluations of contaminant fate result in estimates of the total mass or concentration remaining within the soil profile as a function of space and time. Fate processes can be biotic or abiotic and are closely linked to water and chemical transport mechanisms. In particular, microscale diffusion processes are particularly important in fate considerations and are difficult to separate from the chemical and biological processes.

A conceptual model of contaminant fate in soil is presented in Figure 5-1. This diagram shows that an organic contaminant may exist in 4 phases or compartments in soil: contaminant source (nonaqueous-phase liquid [NAPL], solid), soil solids (sorbed phase), soil atmosphere (gas phase), and soil solution (aqueous phase). The arrows connecting the compartments indicate either equilibrium partitioning or kinetically controlled mass transfer between these 4 compartments or phases. Because of the importance of the mass transfer between the solution phase and soil solids (adsorption and desorption) on the availability for transport and on the bioavailability of the contaminant (Linz and Nakles 1997), interactions between the soil solution and the soil solids is considered in detail in Chapter 7. Contaminant partitioning between the source (NAPL, surface phase) and soil solution (dissolution/solubilization) and between the source and soil atmosphere (volatilization) are considered in the chapter on transport (Chapter 6). Chapter 6 also covers advection and dispersion of contaminants in the soil solution and the soil atmosphere, as well as transport of the soil solids by erosional processes.

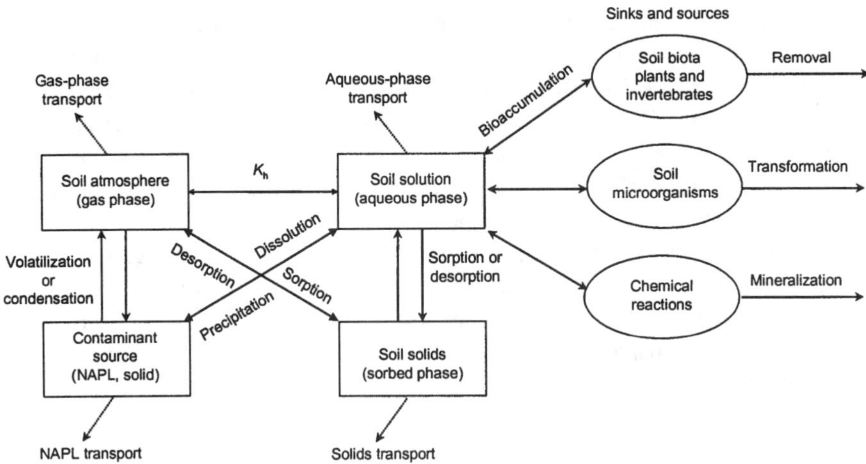

Figure 5-1 Conceptual model of contaminant distribution and fate processes in soil

The focus of this chapter is on the biological and chemical processes that transform the original contaminant to other chemicals or remove a portion of the chemical from the soil by mechanisms such as plant uptake. It generally is assumed that most microbiological reactions, chemical reactions, and bioaccumulation occur from or within the soil solution (aqueous phase) of soil. The arrows on the lower part of the soil solution compartment of Figure 5-1 indicate these processes. It will be apparent that the processes controlling fate of a contaminant in soil can be quite complex and difficult to predict. These processes also are quite variable for different soils and chemicals and often change significantly with space, time, and soil depth. It is also apparent that all the pathways described in Figure 5-1 are not applicable to metals because fate processes such as gas-phase transport and NAPL-phase transport are of little relevance to the fate of most metals in soil. This chapter will describe the major processes affecting the fate of chemicals in soil, indicate gaps in knowledge, and provide ideas on how to increase the accuracy of predictions of chemical fate. It also will discuss some modeling concepts that may be utilized to predict contaminant concentration as a function of time and soil depth as input for exposure assessment (Chapter 2). The uncertainty in soil properties, processes, and model predictions as they relate to exposure assessment is also discussed.

Major Fate Processes

Microbial reactions

Microorganisms mediate the biodegradation of numerous organic contaminants and the transformation of many organic and inorganic compounds. In many reactions involving organic contaminants, organisms obtain energy and C directly from the metabolism of contaminants and release harmless byproducts. Microbial degradation is by far the most important fate pathway for contaminants such as petroleum hydrocarbons, lower molecular weight polyaromatic hydrocarbons (PAHs) and polychlorinated biphenyls (PCBs), explosives, and many pesticides. The soil microbial communities are also essential for the maintenance of other biota as microbes transform plant and animal materials into soil organic matter and play a major role in cycling of C, N, S, and other nutrients.

Microbial populations are both numerous and diverse in soil. The 2 major groups of microorganisms are bacteria and fungi. Total bacterial numbers commonly range from 10^3 to 10^{10} cells/g soil, with higher densities in upper horizons of the soil profile and higher densities around plant roots and in litter than in bulk soil. There is a tremendous diversity in types of microbial metabolic reactions that can affect contaminants. Metabolism is both aerobic and anaerobic and can involve organic and inorganic molecules. Contaminants can serve as electron donors or acceptors or be modified by reactions other than biologically mediated oxidation-reduction reactions.

Microbial reactions in soil, especially biologically mediated reactions involving organic chemicals, usually have a greater impact on contaminant fate than do chemical reactions. Contaminants that provide sources of C and energy result in increased numbers of microorganisms, resulting in increased enzyme levels and reaction rates. Unfortunately, rates of microbial reactions are more difficult to predict than are chemical reaction rates because of the multitude of environmental, soil, chemical, and biological factors that can influence microbial reaction rates. In turn, this can contribute considerable uncertainty to exposure assessment.

There are numerous types of reactions involving contaminants (Alexander 1980). In this section, we will consider 4 general categories of reactions, including complete biodegradation, transformation, cometabolism, and complexation or polymerization. Within each of these categories, there are additional types of specific reactions.

Types of microbial reactions

Complete biodegradation of organic contaminants

Microbial reactions can lead to the complete destruction (also called "mineralization" or "biodegradation") of contaminants, which usually results in the production of innocuous, inorganic compounds (e.g., CO_2, H_2O, ammonium) and cells. Many of these reactions are oxidation-reduction reactions in which the contaminant is used as a source of C and energy, coupled with the utilization of O, NO_3, SO_4, ferric iron, or other compounds as the electron acceptor. These reactions occur most rapidly in the presence of O and can often support growth of the organisms involved. Examples of contaminants that are degraded through their utilization as electron donors under both aerobic and anaerobic conditions include aromatics, alkanes, many pesticides, and some PAHs. In other types of reactions, the contaminant serves as an electron acceptor, and native sources of C or other contaminants supply electrons and C. The latter types of reactions occur under strictly anaerobic conditions and often at very slow rates. Examples of contaminants that may be used as electron acceptors include many of the chlorinated solvents and PCBs.

Transformation of organic contaminants

Transformation reactions result in the alteration of the chemical form of the contaminant. Products of these reactions may be less toxic (e.g., catechol from simple aromatics) or more toxic (e.g., vinyl chloride from trichloroethylene) than the parent compound. Because of the unpredictability of these reaction pathways, recognition of the existence of biodegradation and measurement of metabolites can be essential pieces of information in the assessment of soils polluted with certain contaminants (e.g., chlorinated solvents). For many contaminants, including many of the PAHs, metabolite production has never been measured in soils and remains of unknown significance.

Complexation and polymerization may also result from microbial transformation reactions. Complexation reactions lead to incorporation of the contaminant into organic matter, called "bound residue formation," via strong covalent bonds that are difficult to break. Contaminants containing amine and hydroxyl functional groups can form chemical complexes with regions of soil organic matter substituted with hydroxyl, carboxy, amine, and other functional groups. Some scientists view bound residue formation as a form of bioremediation because research shows that only very small amounts of bound contaminant are bioavailable for uptake by microorganisms, plants, or other biological receptors compared to freshly added contaminants (Alexander 1995). Polymerization reactions, where the same compounds undergo covalent bond formation with one other, may occur between molecules that have been altered by microbial transformations. For example, catechols, which result from biotic transformation of aromatic compounds, undergo rapid polymerization.

Cometabolism

In cometabolism, the contaminant can be degraded only in the presence of another compound (primary substrate), the latter being necessary for enzyme induction and as a C and energy source. The contaminant provides little, if any, benefit to the microorganism. Eventually, the contaminant degradation rate may be decreased by competitive inhibition of the primary substrate, by formation of toxic metabolites, or by running out of reducing equivalents. Some of the most well-studied contaminants subjected to cometabolism are chlorinated solvents and PCBs (Alexander 1985). In some cases, the contaminant is cometabolized to harmless endproducts, and in other cases, toxic byproducts may be formed. An example of a compound that undergoes cometabolism by a variety of microbial groups is trichloroethylene, which can be cometabolized by ammonium oxidizers, methanotrophs, and aromatic and propane oxidizers.

Transformation of inorganic contaminants

Inorganic contaminants may be transformed by several types of reactions catalyzed by microorganisms, including removal from the soil solution and assimilation into microbial tissue through uptake processes. Oxidation reactions occur when the oxidation state of the contaminant increases through its use as an electron donor. Reduction reactions occur when the contaminant becomes reduced through its use as an electron acceptor. These reactions can increase or decrease the toxicity of Cd, As, and many other metals. Complexation of metals with microbially produced organic ligands can increase solubility of many metals such as Fe. Methylation reactions involving inorganic contaminants can increase toxicity and potential for exposure. For example, microbial methylation increases the release of Hg from sediments into the water column where aquatic biota are likely to be exposed. The methylated form of Hg is far more toxic than other forms of the element. Methylation also can decrease exposure when it leads to volatilization of toxic compounds from a contaminated site, as has been found for Se in drainage ponds in California. Microorganisms are also responsible for numerous indirect transformations of inorganic contaminants, such as changes in the speciation or oxidation state through altering the soil pH and O concentration.

Kinetics

In soil, biodegradation is usually characterized by simple kinetic expressions originally developed to describe a single bacterial species degrading a single compound in a well-mixed liquid culture (Alexander and Scow 1989). Such assumptions are unlikely to be valid at contaminated sites. However, simple equations often fit biodegradation data measured in soils and therefore are widely used to calculate contaminant half-lives. The biodegradation kinetics of a contaminant are most often described, when considered at all, by a first-order decay equation. This equation is readily adopted without consideration of the assumptions that must be met for its use because it is mathematically simple and easy to incorporate into more complex fate and transport models. Inappropriate use of first-order kinetics can result in underestimation or, more frequently, overestimation of biodegradation rates (Bekins et al. 1998). Also, problems can result from estimating biodegradation rates for use in exposure assessments independently of considerations of other important fate and transport properties of a contaminant. Biodegradation should be considered within the context of all the environmental processes affecting the contaminant.

Reviews of the kinetics of microbial reactions are given in Simkins and Alexander (1984), Robinson (1985), Alexander and Scow (1989), and Alexander (1994). The following section provides a brief overview of biodegradation kinetics.

The most widely used equation for describing the biodegradation of contaminants by microorganisms over a broad range of contaminant concentrations is the Monod equation. The disappearance of the contaminant is described by

$$dC/dt = -[(\mu_{max} C/(K_s + C)] B (1/Y) \quad (5\text{-}1),$$

where C is the concentration of contaminant (g contaminant/g soil), t is time, μ_{max} is the maximum growth rate (day^{-1}), K_s is the constant equal to concentration at 0.5 maximum rate (g contaminant/g soil), B is the population density of the microorganisms (g cells/g soil), and Y is the yield coefficient (g contaminant/g cells).

The growth of the microorganisms in the presence of the contaminant is described by

$$dB/dt = \mu_{max} C/(K_s + C)B \quad (5\text{-}2).$$

Simkins and Alexander (1984) derived a single expression that describes chemical disappearance by Monod kinetics linked to growth of the microbial population degrading the chemical. The Monod equation for growth is

$$dC/dt = [\mu_{max} C (C_0 + X_0 - C)] \ / \ (K_s + C) \qquad (5\text{-}3),$$

where X_0 (g contaminant/g soil) is the amount of substrate required to produce the population density equivalent to B_0, and B_0 (g cells/g soil) is the initial population density of microorganisms.

Simpler kinetic expressions can be derived from the Monod equation. These equations apply to very specific conditions. For example, the first-order, or exponential decay, equation is appropriate to use when there is no net growth of the microbial population ($dB/dt = 0$) and the contaminant concentration (C) is much lower than K_s. Therefore, as a rule of thumb, first-order kinetics are applicable to locations in the soil where contaminant concentrations are low. The first-order rate equation may be written as

$$dC/dt = -\, k_1 C \qquad (5\text{-}4),$$

and the integrated form for C as a function of time is

$$C_t = C_0 \, (e^{-k_1 t}) \qquad (5\text{-}5),$$

where k_1 is the first-order rate constant (time^{-1}) and C_0 (g contaminant/g soil) is the initial contaminant concentration.

The zero-order, or linear, equation can be used under conditions where there is no net growth of the microbial population and the C_0 far exceeds K_s. The zero-order rate equation is

$$dC/dt = -k_0 \qquad (5\text{-}6),$$

and the integrated form with respect to time is

$$C = -k_0 t + C_0 \qquad (5\text{-}7),$$

where k_0 is the zero-order rate constant.

More complex models than the single substrate models summarized above describe the kinetics of cometabolized contaminants. These models involve the coupling of equations describing growth on the primary substrate as well as degradation of the cometabolized chemical. Cometabolism is described by the

Michaelis-Menten equation for enzyme kinetics that assumes there is no growth of the organisms on the cometabolized chemical (Schmidt et al. 1985). The actual equations used, not considered here, are summarized in Criddle (1993). Equations for predicting the rates of complexation or polymerization of contaminants in soil are not known.

Factors influencing microbial reactions

All microbial reactions are influenced by a common set of environmental and biological factors (Table 5-1) that are well described in any soil microbiology textbook (e.g., Silvia et al. 1998). Reactions are also influenced by properties of the specific contaminant involved; however, such factors will not be considered here. The effect of numerous environmental factors on biodegradation, in particular, is addressed elsewhere (Alexander 1994). Table 5-1 provides estimates of the optimal and limiting values for biodegradation for many of the factors. It is possible, however, that at heavily contaminated sites, biodegradation is very slow at the field scale, but microsites or interfaces supporting pockets of active biodegradation may exist. In addition, extreme conditions may ultimately select for microbial populations over time that are able to biodegrade pollutants outside the ranges cited in the table.

The equations used for biodegradation reactions describe processes occurring under a constant set of environmental conditions and do not reflect many of the factors listed in Table 5-1. A few fate and transport models include coefficients for modifying the rate constant to account for temperature and other factors (Kinerson et al. 1989). These coefficients usually vary from 0 to 1 and are empirical rather than theoretical.

Because biological processes are affected by many more factors than are physical and chemical processes, it is not possible to obtain a biodegradation rate constant for a given chemical in the same manner that vapor pressure or water solubility values can be obtained. It is important to recognize that published rate parameters apply only to the conditions under which they were measured, and extrapolation of these parameters to other situations should be made with caution. Correction for differences in certain factors between soils, such as temperature, is somewhat straightforward, whereas correction for differences in population density or community composition is virtually impossible because of the difficulty in obtaining this information for soils.

Soil factors are such important controllers of microbial reactions that they may prevent degradation from occurring or from proceeding further after some initial activity. Most microorganisms are presumed to be metabolically active primarily in the soil solution phase and have much slower, if any, activity on contaminants sorbed to soil solids or existing as free product. Based on evidence that rates of biodegradation of insoluble and sorbed contaminants sometimes exceed their solubilization or desorption rates, it appears that some microorganisms can use

Table 5-1 Factors controlling biodegradation rates

Factor	Units	Optimal	Limiting[a]
Environmental			
pH	–	5.0–8.0	4.0, >10.0
Water potential	Atm	0.3–15	<0.1, >50
Temperature	°C	15–35	<5, >55[b]
Organic matter	%	–	<0.1
Porosity	%	10.40	<5
Chemical			
Contaminant concentration	mg/kg	Compound-specific	<0.01, >toxic
Inorganic N	mg/kg	5–500[c]	<1, >2000[d]
"Available" P	mg/kg	1–100[e]	<0.2
Oxygen content	% in soil vapor	18–21	<5
Electron acceptors[f]	mg/kg	Compound-specific	Compound-specific
Eh	mV	0–200	<–100[g]
Biological			
Degraders present	Cells/g	10^4–10^7	Must be present
Numbers[h]	–	–	–
Predation[i]	–	–	–
Competition[j]	–	–	–
Mutualism[k]	–	–	–

[a] All values are general guidelines only. Microorganisms can often adapt and be active in severely limiting environments, but degradation rates will often be much slower than under optimal conditions.

[b] An exception is thermophilic organisms, which can withstand temperatures over 70 °C.

[c] Depends on C content. Typical recommended values for biological treatment are C:N:P of 100:5:1.

[d] Ammonia or salt toxicity can occur at high levels.

[e] Varies depending on soil chemistry, particularly pH. Generally reported as *o*-phosphate.

[f] Alternate electron acceptors such as Fe, Mn, NO_3, or SO_4 must be present to support biodegradation in the absence of O.

[g] Anaerobic processes, which occur at redox values from 0 to –400mV, are generally, but not always, slower than aerobic degradation, characterized by positive redox values.

[h] When possible, presence of organisms capable of degrading the target contaminants should be verified, generally by counting organisms that can grow on the target contaminant. In some cases, contaminants can be partially degraded by organisms that cannot use the target compound as sole C and energy source.

[i] Presence of predators can increase or decrease the biodegradation of target compounds.

[j] Competition for nutrients, O, and other requirements by other organisms can impact rates.

[k] Mutualism refers to the fact that several types of organisms (consortia) may be required for degradation. Loss of 1 species may cause loss of degradative ability or impact rates of degradation.

contaminants that are not in solution, either directly or by enhancing the rates of solubilization or desorption through production of biosurfactants or other modifications of the soil–chemical interaction. Approaches for describing biodegradation kinetics of these processes have not yet been developed for use in soil, in part because of the difficulty in studying such phenomena in complex environments. Organic matter content and clay composition and content directly influence the partitioning of the contaminant (considered in Chapter 7) and thus its bioavailability to soil microorganisms (considered in Chapters 8 and 9).

Interactions among the organisms living in soil influence rates of biodegradation, and yet such interactions are rarely considered in contaminated soils. Potential interactions include predation by protozoa and nematodes, competition for resources with other microorganisms, and mutualistic relationships with microorganisms that are involved directly in the biodegradation process or provide nutrients and growth factors to the degrading populations. The presence of contaminants in soil may further complicate these interrelationships by differentially impairing members of the soil food web. These interrelationships are extremely difficult to measure and, though likely to be important, have not been studied sufficiently to be included in exposure assessment of contaminated soils.

Measuring biodegradation in soil

Measuring the rates of biodegradation is a critical parameter for exposure assessments (Larson 1979), but it can be difficult to measure because of field variability, soil environment complexity, and other processes that may be occurring simultaneously. The most significant evidence is the rate of compound disappearance in the field, but apparent disappearance can be due to other processes (e.g., sorption, volatilization, or leaching), and reliable data are difficult to obtain because of the inherent variability within field sites. Other field measures can provide valuable supporting evidence. These kinds of indirect measures include in situ measures of O uptake, production of metabolites (e.g., CO_2, methane, vinyl chloride from trichloroethylene [TCE]), increases in numbers of microorganisms or related organisms, such as protozoa, or reduction of alternate electron acceptors such as Fe, Mn, NO_3, or SO_4 (Borden et al. 1995). These measures all have some limitations (e.g., nonbiological release of CO_2), so the best approach involves combining several lines of evidence. In some cases, metabolites that are produced only through biodegradation can also be measured, such as hydroxylated PAH compounds (Madsen 1991). In some cases, these measures can provide estimates of biodegradation rates based on presumed stoichiometry, although this stoichiometry is not very precise because it is determined for field conditions.

It is often valuable to perform laboratory studies on samples collected from the field to provide evidence under more controlled conditions (Bartha and Pramer 1965). Loss of compounds in active soil samples compared to sterile soil samples is often used as proof of biodegradation and an estimate of biodegradation rates.

However, the environmental conditions in small reactors often differ significantly from the field conditions. Radiolabeled compounds can be added to soil samples to ensure that the native microbiota have the ability to completely degrade (mineralize) the target compounds, to attempt identification of metabolites that may accumulate, and to measure rates of mineralization. The most valuable, and expensive, approach is to perform microcosm studies to measure degradation rates under conditions simulating those observed in the field.

Chemical reactions

Chemical reactions result in the conversion of contaminants into one or several products through abiotic transformation processes involving the cleavage or formation of chemical bonds. Reactions that occur in the absence of light or do not require light include nucleophilic substitution, hydrolysis, elimination, oxidation, reduction, complexation, and precipitation. Reactions in which a compound undergoes transformation due to direct absorption of light or by reaction with highly reactive species, such as free radicals or singlet O, are classified as photochemical transformations. Most chemical reactions that occur in natural systems are not instantaneous, and thus, a brief review of chemical kinetics is appropriate. This is especially relevant for risk assessment purposes because the half-life of a contaminant is frequently used to estimate the decay or disappearance of a contaminant.

Chemical kinetics

Rates of contaminant transformation generally are described by chemical kinetics, often based on the first-order rate law. For the simple transformation of compound A to compound B,

$$A \xrightarrow{k_1} B \qquad (5\text{-}8),$$

a differential rate equation may be written as

$$-\frac{\mathrm{d}[\mathrm{A}]}{\mathrm{d}t} = k_1[\mathrm{A}] \qquad (5\text{-}9),$$

where [A] is the concentration of species A, t is time, and k_1 is the first-order rate constant (t^{-1}). The integrated form of Equation 5-9 may be expressed as

$$[\mathrm{A}] = [\mathrm{A}]_0 \mathrm{e}^{-k_1 t} \text{ or } \ln\mathrm{A} = \ln\mathrm{A}_0 - k_1 t \qquad (5\text{-}10)$$

where $[A]_0$ is the concentration of species A at time zero. The reaction kinetics of a compound exhibiting first-order decay is typically represented by a half-life ($t_{1/2}$), that is defined as the time required for 50% of the compound to be transformed. Setting $[A] = 1/2[A]_0$, the half-life of a compound is defined as

$$t_{1/2} = \frac{\ln 2}{k_1} = \frac{0.693}{k_1} \quad (5\text{-}11).$$

For a slightly more complex reaction such as

$$A + B \xrightarrow{k_2} C \quad (5\text{-}12),$$

the reaction is considered to be second order overall, but first order in terms of each reactant, *A* and *B*. The rate of decay or disappearance of species A may be written as

$$-\frac{d[A]}{dt} = k_2[A][B] \quad (5\text{-}13).$$

To properly solve such an equation, we must know the stoichiometry of the equation and the concentration of the reactants at a particular time. For the simple bimolecular case given above, the integrated form yields a solution of

$$k_2 t = \frac{1}{[A]_0 - [B]_0} \ln\left\{\frac{[A]_t[B]_0}{[A]_0[B]_t}\right\} \quad (5\text{-}14).$$

Solutions to rate equations become considerably more complex when additional reactants and stoichiometry are considered. Fortunately, many reaction rates can be treated as pseudo-first order if 1 reaction step controls the system or if the concentration of 1 species is much larger than another and does not change appreciably over time.

Additional factors that may influence reaction kinetics include temperature and molecular diffusion. The dependence of a reaction rate (k) on temperature may be represented using the Arrhenius equation

$$k = Ae^{-E_a/RT} \quad (5\text{-}15),$$

where A is a constant for the particular reaction, E_a is the activation energy, R is the gas constant, and T is temperature in degrees Kelvin. This equation may be rewritten in natural log form:

$$\ln k = \ln A - {}^{E_a}\!/_{RT} \qquad (5\text{-}16).$$

If the plot of ln$k$ versus $1/T$ yields a straight line, the values of A (intercept) and E_a/RT (slope) can be readily obtained. When a reaction occurs so rapidly that 2 species immediately react upon contact, then the reaction rate may be limited by diffusion of the species. This means that the reaction rate (k) can never exceed the diffusion rate of the species in solution. This situation may be further exacerbated in soil systems because the concentration of a contaminant in solution is often limited by the rate of desorption from the solid phase. Thus, one may need to account for diffusion limitations within the solution phase as well as rate-limited desorption processes. These issues are discussed further in Chapter 7.

Nucleophilic substitution

Nucleophilic substitution reactions typically involve covalent bonds of an organic molecule, where a nucleophilic species (positive charge-liking species) is attracted to the electron-deficient atom of the bond (e.g., C) or an electrophilic species (negative charge-liking species) is attracted to the electron-rich atom of the bond (e.g., halogen). In natural systems, the majority of chemicals that react with organic compounds by nucleophilic substitution are inorganic anions such as F^-, Cl^-, Br^-, NO_3^-, and SO_4^{2-}. These nucleophilic species, which possess a net negative charge, may displace the original atom (commonly referred to as the "leaving group") and form a new covalent bond with the electron-deficient atom of the organic compound.

Two reaction mechanisms, S_N1 (substitution-nucleophilic-unimolecular) and S_N2 (substitution-nucleophilic-bimolecular), have been proposed to describe the nucleophilic substitution process (Schwarzenbach et al. 1993; Larsen and Weber 1994). The S_N1 reaction is considered to be an ionization process involving the dissociation of the organic compound to form a sp^2 hybridized carbocation, referred to as a "carbonium ion." Formation of the carbonium ion is often the rate-limiting step in this process, and thus, the transformation of organic molecules subject to an S_N1 mechanism generally conforms to first-order kinetics. In aqueous systems, the formation of the carbocation is typically followed by reaction with water (hydrolysis), although rearrangement of substituents may occur prior to reaction with water. The S_N2 reaction mechanism is considered to be a direct replacement process in which both the organic compound and the hydrophile exist in a transition state. In this reaction, bond cleavage and formation occur simultaneously, with the nucleophile attacking the organic compound 180° from the leaving group. Because the S_N2 mechanism is dependent on the concentration of both the organic compound and the nucleophile, the reaction is bimolecular and exhibits second-order rate kinetics.

Hydrolysis

Hydrolysis is a specific type of nucleophilic substitution reaction in which water reacts with an organic compound (RX). This process results in cleavage of the covalent bond between the organic compound, R, and the nucleophile, X^-, and formation of a new covalent bond with hydroxide ion (OH^-). The general form of the hydrolysis reaction is

$$RX + H_2O \rightarrow ROH + X^- + H^+ \qquad (5\text{-}17).$$

Based on the reaction give above, the hydrolysis of simple halogenated aliphatic compounds is pH dependent and results in the formation of an alcohol. The products of hydrolysis are generally more polar and of less environmental concern than the parent compound. An example of a hydrolysis reaction is the transformation of methyl bromide to methanol,

$$CH_3Br + H_2O \rightarrow CH_3OH + H^+ + Br^- \qquad (5\text{-}18).$$

The half-life of this reaction is approximately 20 days at pH 7 and 25 °C (Maybey and Mill 1978). Note that the reaction half-life becomes greater as the strength of the C-halogen bond increases. For example, the hydrolysis half-life for CH_3Br is 20 days, while that of CH_3Cl is 0.93 years and CH_3F is 30 years.

Although the above discussion focused on nucleophilic reactions involving water (neutral hydrolysis), acid-catalyzed hydrolysis (H^+) and base-catalyzed hydrolysis (OH^-) may contribute significantly to the overall hydrolysis rate. Even when the concentrations of H^+ and OH^- are small (e.g., 10^{-7} M at pH 7.0), acid and base catalysis can accelerate hydrolysis kinetics. Assuming that the individual reactions obey first-order kinetics, the overall hydrolysis rate (k_{hyd}) may be expressed as

$$k_{hyd} = k_a[H^+] + k_n + k_b[OH^-] \qquad (5\text{-}19),$$

where k_a, k_n, k_b are the acid-catalyzed, neutral, and base-catalyzed rate constants, respectively. Detailed discussions and examples of hydrolysis reactions involving carboxylic acid esters, carboxylic acid amides, carbamates, and phosphoric acid esters can be found in Larsen and Weber (1994) and Schwarzenbach et al. (1993).

Elimination

Under certain environmental conditions, polyhalogenated alkanes have been shown to react by a second mechanism, β–elimination. In the elimination reaction, a C-halogen bond (C–X) is broken, a proton (H) is lost from the adjacent C atom (β–C), and a double bond is subsequently formed between the 2 C atoms. The overall process, in which a halogen atom (X) and a hydrogen atom (H) are removed from the compound, is referred to as "dehydrohalogenation." This process may result in the formation of halogenated alkenes that are more recalcitrant and toxic than the products of nucleophilic substitution.

The importance of elimination reactions depends, in part, on the rate at which comparable nucleophilic substitution reactions occur. In general, nucleophilic substitution involving polyhalogenated ethanes and propanes occurs extremely slowly, due to steric hindrances arising from the relatively large size of halogens. For example, the nucleophilic substitution half-life for CH_2Cl_2 is approximately 700 years (Maybey and Mill 1978). In addition, elimination is favored for base-catalyzed reactions (e.g., reactions with OH^-) and thus is pH dependent. Examples of environmentally relevant elimination reactions include the transformation of 1,1,2,2-tetrachloroethane to TCE and the transformation of 1,1,1trichloro-2,2-bis-(*p*-chlorophenyl)ethane (DDT) to 1,1,-dichloro-2,2-bis-(*p*-chlorophenyl)ethylene (DDE).

Reduction and oxidation

Chemical reactions involving the transfer of electrons either from (oxidation) or to (reduction) a contaminant constitute a major transformation pathway in environmental systems. This section will focus on abiotic reactions, although most "abiotic" reduction–oxidation (redox) reactions occurring in soils are strongly influenced by microbial activity. Photochemical redox reactions will be addressed in the following section of this chapter.

Redox reactions can be distinguished from chemical reactions that do not involve electron transfer (e.g., hydrolysis) by evaluating the oxidation state of the compound of interest. Assuming that elemental C is in a zero oxidation state, a bond to an atom that is more electronegative than C (e.g., Cl) is assigned a value of +1, while a bond to an atom that is less electronegative than C (e.g., H) is given a value of –1. Based on this convention, methane (CH_4) represents the lowest possible oxidation state of C (–4), whereas CO_2 represents the highest oxidation state of C (+4). A reduction reaction occurs when electrons are transferred from an electron donor (reductant) to an electron acceptor (oxidant). A classic example of a reduction reaction is the transformation of DDT to 1,1-dichloro-2,2-bis (*p*-chlorophenyl)ethane (DDD) by Fe^{+2} where a chlorine atom is replaced by a hydrogen atom (Glass 1972). This general process is referred to as "reductive dehalogenation." It should be noted, however, that reductive dechlorination can lead to the formation of metabolites that are more toxic than the parent compound, as is the case for the sequential reductive dechlorination of tetrachloroethene or trichloroethene to vinyl chloride, a suspected carcinogen.

Although the above discussion of redox reactions focused on the transformation of organic compounds, redox reactions play a critical role in the solubility, mobility, toxicity, and speciation of metals in soils. For a reaction of the general form,

$$Oxidant + mH^{+} + ne^{-} = Reductant \quad (5\text{-}20).$$

The redox potential (E_h) of the reaction may be represented by the Nernst equation:

$$E_h = E_h^0 + \frac{RT}{nF}\ln\frac{[Ox][H^+]^m}{[Red]} \quad (5\text{-}21),$$

where E_h^0 is the standard state redox potential of the reaction (volts), R is the gas constant, T is temperature in degrees Kelvin, n is the number of moles of electrons, F is the Faraday constant, $[Ox]$ is the concentration of oxidant, $[Red]$ is the concentration of reductant, $[H^+]$ is the concentration of hydrogen ions, and m is the moles of H^+. Standard state redox potentials for half reactions of importance in soils are given in Table 5-2. As is evident from these half reactions, redox reactions can alter the pH of the soil solution. It is often useful to develop plots of E_h versus pH, as well as plots of pH versus pE (stability field diagrams), where pE = – log (e^-). Such diagrams allow one to estimate metal speciation under various system conditions. The reader may wish to consult texts by McBride (1994) and Stumm and Morgan (1981) for a more detailed treatment of this subject matter.

Table 5-2 Standard-state redox potentials for representative half reactions

Half reaction Oxidized species (reductant) → Reduced species (oxidant)	E_h^0 (volts)
$O_2 + 4H^+ + 4e^- \rightarrow 2H_2O$	+1.229
$MnO_2(s) + 4H^+ + 2e^- \rightarrow Mn^{2+} + 2H_2O$	+1.229
$Fe(OH)_3(s) + 3H^+ + e^- \rightarrow Fe^{2+} + 3H_2O$	+1.057
$NO_3^- + 2H^+ + 2e^- \rightarrow NO_2^- + H_2O$	+0.834
$SO_4^{2-} + 10H^+ + 8e^- \rightarrow H_2S + 4H_2O$	+0.303
$CO_2 + 8H^+ + 8e^- \rightarrow CH_4 + 2H_2O$	+0.169

In well aerated soils, E_h values typically range from 400 to 500 mV. In contrast, water-saturated or flooded soils have E_h values in the range of –100 to –300 mV, which are considered to be highly reduced (McBride 1994). Under such conditions, species such as NO_2^- and Mn^{2+} are often present in the soil solution. In general, reduced forms of polyvalent cations are more soluble than their oxidized counterparts. Because metal ions may serve to catalyze chemical reactions, the reduction and subsequent dissolution of metals may enhance the abiotic transformation of contaminants.

Photochemical transformation

When organic molecules absorb light from the near-ultraviolet (UV; 200 to 400 nm) and visible (Vis; 400 to 800 nm) regions of the electromagnetic spectrum, sufficient energy may be provided to promote electrons from their ground state to a higher energy or "excited" state. This process typically involves the excitation of pi (π) and nonbonding (n) electrons of organic molecules containing a chromophore. A chromophore is a functional group or conjugated system of double bonds (e.g., aromatic ring) that absorbs UV/Vis light. For example, the UV absorption spectrum for benzophenone contains 2 maxima, one at a wavelength of 245 nm and the second at a wavelength of 345 nm (Pine et al. 1980). The 245 nm maximum corresponds to electron excitation from a bonding π orbital to an antibonding π^* orbital ($\pi \rightarrow \pi^*$), while the 345 nm maximum represents electron excitation from a nonbonding n orbital to an antibonding π^* orbital ($n \rightarrow \pi^*$). For most unsaturated organic molecules, the π orbital is lower energy than the n orbital, and thus, more energy is required to excite or move electrons from π to π^* orbitals than from n to π^* orbitals.

In natural systems, most of the UV light at wavelengths less than 285 nm is absorbed in the atmosphere. Thus, only organic compounds that absorb light at wavelengths greater than 285 nm will be susceptible to direct photochemical transformations in the environment. Many photochemical reactions involve a single molecule (unimolecular), which can be treated mathematically as a simple first-order decay process. Photolysis (bond cleavage) and intramolecular rearrangement are examples of common unimolecular photochemical transformations. For wastewater treatment applications, photolysis has been investigated as a means to break recalcitrant C-halogen bonds in chlorophenols that inhibit conventional biological treatment (Miller et al. 1988). In bimolecular photochemical reactions, a molecule in an excited state reacts with a ground-state molecule of the same species, or with some other constituent in the reaction matrix. Such reactions are also referred to as "indirect" or "sensitized" photochemical transformations. In natural water bodies, light absorption by dissolved organic matter is one of the most important contributors to indirect photolytic reactions. Experiments conducted in aquatic systems show that the presence of dissolved organic matter enhances the photolytic transformation of many organic contaminants (Zepp et al. 1985).

In natural bodies of water, light that penetrates the air–water interface will be subject to scattering by nonabsorbing species, absorption by organic particulate matter, and absorption by dissolved species, primarily natural organic matter. The intensity (W) of a particular wavelength of light (λ) at a particular depth (z) in a water column can be estimated using the Lambert-Beer equation,

$$W(z,\lambda) = W(\lambda) \times 10^{-\alpha_D(\lambda)z} \quad (5\text{-}22),$$

where $W(\lambda)$ is the intensity of the light at the water surface and $\alpha_D(\lambda)$ is the apparent or diffuse attenuation coefficient. In soil systems, the photochemical reactions are generally limited to the upper few mm of soil because soil minerals strongly absorb light (Kieatiwong et al. 1990). Photolysis experiments conducted with thin layers of soil suggest that indirect photolysis can occur in soil. In the presence of UV light, some substances (e.g., hydrogen peroxide) form free radicals that readily attack and transform many types of organic compounds. The rate of the reaction will depend on the soil moisture content and the incident angle of solar radiation. Photolysis may continue to be a significant degradative pathway over long periods of time if chemicals are transported to the soil surface during soil water evaporation. Generally, the products of photolysis are similar to those of enzymatic reactions; however, some unique products have been isolated (e.g., photo-dieldrin), which may be as toxic as the parent chemical. For very dry soils, and in the presence of ozone and sunlight, phosphorothioate insecticides may be oxidized to toxic phosphoraotes or oxons (Spencer et al. 1980).

Complexation and precipitation

Complexation and precipitation reactions typically involve metal ions, although organic compounds may influence or take part in these reactions. From a risk assessment or toxicology perspective, the formation of complexes and precipitates can dramatically reduce the free metal ion concentration in solution. This may not only influence the bioavailability of a contaminant but will also impact contaminant mobility in soils.

Complexes (also referred to as "coordination compounds") consist of one or more central atoms (typically a metal ion) surrounded by ions or molecules, referred to as "ligands." In general, the ligand forms a covalent bond with the central atom, where the central atom of the complex serves as the electron acceptor (a Lewis acid) while the ligand is considered to be the electron donor (a Lewis base) (Snoeyink and Jenkins 1980). Ligands that bond with the central atom at only one point, such as OH^- and CN^-, are referred to as "monodentate ligands." However, if the electron donor consists of more than one ligand atom and thus can occupy more than one coordination position (multidentate ligand), the resulting complex is referred to as a "chelate." Ethylene diamine tetraacetate (EDTA) is frequently used as a chelating agent and can attach to the central atom at 6 points (hexadentate ligand). In soils, functional groups associated with organic matter may interact with metal ions or metal hydroxide particles to form complexes or chelates. The strong affinity exhibited by certain metal ions for soil organic matter indicates that some metals form inner-sphere complexes as a result of strong ionic or covalent bonding with functional groups (McBride 1994). Such strong bond formation tends to be associated with irreversible chemisorption, or at the very least, the metal ion exhibits a desorption rate that is orders of magnitude slower than the adsorption rate.

The tendency to form a stable complex can be expressed in terms of the stability constant (K). For a simple amine-Cu complex, such as (Snoeyink and Jenkins 1980)

$$NH_{3(aq)} + Cu^{2+} \leftrightarrow Cu(NH_3)^{2+} \qquad (5\text{-}23),$$

the stability constant may be expressed as (Snoeyink and Jenkins 1980)

$$K = \frac{[Cu(NH_3)^{2+}]}{[Cu^{2+}][NH_{3(aq)}]} = 10^{+4} \qquad (5\text{-}24).$$

The stability of the chemical complex is directly proportional to the value of the stability constant. As one might anticipate, the stability of a chelate increases as the number of points of attachment between the chelating agent and the central atom increases. Thus, very large values of K are associated with complexes or chelates that may, for practical purposes, be irreversible.

In general, precipitation will occur if the aqueous solubility of a dissolved species is exceeded (super saturation) and the ions or molecules in solution are able to form a solid phase. The process of precipitation is thought to involve 3 sequential steps: nucleation, crystal growth, and agglomeration of small crystals into larger crystals (Snoeyink and Jenkins 1980; Stumm and Morgan 1981). In order for nucleation to occur, the concentration of a species in solution must exceed the solubility product of the corresponding solid phase, with stable nuclei forming only after a certain threshold or activation energy has been overcome. Subsequent crystal growth is governed by either diffusion of new ions to the existing crystal or by processes occurring at the interface, such as adsorption or incorporation of ions into the crystal. Accordingly, the rate of crystal growth has been described by first- or second-order linear driving force expressions (Stumm and Morgan 1981).

The equilibrium solubility product (K_{so}) for the dissolution/precipitation of a generic solid (A_xB_y) may be written as

$$A_xB_{y(s)} \leftrightarrow xA^{y+} + yB^{x-} \tag{5-25}$$

$$K_{so} = \frac{\{A^{y+}\}^x \{B^{x-}\}^y}{\{A_zB_{y(s)}\}} \tag{5-26}$$

If the activity of the solid phase is assumed to be unity, Equation 5-26 can be rewritten to yield a concentration solubility product (${}^cK_{so}$):

$${}^cK_{so} = [A^{y+}]^x[B^{x-}]^y \tag{5-27}$$

Furthermore, if the components of the solution have activity coefficients of 1, then ${}^cK_{so} = K_{so}$, otherwise ${}^cK_{so}$ will be a function of ion strength. Using these equations, one can determine the solubility of each species in solution based on the solubility product of the reaction. In soil systems, important species that undergo precipitation and dissolution reactions include carbonates, hydroxides, and phosphates. In closing, it should be recognized that rates of precipitation or dissolution, which are often difficult to estimate, frequently control the distribution of a species between the solid (precipitate) and solution phases. Thus, reliance solely on equilibrium solubility product data can lead to erroneous predictions of ion concentrations in solution.

Bioaccumulation

Plants and soil biota can be conceptualized as another phase to which transfer of contaminants may occur, analogous to the transfer between soil water and soil solids. Thus, movement of contaminants into the biotic phase represents a sink from the perspective of the soil as a whole. Given the ratio of total soil mass to soil organism biomass, however, it is improbable that the biota could remove enough contaminant to appreciably alter the total soil concentration. Plants capable of hyperaccumulating metals have been tested as a bioremediation technique, but the low rates of uptake do not appreciably alter concentrations except over very long time periods (Huang et al. 1997). Thus, contaminant transfer or uptake by plant and soil animals is of immediate relevance to exposure assessment for organisms occupying trophic levels above primary producers and consumers. In this case, the soil animals of most concern would be the invertebrates. Once transfer occurs from soil into soil-dwelling animals and plants, qualitative and quantitative transformations are fundamental to calculating doses to organisms at the next trophic level. Both the process of uptake (i.e., phase transfer) and metabolic fate are considered herein as bioaccumulation.

Accurate predictions of contaminant bioaccumulation are complicated by the necessity to consider the interactions of soil sorption, the differential uptake among species, and the toxicodynamic disposition within an organism. Thus, bioaccumulation of contaminants in plants and soil invertebrates is a fate process controlled as much by environmental conditions as by physiology and biochemistry.

Risk management decisions for remediation of a contaminated area can be aided by an understanding of the potential for bioaccumulation. Such understanding will be enhanced by consideration of the principles affecting uptake and disposition of the contaminant within the receptor organism. This section gives consideration to the processes affecting uptake of contaminants by soil invertebrates and plants, and a brief overview of the distribution and metabolism within these organisms (i.e., toxicodynamics). Special attention is paid to the possibility that contaminants may be metabolized to equally toxic or more toxic products in addition to the normal detoxification and excretion mechanisms. Finally, some insights are offered regarding the prospects for prediction of bioaccumulation.

Contaminant uptake by soil invertebrates

Processes affecting uptake

Sorption equilibria

The dependence of uptake upon the presence of a contaminant in soil solution was suggested by early studies that showed an inverse correlation with soil organic matter and toxicity of lindane and dieldrin to *Drosophila* (Edwards et al. 1957; Hermanson and Forbes 1966). The fact that soil solution concentration is a limiting factor for contaminant uptake can be inferred from many experiments showing significant correlations between sorption of neutral organic compounds and soil organic matter content (Chiou 1989). Evidence that desorption influences contaminant uptake was demonstrated by positive correlations between increasing percentages of soil desorbable, radiolabelled, nonionic insecticides, and increasing levels of radiolabel uptake by corn rootworm larvae exposed in soils differing in organic C content (Felsot and Lew 1989). The relationship between desorbability and uptake of contaminants from soil water may be generalized among invertebrates, as evidenced by studies with earthworms (Ma et al. 1998).

Diffusion across the invertebrate surface

A mechanistic hypothesis that explains the uptake of contaminants from soil solution is tied to the biochemical nature of the exterior surface of important soil invertebrate phyla. The surface or cuticle of arthropods, annelids, and nematodes is to varying degrees lipophilic by virtue of the exterior deposition of different hydrocarbons or waxes. Thus, the cuticle functions as a matrix similar to the hydrophobic regions of soil organic matter, enhancing phase transfer at least to the outer surface of the organism.

Once sorption of the contaminant to the invertebrate cuticle has occurred, absorption into the body will be limited by diffusion across the cuticle and through the underlying epidermis. In the case of insects, diffusion across the cuticle might be enhanced via movement into spiracles and trachea or by diffusion across less lipophilic intersegmental membranes.

Compounds of extremely high log K_{ow} (>5) may not diffuse sufficiently out of the lipophilic cuticular layers to significantly influence internal body burdens. For example, early work with isolated insect cuticle showed that penetration was inversely related to hydrophobicity (Olson and O'Brien 1963). In addition to the possibility that highly hydrophobic compounds may diffuse only slowly through outer waxy layers, the comparatively more polar endocuticular layers may serve as an additional diffusional barrier. In these cases, dietary exposure to contaminated particles may be most influential in ultimate body burdens, as has been shown for earthworms exposed to very hydrophobic polycyclic aromatic hydrocarbons (Ma et al. 1998).

Food intake

In addition to bioconcentration via surface sorption and diffusion, invertebrates will acquire contaminants through the food chain and/or through the ingestion of contaminated soil and litter particles. Accumulation within the body then will depend on the efficiency of transfer across the gut epithelium. For organic compounds, diffusion across the gut lining is probably the main process controlling uptake. Heavy metals, on the other hand, may require ligand exchange processes, facilitated diffusion, or even active transport.

Studies of various earthworm species indicate that some heavy metals are preferentially accumulated by ingestion of soil particles, whereas others are largely taken up and stored in the surface integument. A comparison between whole body concentrations of a large variety of metals with earthworm casts showed that Cd and Hg are readily accumulated via food ingestion, but Cr is largely associated with the earthworm surface (Helmke et al. 1979). In experiments where metal salts were added to sewage sludge, only Cd was accumulated to a whole body concentration that exceeded the sludge concentration (Hartenstein et al. 1980). Ni, Pb, and Zn body concentrations remained nearly 2 orders of magnitude below the levels added to the sludge.

Prospects for prediction of contaminant uptake by invertebrates

Organic compounds

Equilibrium partitioning theory has been used to explain the uptake of organic contaminants from sediments and, to a lesser extent, from soil (Di Toro et al.

1991). The theory suggests that uptake of contaminants is from interstitial or pore water. Knowing the concentration of contaminant in water or soil, the fraction of organic matter in the matrix, and the lipid-adjusted residues in the organism of interest, a biota–soil accumulation factor (BSAF, ratio of concentration in organism to concentration in soil, at steady state) that is independent of K_{ow} (octanol–water partition coefficient) can be calculated (Di Toro et al. 1991; Ma et al. 1998). For example, the average BSAF (expressed as ([contaminant]/kg SOM)/([contaminant]/kg lipid) for a diversity of PAHs in *Lumbricus rubellus* exposed to contaminated soil was 0.1 with a range from 0.03 to 0.26 (Ma et al. 1998). The average BSAF for PCBs in the same species was reported as 0.75 (Hendriks 1995). With this information, if shown to be generally true, the concentration in *L. rubellus* could be predicted from soil chemical monitoring alone. However, the magnitude of BSAF likely will be species specific. For the earthworm *Eisenia andrei*, the BSAF (soil organic matter/lipid basis) ranged from 5.3 to 7.4 for a series of PCBs (Belfroid et al. 1995). In the same study, BSAF for a series of chlorobenzenes ranged from 0.44 to 10.1.

The limitations of equilibrium partitioning theory for incorporating fate processes into exposure assessment suggest a need for empirical studies of different classes of compounds and a diversity of indicator species. In determining the BSAF, the first step is determination of the uptake of soil contaminants when steady state is reached. A corresponding measurement of contaminant concentration must be determined in the matrix. Only at this point will the BSAF have relevance for exposure assessment. However, if organisms will be extracted from soil sampled from a site that has been contaminated for a long period of time, steady state is likely to exist and this assumption is valid.

Certain anomalies among compounds within a class are unpredictable. For example, high molecular weight PAHs do not seem to behave as the lower molecular weight PAHs (Ma et al. 1998). Compounds that are biotransformed will yield a lower BSAF than those substantially recalcitrant to transformation. It is not clear if low hydrophobicity neutral organic compounds will behave as the PAHs and PCBs. Furthermore, contaminants are just as likely to be somewhat polar or even ionizable, and too few studies have applied equilibrium partitioning theory to these types of compounds to make it generally useful for a priori risk assessments.

Heavy metals

Prediction of body burdens of metals by terrestrial species is even more precarious than for organics. Sorption to the cuticle and diffusion across the epidermis may require specific ligand interactions. Uptake of metals that are essential elements are regulated at the cellular level (Chapman et al. 1996). Thus, uptake may be quite limited in the presence of high metal concentrations.

Specific tissues or organs of some groups of invertebrates can store and release metals, thereby making generalizations about uptake and excretion kinetics difficult (Eijsackers 1994). Furthermore, large differences in potential for uptake exist among the myriad of metal contaminants. As mentioned previously, Cd seems to be the metal most easily bioaccumulated by earthworms to levels that are orders of magnitude above the concentrations present in soil or sludge.

In one experiment exposing earthworms to soil from a contaminated industrial site, steady-state uptake of Ni and Cu occurred within 1 and 14 days, respectively (Marinussen et al. 1997). Excretion of metals after removal of worms to uncontaminated soil was modeled as a 2-compartment process. The metals were very rapidly eliminated within a day, followed by a very slow decline over the next 2 months. This experiment illustrates a problem that will be encountered when trying to model metal bioaccumulation by invertebrates. The animals will move around from areas of high concentration to low contamination and back again. Prolonged time spent in an uncontaminated area could quickly cause a drastic reduction in body burden, but upon reentry into a contaminated zone, contaminants will quickly accumulate again.

Contaminant uptake by plants

Processes affecting uptake

Sorption equilibria

Early studies of herbicide phytotoxicity in soils differing in organic matter content generated the hypothesis that sorption was rate limiting for plant uptake (Harris 1966). Taking advantage of the dependency of uptake on sorption, researchers recommended additions of activated C to herbicide-contaminated soils in an effort to reduce bioavailability and thus damage to desirable crops (Burnside 1974). Amendment of soil with high organic matter wastes like municipal sewage sludge and manure simultaneously increases K_d (equilibrium distribution coefficient) and decreases phytotoxicity (Guo et al. 1991, 1992). Studies with uptake of insecticide residues by pea plants in different soils also showed a dependence on organic matter content and, by implication, equilibrium distribution of insecticide between solid and liquid phases (Getzin and Chapman 1959).

Movement into roots

Studies with herbicides lead to the conclusion that the K_{ow} of a chemical is an additional factor, albeit related to sorption, influencing plant uptake (Devine et al. 1993). Chemicals of comparatively higher K_{ow} will readily sorb to root surfaces, whereas ionized organic chemicals will not. It is assumed that neutral organics move from soil water in a phase transfer process to the exterior surface of the root epidermal cells. The surface of younger roots is covered with cutin and a mucilaginous film, which enhances intimate contact with soil particles (Kramer 1969), potentially facilitating interphase mass transfer. Plant roots often have mutualistic associations with fungi, known as "mycorrhizae," that also take up contaminants and affect mass transfer into the root itself (McFarlane 1995).

Water moves into the interior of the root through the cortex towards the endodermis, the cell layer surrounding the vascular elements (xylem and phloem). Water movement may occur mostly by diffusion mostly through the cell walls (i.e., apoplastically) of the endodermis and cortex (Kramer 1969; Oertli 1996), but it also can occur via diffusion through the cytoplasm (symplastically), vacuoles, and plasmodesmata (narrow channels connecting adjacent cells). Any contaminant in solution should move via mass transfer along with the water or by diffusion toward the interior of the root. However, adsorption to the cell walls or membranes would reduce the concentration of chemical reaching the endodermis.

The endodermis associated with the section of roots distal to the meristematic region contains the Casperian strips, a hydrophobic barrier to water and solute movement. Water and solutes can cross this barrier via diffusion in the cytoplasm, vacuoles, and plasmodesmata to enter the xylem. Thus, sorption to cell constituents, as well as rates of diffusion, can limit the movement of comparatively hydrophobic compounds across the endodermis. In the apical part of the root just above the meristematic region, the Casperian strip is least developed, and the movement of water and solutes will be comparatively greater than in more suberized root regions at increasing distances from the apex (Moreshet et al. 1996).

The efficiency of uptake of chemicals from soil into roots has been characterized by calculating a root concentration factor (RCF) (Bromilow and Chamberlain 1995). The RCF is a measure of the ratio of the concentration of a solute in the roots relative to the concentration in the external soil solution. The mechanism of movement into the root may be due to aqueous diffusion or to vapor-phase diffusion depending on Henry's Law constant. For example, compounds having K_H (dimensionless) of 10^{-7} or less would enter roots via diffusion in the aqueous phase, and those having K_H above 10^{-4} would likely move to roots via air diffusion. Compounds with K_H in between could move in either phase (Bromilow and Chamberlain 1995).

Experimental measurements of the RCF are usually made by bathing young plants in nutrient solutions rather than determining chemical uptake from soil. The potential for concentration in the roots is related to the hydrophobicity of a chemical, with RCF increasing nearly exponentially as K_{ow} increases. This relationship may be functional, however, only if RCF is measured when roots are in a nutrient solution. In soil, sorption (as measured by K_{oc}) would increase as K_{ow} increased, effectively reducing or slowing movement toward the roots by both aqueous and vapor-phase diffusion.

Solution pH is an important factor influencing the RCF, especially for weak acids and metals. For phenoxyacetic acids, RCF decreases with a change in solution pH from 4 to 8 (Briggs et al. 1987). On the other hand, decreases in solution pH would tend to favor speciation of certain heavy metals (e.g., Pb, Cd) to more soluble forms, causing an increase in uptake. Liming, which raises soil pH and reduces solubility of Pb, has been associated with a reduction in uptake of added Pb salts by lettuce and oats (Cox and Rains 1972; John 1972; John and Van Laerhoven 1972). Liming may cause the formation and precipitation of relatively insoluble Pb carbonate (Zimdahl and Skogerboe 1977). While as a general principle, liming can reduce uptake of metals, it may not universally reduce bioavailability, as shown in some studies (Sterritt and Lester 1980; Sims and Kline 1991).

Plants can influence the efficiency of metal uptake by exudation of organic acids and phenols. This process is particularly useful for increasing uptake efficiency of essential elements like P or Fe when plants are deficient (Jungk 1996; Marschner and Romheld 1996). However, mobilization of Al can have adverse consequences (Jungk 1996). While exudation of organic acids can alter soil pH, little is known about whether such changes will alter uptake of ionizable organics. On the other hand, several plants in the family Cucurbitaceae (cucumbers, squash, melons) are extraordinarily proficient at mobilizing and taking up dioxins, which are normally considered immobile in soil and nonbioaccumulative due to their extremely high K_{oc} (Huylster et al. 1994).

Bound residues are usually not considered to be bioavailable, but years of pesticide research has shown that small percentages of the total bound material can be taken up by plants and animals (Khan 1982). The proportion of bound residues typically reported to have been bioaccumulated is 5% or less of the bound pool. The problem with characterizing the risk associated with these small amounts of bioaccumulated bound residues is lack of information concerning their nature. Studies with various pesticides have shown that some of the bound residues are actually parent compound, but a portion may also be present as metabolites. The nature of bound residues of contaminants other than pesticides has been neglected.

Movement into shoots

Movement from the root into the vascular elements is prerequisite for further distribution to the leaves and potential for extensive metabolism. The efficiency of movement from the roots into the xylem is expressed as the transpiration stream concentration factor (TSCF). The TSCF represents the ratio of the concentration in xylem sap relative to the concentration in the external solution. Measurement of the concentration of a chemical in the xylem sap is impractical, but the TSCF can be estimated from the total residues in the aerial plant parts accumulated for a known time. If a compound moves into the stem with the same efficiency as water, then the TSCF is 1.0, the maximum value for passive uptake. A few experiments suggest that translocation of contaminants into the stem is influenced by the transpiration rate. For example, within 1 day, the explosive cyclonite (RDX) accumulated in bush bean stems to levels similar to those present in a hydroponic culture medium (Harvey et al. 1991). Bromacil herbicide, phenol, and nitrobenzene translocation into soybean stems and leaves increased as transpiration rate increased (McFarlane et al. 1987).

Chemicals with a log K_{ow} of approximately 1.8 tend to have the highest potential for movement into the xylem (Briggs et al. 1982). These seem to cross the endodermis more readily than either very low or very high K_{ow} compounds. For example, very little of the dinitroaniline herbicide trifluralin passes into the vascular cells; the majority of the residues are recovered from the roots. Dioxins have even higher K_{ow} values and are not absorbed by plants into the vascular elements. Part of this limited translocation is related to high soil sorption potential, but part is related to the inability to cross the cortex and move into the xylem (Jones and Sewart 1997). Acetanilide and triazine herbicides have comparatively lower soil sorption potential but also easily diffuse into the xylem and are distributed to apical regions.

The relationship between K_{ow} and movement from the root into the xylem stream was developed from experiments on several groups of herbicides. The relationship may not hold for all compounds, however. Bromacil, phenol, and nitrobenzene have similar K_{ow} values. Nearly 10% of the bromacil added to a hydroponic solution was taken up by soybean shoots, while less than 2% of added phenol and nitrobenzene was absorbed. Phenol and nitrobenzene were rapidly absorbed by the roots (15% to 20% of the concentration present in solution) but did not translocate into the stem.

Both the pK_a and the K_{ow} of the chemical influence translocation of ionizable chemicals. The pH of the xylem fluid is approximately 5, while the phloem is approximately 8. Thus, acidic compounds with $pK_a < 5$ and log $K_{ow} < 3$ will have optimum phloem mobility, in addition to being readily translocated in the xylem (Bromilow et al. 1990). Compounds that are phloem-mobile may ultimately attain a greater distribution throughout the plant than those entering only the xylem. Xylem-only mobile contaminants, however, may concentrate in distal, rapidly growing regions to effectively higher concentrations than phloem mobile compounds. These regions of new, tender growth might be more palatable to herbivores and thus serve to increase potential exposure.

Highly sorbed contaminants like PCBs have been found in plant foliage, suggesting root uptake and translocation in the xylem (Moza et al. 1979). The presence of these compounds, however, may not be due to direct absorption from soil. PCBs and similar low-solubility compounds can volatilize from the soil surface and be redeposited and retained by plant foliage. For example, after 52 days of growth in a sandy loam soil, soybean foliage contained up to 18% of the radiolabel from surface-applied ^{14}C-PCBs (Fries and Marrow 1981). However, plants grown in treated soil covered with a vapor barrier contained no radioactivity.

Uptake of metals by plants differs greatly among species and among plant parts. For example, Pb content in corn stover growing in soil containing over 1400 mg/kg Pb acetate was significantly greater than corn grown in uncontaminated soil, but Pb levels in corn kernels were unaffected (Baumhardt and Welch 1972). The peels of beet roots growing in a contaminated soil accumulated about 4X more Pb than the fleshy tissue; however, the reverse was true for turnip peels and roots (Nicklow et al. 1983). Obviously, metal uptake by plants, in addition to being modulated by soil factors, is difficult to predict by assuming a linear relationship between concentration in soil and concentration in the plant. This uncertainty thus affects predictions of contaminant food-chain transfer among foraging vertebrates.

Prospects for prediction of plant uptake

Current models generally rely on assumptions of equilibrium between the plant surface and soil solution (Briggs et al. 1982; Travis and Arms 1988). Thus, K_{ow} tends to be frequently correlated with uptake. While this simple assumption may hold useful for chemical homologs, extension of this approach to a wide diversity of compounds will not produce valid predictions of uptake. Nevertheless, regression equations have been developed with a range of very different chemicals for bioconcentration in vegetation (Travis and Arms 1988). However, the regression model suffers from reliance on limited representation of specific properties and overreliance on a representative group of chemicals. The applicability of the results is uncertain because of the variability of the data chosen and the weighting of the regression slope by 1 or 2 extreme data points (Ferguson et al. 1998).

A detailed mathematical model for plant uptake of organic chemicals and translocation within the plant has been developed and calibrated with the herbicide bromacil in soybean plants (Boersma et al. 1991; Lindstrom et al. 1991). The model was based on the principles of concentration of mass and xylem and phloem transport. Equations were reported for transfer of chemical between soil and root xylem and all possible combinations of xylem and phloem in stem and leaves. The model used transpiration rates as a driver for water flux within the plant. The model estimated the concentration of contaminant in roots, individual stem parts or whole stem, and individual leaf clusters or whole leaf.

A fugacity-based model has been developed for modeling the bioconcentration of organic chemicals in plants (Trapp et al. 1990). The model specifically considers as compartments soil, roots, shoots, and air. Transfer of chemical between compartments is considered for soil–air, soil–roots, roots–shoots, and air–foliage. The latter compartmental transfer relies on assumptions about plant growth and total volume. The model has been validated in a series of experiments with radiolabelled pesticides (atrazine, DDT, dieldrin, hexachlorobenzene) and PCBs (tri-, tetra-, and pentachlorobiphenyl). Output from the model includes a calculation of the bioconcentration factor (BCF, concentration in fresh plant or concentration in dry soil). Despite differences among K_{ow} values for compounds, BCFs differed little, and when corrected for metabolism, the values differed by less than a factor of 2. The model was sensitive to the inputs for K_{ow} and K_{oc}, with the greatest uncertainty introduced by inaccuracies in K_{oc} values.

The fugacity-based model of Trapp et al. (1990) was improved and tested specifically with the herbicide bromacil (Trapp et al. 1994). In the improved model, known as PLANTX, more possible compartmental transfers were modeled. The following processes were considered:

- diffusive exchange in soil water and air pores to roots,
- transfer into roots with the transpiration stream,
- translocation into stems and leaves via the transpiration stream,
- partition into the stem,
- transport into fruits via the assimilation stream,
- diffusive exchange between air and leaves via stomata and cuticle,
- metabolism, and
- dilution by growth.

The model is purportedly applicable to different plant species and most nonionizable organic compounds. As expected, operation of the model would require specific inputs for individual plant species as well as phase transfer characteristics and a biological degradation half-life for the chemical.

Metabolic transformations in plants and invertebrates

Once a chemical has been absorbed into the tissues of plants and invertebrates, its fate is determined by the affinity for enzyme systems evolved ostensibly for primary metabolism and secondarily for detoxification. Two kinds of metabolic detoxification reactions, commonly known as "Phase I reactions" and "Phase II reactions" (Dauterman and Hodgson 1978), can occur in both plants and animals. The nature of these reactions has great relevance to the toxic dose transferred to the next trophic level as well as for the survival of the receptor organism itself. The typical metabolites produced are more water soluble than the parent molecule, and thus should be more readily excretable, at least in animals.

Phase I reactions involve enzymes capable of oxidations, hydrolyses, and reductions. Oxidation reactions are carried out by the microsomal oxidases dependent on different forms (isozymes) of cytochrome P450. Typical reactions may include hydroxylation leading to ring cleavage, dealkylation, and deamination. Hydrolysis reactions are carried out by specialized esterases known also as "hydrolases." These enzymes are active on esters, resulting in alcohol and acid products. The most typical reductive reactions include dehydrochlorination, or the removal of one chlorine from an alkyl or aromatic structure, and denitrification of nitro substituents to amines.

Phase II reactions involve conjugation of a glucose, galactose, or glutathione molecule to a partially oxidized parent compound. Thus, a typical reaction might be hydroxylation of molecule followed by conjugation. The conjugate formed is more water soluble than the parent and is more easily excreted by animals. Plants will store the conjugated metabolites in vacuoles or in the cell wall (Devine et al. 1993).

Despite the generic use of detoxification to describe enzymatic transformations of contaminants, some reactions are actually metabolic activations from the viewpoint of toxicity, even though the products are more water soluble than the parent compound. Well-known examples are the P450-mediated oxidations of low toxicity phosphorothioate insecticides to phosphoroates (oxons), which have greatly enhanced anticholinesterase activity. For example, probably one of the most studied insecticide activations was the transformation of ethyl parathion to ethyl paraoxon. PCBs can be hydroxylated via P450 reactions to yield polychlorinated biphenylols that are responsible for the reported estrogenic activity of PCBs (Fielden et al. 1997).

What is not well known is the rate of elimination of these toxic metabolites from plants and invertebrates. In animals, the more water soluble phosphoroates and hydroxylated PCBs are excreted at enhanced rates but how quickly will be critical to predicting doses at the next trophic level.

Studies of pesticide residues in plants have shown that metabolites may take 3 possible forms—freely extractable, extractable conjugates with sugars or peptides, and unextractable or bound residues incorporated into plant constituents (Khan 1982). Thus, the term "bound residues" has been applied to contaminant residues in plants in an analogous manner to bound residues in soil. A large fraction of pesticide absorbed by the plant ends up as a bound residue, mainly associated with lignin in cell walls. A general consensus is that bound plant residues have low bioavailability (although not zero), as shown in animal feeding studies with elimination in the feces. Little work has been done with invertebrates on processing of bound plant residues. Furthermore, research on contaminants other than pesticides is virtually nonexistent.

Absorbed metals will tend to have a uniquely different metabolic fate than organic chemicals. First, metals might be reduced or oxidized to form either more or less toxic species. Some metals will form ligands with organic acids or proteins, keeping them out of general circulation. In general, however, animals and plants tend to sequester metals in physiologically unavailable pools. For example, insects have the capability to shunt metals to the cuticular exoskeleton. Plants may shunt metals to the cell walls. Few studies have been conducted to determine the bioavailability of these sequestered pools of metals to organisms in higher trophic levels.

Mathematical Modeling of Contaminant Fate in Soils

Introduction

Mathematical modeling can provide site managers, risk assessment specialists, and regulators with a cost-effective tool to assess and predict the fate of soil contaminants. From a practical perspective, contaminant fate models can be used to estimate the concentration of a contaminant within each phase of a soil system (gas, liquid, and solid) and to predict the behavior and dissipation of contaminants over time. Such models are based on many of the concepts and parameters discussed in previous sections of this chapter and serve to integrate these principles within a framework that can be readily applied to environmental risk assessment. In addition to providing estimates of environmental concentrations, fate models can also be used to identify important processes governing fate, provide information needed for the design of environmental monitoring and remediation activities, and analyze the sensitivity of contaminant fate to changes in site conditions and contaminant properties.

Two types of fate models will be discussed in this section: multimedia models and site-specific models. The primary focus will be on multimedia models, while the discussion of site-specific models will lead into the contaminant transport chapter (Chapter 6). In general, multimedia models are much less complex than site-specific models and are best suited for predicting average contaminant concentrations on large (e.g., regional) scales. Site-specific models are the most appropriate for predicting contaminant fate and transport at or near the contaminant source and can be used for site-specific risk assessment. Nevertheless, multimedia models are very useful for first-order approximations of contaminant distributions and as qualitative tools to identify fate and transport processes that should be included in more detailed site-specific models.

Multimedia fate models

As the name implies, multimedia models describe the distribution of a contaminant between multiple media or phases in a system. In general, these models are based on the concept of a compound's fugacity (f, Pa = kg/m s^2) in each phase as described by Mackay (1991). The fugacity of a compound is linearly related to concentration and may be defined as

$$\mathrm{f} = C / Z \text{ or } C = Z \times \mathrm{f} \tag{5-28}$$

where C is the molar concentration of the compound in a particular phase (mol/m^3) and Z is the capacity of the phase for the contaminant (mol/m^3 Pa^{-1}). If equilibrium is reached between 2 phases, then the contaminant fugacity in phase 1 (f_1) will be equal to the contaminant fugacity in phase 2 (f_2). This concept can be expressed as

$$\mathrm{f}_1 = \mathrm{f}_2 \text{ or } C_1 / Z_1 = C_2 / Z_2 \tag{5-29}$$

Equation 5-22 may be rearranged as

$$C_1/C_2 = Z_1/Z_2 = K_{12} \quad (5\text{-}30).$$

Notice that the ratios of C and Z given in the above equation are equal to the conventional partition or distribution coefficient between Phase 1 and Phase 2 (K_{12}). If the units for C_1 and C_2 are identical, then K_{12} will be dimensionless. In practice, however, the units of C for each phase are often different. For example, if the concentration in water, C_{water}, is expressed as mg/L and the concentration in solid phase, C_{solid}, is expressed as mg/kg, then K will have units of L/kg.

Unfortunately, the concept of fugacity is somewhat ambiguous and requires the user to learn new terminology and relationships. Thus, it is instructive to express fugacity as the "effective solubility" of a contaminant in each phase (Mackay and Shiu 1992). This method involves computing the amount or concentration of a compound in each phase based on chemical properties such as aqueous solubility. Based on this approach, the total concentration of a contaminant in the soil (C_T, mg/m^3) can be expressed as the sum of the concentration in each phase:

$$C_T = C_{water}\, n\, s_{water} + C_{solid}\, \rho_b + C_{gas}\, n\, s_{gas} + n\, s_{NAPL}\, \rho_{NAPL} \quad (5\text{-}31),$$

where C_{water} is the concentration in the aqueous phase (mg/L), n is the soil porosity (cm^3 voids/cm^3 total), s_{water} is the saturation of water (cm^3 water/cm^3 voids), C_{solid} is the sorbed-phase concentration (mg/g), ρ_b is the soil bulk density (g/cm^3), C_{gas} is the concentration in the gas phase (mg/L), s_{gas} is the gas-phase saturation (cm^3 gas/cm^3 voids), s_{NAPL} is the NAPL saturation (cm^3 NAPL/cm^3 voids), ρ_{NAPL} is the density of the NAPL (g/cm^3). If the contaminated zone does not contain a pure organic liquid or the contaminant is dissolved in the NAPL (e.g., PCB dissolved in transformer oil), then the ρ_{NAPL} term is replaced with the concentration of contaminant in the NAPL (C_{NAPL}, mg/L NAPL).

To illustrate the practical utility of this approach for estimating the distribution of a contaminant among solid, liquid (aqueous), and gas phases, consider a source area contaminated with TCE. Based on laboratory analysis of soil cores collected at the site, the average amount of TCE extracted from a 50 mL core sample was 200 mg TCE, or 4 mg TCE/cm^3. A regulator or site manager might then ask whether or not this concentration is indicative of entrapped free product or NAPL in the source zone. Assuming that local equilibrium conditions exist among the phases, and using the chemical and soil properties given in Table 5-3, one can compute the maximum amount of TCE that could exist in each phase. Starting with the aqueous phase, the maximum value of C_{water} is 1100 mg/L, the aqueous solubility of TCE. The porosity (n) is then calculated using the relationship, $n = 1 - \rho_b/\rho_s$, to yield a value of 0.351 cm^3 voids/cm^3 total. The water saturation (s_{water}) is computed by

Table 5-3 Contaminant (TCE) and soil properties necessary for estimating contaminant distribution within the source zone

Contaminant (TCE) properties	Soil properties
Molecular weight (MW) = 131.4 g $mole^{-1}$	Organic carbon content (OC) = 1% (wt)
Aqueous solubility = 1100 mg L^{-1}	Water content (WC) = 8% (wt)
Saturated vapor pressure = 68.1 mm Hg	Bulk density (ρ_b) = 1.72 g cm^{-3}
TCE liquid density (ρ_{NAPL}) = 1.43 g cm^{-3}	Particle density (ρ_s) = 2.65 g cm^{-3}
K_{oc} = 126.0 ml g^{-1}	Soil porosity (n) = 0.35
K_H = 10.72 L atm $mole^{-1}$	Temperature (T) = 20 °C

converting the gravimetric water content to volumetric water content, $\theta_w = WC \times \rho_b \times 1/\rho_w = 0.138$ cm^3 water/cm^3 total, and using the relationship between volumetric water content and water saturation, $s_w = \theta_w/n$. This yields a water saturation of 0.392 cm^3 water/cm^3 voids. These values can then be used to calculate the amount of TCE in the soil core that is attributable to the aqueous phase:

$$C_{water}\, n\, s_{water} = (1100 \text{ mg L}^{-1})(0.351 \text{ cm}^3 \text{ cm}^{-3})(0.392 \text{ cm}^3 \text{ cm}^{-3})(1\text{L } 1000 \text{ cm}^{-3}) = 0.151 \text{ mg cm}^{-3} \qquad (5\text{-}32).$$

Assuming linear sorption of the TCE by the soil, the solid-phase concentration can be estimated using the following relationship: $C_{solid} = C_{water} \times K_{oc} \times OC$, where K_{oc} is the organic C–water distribution coefficient, and OC is the organic C content of the soil. Based on the data given in Table 5-3, one would compute a value for C_{solid} of 1.386 mg TCE/g solid. Converting this value into a total volume concentration, the amount of TCE attributable to the solid phase would be

$$C_{solid}\, \rho_b = (1.386 \text{ mg g}^{-1})(1.72 \text{ g cm}^{-3}) = 2.38 \text{ mg cm}^{-3} \qquad (5\text{-}33).$$

The pressure (P) of TCE in the gas phase can be based on the saturated vapor pressure or computed using the Henry's Law constant and the aqueous-phase concentration, $P = K_H \times C_{water} \times$ molecular weight (MW). Since we are assuming that the aqueous-phase concentration is equal to the solubility value, the pressure calculated using the Henry's law constant would be equal to the saturated vapor pressure of TCE. The pressure can then be converted to a gas-phase concentration using the ideal gas law, PV = nRT. The ideal gas law can be rearranged such that $C_{gas} = P/RT \times MW$.

Expressing the gas-phase concentration in terms of the total volume of the soil core gives

$$C_{gas}\, n\, s_{gas} = (489.1 \text{ mg L}^{-1})(0.351 \text{ cm}^3 \text{ cm}^{-3})(0.392 \text{ cm}^3 \text{ cm}^{-3})(1\text{L } 1000 \text{ cm}^{-3}) \\ = 0.067 \text{ mg cm}^{-3} \quad (5\text{-}34).$$

Summing the water-phase, soil-phase, and gas-phase concentrations together gives a total contaminant concentration, C_T, of 2.59 mg/cm^3. Because the measured total concentration of TCE in the soil core (4 mg/cm^3) is greater than 2.59 mg/cm^3, one would anticipate that NAPL was present in the soil core, assuming that equilibrium had been reached between all phases.

These simple calculations serve to illustrate the principles that may be used to estimate the distribution of a contaminant among different soil phases. Multimedia models expand upon this basic approach to include degradation and nonequilibrium conditions over much larger scales, accounting for contaminant distributions between the atmosphere, water bodies, biota, and soil.

Multimedia models vary in their level of complexity, from simple steady-state, equilibrium partitioning of a contaminant between phases in a closed system (Level 1) to nonsteady-state, nonequilibrium partitioning in open systems (Level IV). A summary of the conceptual distinctions between different levels of multimedia model is given in Table 5-4. Level I and II models are considered to be oversimplifications of natural systems and are recommended for use only in determining first-order approximations of contaminant fate or to identify chemical properties or transformation processes that may be important in determining contaminant fate. It is generally agreed that Level III models are necessary to adequately assess the fate of contaminants in soil (Cowen et al. 1995). A conceptual model of a Level III multimedia model is shown in Figure 5-2.

Table 5-4 System parameters that can be accommodated by different levels of multimedia contaminant fate models

Model level	Closed system	Steady state	Equilibrium	Open system	Non-equilibrium	Nonsteady state
I	✓	✓	✓	✓	✓	✓
II	✓	✓	✓	✓	✓	✓
III	✓	✓	✓	✓	✓	✓
IV	✓	✓	✓	✓	✓	✓

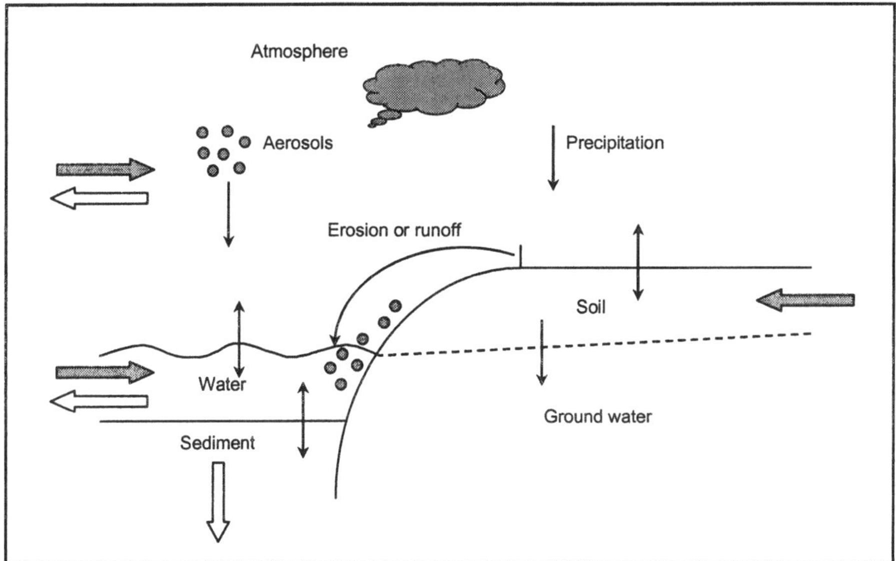

Figure 5-2 Schematic diagram of multimedia model compartments with inputs, outputs, and inter-compartmental mass transfer processes indicated by arrows

Several Level III multimedia models are readily available, including CalTOX, ChemCAN, HAZCHEM, and SimpleBox 2.0. CalTOX was developed for the California Environmental Protection Agency to estimate chemical fate near hazardous waste sites (McKone 1992, 1994). ChemCAN was developed for Health Canada to predict regional chemical fate, and is available as a Windows application (MacLeod and Mackay 1999). HAZCHEM was developed as a regional scale model for European Community member states (ECETOC 1994). SimpleBOX is slightly more complicated than the other models, capable of estimating dynamic (nonsteady-state) contaminant behavior (Level IV), and uses contaminant concentration rather than fugacity for computing mass balances. This model was developed in the Netherlands as a regional scale model (Brandes et al. 1996).

All of the multimedia models introduced above use similar mathematical expressions (either fugacity or concentration) to represent contaminant distribution between phases and transformation processes and can yield estimates of nonequilibrium concentrations under steady-state conditions. Important assumptions of these models are these: 1) media within a phase are homogeneous and "well mixed," 2) partition equilibrium is assumed within each compartment or phase, and 3) rates of intermedia mass transfer are proportional to the concentration difference. A comprehensive description of these models, along with detailed comparisons of model performance, is presented by Cowen et al. (1995).

In general, the models require the area or volume of each system compartment, the compartment properties such as organic C content, the environmental conditions such as temperature, and the contaminant properties. Perhaps the most difficult parameters to obtain include reaction half-lives, which are often extrapolated from laboratory data, and mass transfer rates or residence times in each phase. Model outputs may vary considerably based on the selection of input parameters, and it is advisable to perform some type of sensitivity analysis in which the parameters are varied within reasonable ranges. More detailed discussions of sensitivity analysis and model validation procedures can be found in Cowen et al. (1995). The input requirements for these Level III models are similar to those described above, although it should be noted that the collection of appropriate input data is often a time-consuming and expensive process. It should also be remembered that these models only describe chemical transport and potential contaminant levels in various environmental compartments, but none of them link these concentrations to biological responses.

Site-specific models

Site-specific models typically are used to estimate contaminant fate and transport near the source of contamination and to support detailed, site-specific risk assessments. Models developed for soil-based or near-surface contamination scenarios are often identified as unsaturated or vadose zone models. In general, these models account for water infiltration and flow through the unsaturated zone, advective and dispersive transport of contaminants in soil water, sorption of contaminants by the solid phase, and chemical and biological transformation processes. More sophisticated models may incorporate equations to describe surface runoff and erosion, plant uptake, preferential flow paths, volatilization and gas-phase transport, as well as the fate and transport of daughter products or metabolites.

By their very nature, site-specific models are designed to integrate many detailed soil processes, thereby providing a mechanistic prediction of contaminant fate in soil. This deterministic modeling approach requires that input parameters be provided for each process considered in the model. More specifically, these models frequently require rainfall and evapotranspiration data, sorption coefficients, transformation rate constants, vegetative properties, and transformation pathways. In addition, detailed soil property data usually are needed as a function of depth, including soil bulk density, residual and saturated water content, and organic C content. Due to this level of complexity, collaboration with soil scientists or geohydrologists specializing in this area is often necessary to obtain appropriate input data to operate the model and to correctly interpret model output.

A summary of site-specific models designed to predict the fate and transport of contaminants in unsaturated soils is given in Table 5-5. This listing of models is not intended to be exhaustive, but rather to provide the reader with a useful reference of unsaturated zone fate and transport models. The site-specific models included in

Table 5-5 Summary of site-specific models capable of simulating contaminant fate and transport in the unsaturated zone

Model (version)	Model description	Fate and transport processes considered	Reference and distribution source
2DFATMIC (Version 1.0)	2-D water flow and chemical transport with transformations under both unsaturated and saturated conditions. Solves Richard's equation by finite element method and transport equation using a Lagrangian-Eulerian numerical procedure.	Advection, dispersion, microbial growth and degradation, density effects, linear equilibrium sorption.	Yeh et al. 1997
CHEMFLO (Version 1.3)	1-D water flow and chemical transport in the unsaturated zone. Solves Richard's equation and advection-dispersion transport using finite difference methods.	Advection, dispersion, 1st-order decay, linear equilibrium sorption.	Nofziger et al. 1989 USEPA Kerr Lab www.epa.gov/ada/kerrlab.html
CMLS94 (Version 95.09.18)	1-D water flow simulated using Green-Ampt approach with solute transport and transformation. Tabular and graphical output.	Advection, 1st-order decay, linear equilibrium sorption, evapo-transpiration, Monte Carlo capability.	Nofziger and Hornsby 1986 http://soilphysics.okstate.edu/software1/cmls94a.htm
GLEAMS	1-D simulation of water flow based on Green-Ampt approach, with chemical transport through root zone.	Advection, 1st-order decay, linear equilibrium sorption, evapo-transpiration, runoff, and erosion.	Knisel et al. 1994 USDA/ARS
HELP (Version 3.0)	Water budget simulation model designed for landfill evaluations. Water inputs are partitioned into different compartments with flow based on exceeding compartment capacity to store water.	Water storage, steady-state unsaturated flow, evapo-transpiration, runoff, lateral flow.	Schroeder et al. 1994 U.S. Army WES

-continued-

Table 5-5 continued

Model (version)	Model description	Fate and transport processes considered	Reference and distribution source
HYDRUS (Version 6.0)	1- and 2-D simulation of water, heat and multiple solute transports in variably saturated porous media. Solves Richard's equation for water flow and Fickian-based equations for heat and solute transport. Includes a high-resolution graphical interface.	Advective, dispersion, nonlinear and nonequilibrium sorption, zero- and 1st-order degradation, linear equilibrium between liquid and gas phases, soil water hysteresis, and root growth.	Simunek et al. 1994 USDA-ARS Salinity Lab www.ussl.ars.usda.gov
PESTAN (Version 4.0)	1-D water flow and chemical transport through homogenous unsaturated soil. Analytical solution to the transport equation.	Advection, dispersion, 1st-order decay, linear equilibrium sorption.	Ravi and Johnson 1994 USEPA Kerr Lab www.epa.gov/ada/kerr lab.html
PRZM-2 (Version 1.02)	1-D water flow and chemical transport in the root zone and unsaturated zone. Couples PRZM and VADOFT codes.	Advection, dispersion, transformation, linear equilibrium sorption, volatilization, and gas-phase transport, daughter products, Monte Carlo capability	Mullins et al. 1993 USEPA ERL
RITZ (Version 2.12)	1-D water flow and contaminant transport, specifically designed for predicting the fate of immobile oily wastes applied to land surface.	Advection, transformation, volatilization, linear equilibrium partitioning among phases.	Nofziger et al. 1988 USEPA Kerr Lab www.epa.gov/ada/kerr lab.html
RUSTIC (1989)	1-D water flow and chemical transport in the unsaturated zone linked with 2-D saturated zone flow and transport. Couples PRZM, VADOFT, and SAFTMOD codes.	Advection, dispersion, transformation, linear equilibrium sorption volatilization, and gas-phase transport, daughter products, Monte Carlo capability.	Dean et al. 1989 USEPA ERL

-continued-

Table 5-5 continued

Model (version)	Model description	Fate and transport processes considered	Reference and distribution source
SESOIL	1-D soil compartment model solved by finite difference method. Hydrologic cycle, sediment cycle, and chemical fate and transport.	Advection, diffusion, volatilization, decay, linear equilibrium sorption, transformation.	Kinerson et al. 1989 Oak Ridge Nat'l Lab
SUTRA	2-D model capable of simulating both transient and steady-state flow under saturated and unsaturated conditions. Utilizes hybrid finite element and finite difference methods.	Advection, dispersion, diffusion, linear equilibrium sorption, zero- and 1st-order decay.	Voss 1984 USGS http://water.usgs.gov/software
SWMS-2D (Version 1.21)	2-D model designed for simulating water and solute transport in variably saturated soils. Solves Richard's equation and advective-dispersive transport equation.	Advection, dispersion, linear equilibrium sorption, zero-order source term, 1st-order decay, water uptake by plants.	Simunek et al. 1994 USDA/ARS
VLEACH (Version. 2.2)	1-D finite difference model for simulating water and contaminant transport in the unsaturated zone. Assumes homogeneous soil, no dispersion, no decay or source term.	Advection, linear partitioning between phases, linear equilibrium sorption, vapor-phase diffusion.	Ravi and Johnson 1997 USEPA Kerr Lab www.epa.gov/ada/kerrlab.html

Table 5-5 are readily available; several can easily be downloaded from the USEPA R.S. Kerr Environmental Laboratory, Center for Subsurface Modeling Support (CSMOS) web page: www.epa.gov/ada/csmos.html. A more detailed compilation of unsaturated zone models has also been developed by the USEPA (1994). All of the models considered herein couple water flow with chemical or contaminant fate and transport processes. The level of model complexity varies, from simplifications of water infiltration and flow based on steady-state conditions or piston displacement to numerical solutions of Richard's equation for transient water flow. In a similar manner, the complexity of the soil matrix varies from a single, homogeneous porous medium (i.e., no change in soil properties with depth) to layered soil systems that account for surface runoff and plant uptake of water and contaminant. In the simplest cases, contaminant fate and transport is based solely on advective transport in water, while more complex models incorporate transformation processes, formation of daughter products, source–sink terms, sorption, volatilization, and gas-phase transport processes. Therefore, it is advisable to select a model that accounts for the fate and transport processes that are relevant to the particular contaminant and conditions of the site under consideration. As noted in the previous section, multimedia models are often useful for identifying the major compartments and mass transfer processes that need to be accounted for, particularly when assessing larger-scale implications of contaminant distribution in the environment. A detailed description of the mathematical equations used to describe contaminant release and transport from source zones existing in the unsaturated zone is provided in the transport chapter (Chapter 6).

Implications for Exposure Assessments

The interrelated fate processes described in earlier sections can have significant impacts on potential exposures in several ways: 1) removal or destruction of contaminants, 2) chemical or physical changes affecting the toxicity of the compound (or its metabolites), and 3) changes in the bioavailability of the compound or its availability for leaching and other transport processes. All of these impacts need to be considered in human and ecological risk assessments.

Although the need for considering environmental fate in risk assessments seems obvious, it is often ignored or treated in a very simplistic manner. The following sections discuss implications of some of the key fate processes on risk assessments, in hopes that risk managers will give greater attention to this area. Particular emphasis is given to biodegradation because of its strong influence on organic contaminant persistence and attenuation during transport and the high degree of uncertainty surrounding the use of degradation parameters (e.g., half-life).

Contaminant concentrations and availability

The removal and destruction of organic contaminants is the most commonly considered impact of fate processes, particularly through biodegradation. Degradation rates can have a significant impact on the mass and average concentration of chemicals at a contaminated site and on the attenuation of contaminants during transport (Larson and Cowan 1995). For example, if environmental conditions are favorable, some lighter hydrocarbons (e.g., benzene, toluene, ethylbenzene, and xylene [BTEX]) can have half-lives in soil of a few days or weeks. On the other hand, persistent chlorinated pesticides such as DDT and its metabolites can persist for decades. For most organic compounds, the assumption that no degradation occurs results in unrealistic and extremely conservative exposure assessments, which can in turn lead to overly expensive and unnecessary remediation requirements.

However, risk assessment models typically either ignore degradation or use simple models incorporating half-life estimates based on laboratory studies. The importance of incorporating biodegradation into risk assessments was clearly shown for petroleum hydrocarbons in California soils by Odencrantz et al. (1992). In sensitivity testing with the widely used SESOIL model, using the range of typical degradation rate estimates for dissolved hydrocarbons, biodegradation had more impact than any other model parameter in calculating the soil total petroleum hydrocarbon (TPH) and volatile organic chemical (VOC) concentrations protective of ground water for most sites. For example, in the base case, peak benzene concentration in ground water beneath the impacted soil was 600 mg/L without biodegradation and only 33 mg/L assuming a half-life of approximately 1 month (0.02 day^{-1}). Similar sensitivity analyses suggest biodegradation is the predominant attenuation mechanism for organic vapors, which is a particularly important exposure route for assessing risks to ecological receptors such as burrowing mammals.

Degradation can also reduce source concentrations, although the rates are often slow, and there is considerable uncertainty surrounding current model predictions (Madsen 1991). Also, most assessments assume an infinite source, which may be realistic for extremely recalcitrant compounds or residual NAPL, but in many cases, site-specific data regarding soil conditions can be used to estimate source depletion rates. Including source degradation can yield more realistic risk assessments, although long-term monitoring will be needed to validate and refine the model predictions.

Fate impacts contaminant availability as well as total concentration. Precipitation, complexation, and polymerization reactions can markedly reduce the availability of metals and organics, even though total extractable concentrations may not change. This reduced availability should be incorporated into exposure assessments because any reversion of the contaminants to more available forms is generally very slow. Of course, biological and chemical processes can also increase availability, most notably in the case of the increased solubility of metals in oxidized or acidic environments. These kinds of redox-induced or pH-induced changes in solubility need to be considered when future land uses may change these conditions (e.g., conversion to wetlands). Fate processes can also produce more toxic and/or mobile products, such as the production of vinyl chloride from perchloroethylene (PCE) and TCE.

Fate processes can also have a more subtle effect on exposure potential by altering the availability of the residual compounds. Biodegradation and other fate processes act primarily on the soluble or readily solubilized fraction of the total contaminant concentration (Figure 5-2). Because these fractions are also the most available for leaching and biological uptake, it is not surprising that degradation will often reduce the toxicity and mobility far more than the reductions in total concentrations would suggest (Loehr and Webster 1996). However, risk assessments rarely consider this preferential loss of the highest-risk fraction.

Finally, processes such as plant and invertebrate uptake can lead to redistribution of metals within the soil. In some cases, this may result in concentrating metals within the upper litter layer where they may pose an increased potential for direct exposure. In other cases, such as earthworm uptake, contaminants may be further dispersed through the soil profile.

Reliability of half-life values for biodegradation

By far the most commonly used equation for estimating the rate of biodegradation is the first-order kinetic model, which yields a half-life value. These values are often used without consideration of their scientific basis. Half-lives measured directly in the field may reflect overall contaminant attenuation due to transport processes, such as volatilization and dispersion, as well as simultaneously occurring fate processes, such as chemical reactions. Thus, the half-life estimate will be specific to the array of soil and environmental conditions existing at the site where it is measured, and extrapolation to other sites may be misleading.

Half-life estimates for isolated processes, such as biodegradation, are more useful than field-measured half-lives. However, use of a half-life assumes that first-order kinetics is the proper rate expression for biodegradation. As discussed earlier, first-order kinetics apply to conditions where there is no net growth of microorganisms and the contaminant concentration is the only factor governing the rate.

Despite its limitations, most lab and field studies have reported the results of degradation studies as half-lives because the first-order model is easy to use, the decay rates are easy to calculate, and the data for a more complex kinetic analysis are rarely enough. Therefore, common fate and transport models use half-lives as input, even when it is clearly not appropriate. Half-lives based on laboratory studies should be used with caution because degradation under controlled conditions is often much faster than under dynamic field conditions.

The use of half-life estimates, which are available for many compounds for both aerobic and anaerobic soils (Howard et al. 1991), is most appropriate for modeling disappearance of low concentrations of contaminants in areas of the site that do not have other limitations. Literature values provide a useful first approximation regarding relative biodegradation rates and the likely impact of incorporating biodegradation into risk assessments. But they must be viewed with considerable caution because in situ biodegradation rates are generally governed by several environmental factors, and can vary over several orders of magnitude between sites or even within a given site. Half-lives may be most valuable for 2 primary uses: 1) estimating the persistence of chemicals added to soils in relatively low concentrations (such as pesticide applications) and 2) estimating attenuation as contaminants migrate through soil because the concentrations are much lower than in the source areas and environmental limitations are generally less severe.

Half-lives are much less valuable for estimating depletion of high-concentration source areas such as residual saturation with NAPLs because degradation in these areas is generally limited by the availability of O and, in some cases, water or nutrients. As a result, degradation will occur primarily at the fringes of source areas and along transport pathways because the concentrations are not so high as to be toxic to microorganisms or lead to consumption of available electron acceptors. This natural attenuation can rapidly reduce concentrations in the vapors and leachates migrating away from the source area, in some cases providing effective containment so that receptors are not exposed. Over time (several years or even decades), degradation will move inwards to reduce the overall mass in the source area.

Models are available to predict fate over time for situations where a half-life estimate is not appropriate. Such models are more commonly used for ground water but are more complex for unsaturated soil. Using these models may require collecting much more data than is typically collected at contaminated sites, such as the availability and spatial distribution of contaminants, electron acceptors, water, and nutrients.

Production of toxic metabolites

Degradation can also result in the formation of toxic metabolites that will need to be considered in risk assessment. Photoactivation of PAHs is one example, as is the anaerobic biodegradation of TCE and PCE to form vinyl chloride. Other examples include the biological methylation of Hg and the influence of oxidation state on the toxicity of Cr. Degradation of organics will often result in consumption of O and a consequent lowering of the redox status so that previously precipitated metals may be mobilized.

In some cases, we still do not understand the potential for production of toxic metabolites in soil, introducing some uncertainty into exposure estimates. For example, the metabolites formed during petroleum biodegradation have never been fully characterized.

Temporal and spatial variability

The concentrations of contaminants vary tremendously within a site, and the rates of various fate processes also vary in time and space. It is convenient to assume that contaminant concentrations and distributions remain static over time, but the concentrations and availability will change over time and space. As fate processes occur, the potential for exposure in some areas will generally decrease, adding a further degree of complexity to risk assessments.

One common example of spatial heterogeneity occurs in soils containing oil at levels of residual saturation (concentrations at which the oil no longer moves because of gravitational or matrix forces, approximately 1% to 5% oil depending on the soil composition). In these situations, degradation can occur at the fringes of the saturated area, where O is available. But, virtually no degradation will occur within the source area because O is depleted. Complex gradients of the concentrations of contaminants, electron acceptors, and microorganisms will result. As a result, the contaminated area may slowly shrink, but the average concentrations (and particularly the upper confidence limits) may not change for several years.

Degradation rates also may vary with time, for example, slowing as the contamination is depleted. Rates also may increase as microorganisms become acclimated to the contaminant or the site environmental conditions. The rates of fate processes also vary seasonally with temperature and moisture fluctuations. The rates also can vary over much shorter time periods. For example, biodegradation actually may occur in "bursts" following rainfall events, followed by long periods of dormancy. As a result, most of the losses may occur over short periods of time.

Uncertainty of exposure estimates

The preceding discussion has pointed out some of the sources of uncertainty in attempting to quantify potential exposure, particularly when the assessments incorporate environmental fate processes. The rates of many of the fate processes described are governed by numerous site-specific factors and vary over several orders of magnitude depending on these factors. Further, these factors are not constant in soil, but vary themselves over time and space within a given site. Soils are generally very complex systems with a high degree of spatial variability, both laterally and vertically. In addition, contaminant concentrations and distribution will often be highly heterogeneous. A final source of uncertainty is the limitation of most site characterization data in providing a complete picture of the distribution of contaminants and reactants.

The uncertainty in estimating the rates and extents of fate processes is greatest at small scales. At the larger scales that are most significant for risk assessments (a few meters to a few hectares), the uncertainty is reduced, and predictions regarding fate and transport often are accurate within an order of magnitude. Such models may be adequate for screening-level predictions in many cases, but the limitations of these models need to be recognized. Continued monitoring will often be required to be able to define more clearly the reductions in source concentrations over time, the potential for biological uptake, and particularly the actual attenuation occurring during transport.

Methods to evaluate fate processes

Despite all of the uncertainties, adequate models are generally available and useful for initial predictions. There is a long history of the use of several of the models discussed in earlier sections, providing some confidence that adequate initial predictions can be made. Calibration of models to site-specific data is probably the most common method to evaluate the rate and extent of critical fate processes. In most cases, long-term monitoring will be needed to validate the site-specific model predictions, but this monitoring may be far less expensive and disruptive to the environment than active remediation.

However, lab and field evaluations are very important in developing and defending any site-specific model. A valuable example is the general guidelines that have been developed to demonstrate and estimate natural attenuation rates (e.g., ASTM 1996). These guidelines detail the data needs and appropriate test methods for using natural attenuation as part of the overall remedial strategy at a site. Similar standardized lab and field tests are available for assessing other biological and chemical fate processes.

Site evaluations can be expensive. The simplest laboratory tests designed to ensure that a site is not sterile, and that the potential for biodegradation is present, can cost $2000 to $10,000. Microcosm studies to ensure that the soil is biologically active, to develop initial biodegradation rate estimates, and to identify potential toxic metabolites typically cost $10,000 to more than $50,000, depending on the contaminants and the complexity of the site. More reliable field studies to evaluate site-specific degradation rates, and to monitor progress over time, can typically cost $50,000 to $250,000 or more and require a period of several years.

Future Directions and Needed Research

In the area of biodegradation in soils, one of the largest challenges is to realistically predict long-term degradation rates in situ. Predicting rates is intimately connected to identifying and measuring the population density of microbes involved in biodegradation, an endeavor that is still in its infancy. Another challenge is to be able to extrapolate laboratory data to the field environment. Our relatively poor understanding of the kinetics for biodegradation in situ is also confounded by factors such as sorption and transport and by local factors that cannot be measured at the scale where microorganisms live. Thus, research is needed to understand the complex processes affecting rates in field environments and at the same time to develop simple relationships or predictive capabilities that can be used to estimate long-term, field-scale biodegradation for useful exposure assessments.

Reliable estimates of contaminant bioaccumulation depend partly on quantitative descriptions of residues in food sources. Presently, these estimates are very difficult to predict because available information is essentially site specific or situation specific. The database for pesticides is large because of the regulatory mandate for prospective risk assessment. Currently registered pesticides tend to be moderately hydrophobic to moderately hydrophilic and do not bioaccumulate to high levels in lipid phases of biota. Thus, any empirical equations relating bioaccumulation of pesticides to residues in plants will be unreliable if applied to persistent, highly hydrophobic compounds.

Of the contaminants other than pesticides, PCBs, and secondarily PAHs, have probably been most studied. Despite similarities in physicochemical properties like K_{ow} and water solubility, any bioconcentration factors developed for one group of compounds will not necessarily be applicable to different groups. While rate-limiting processes like desorption may predict sorption to roots or invertebrate cuticles, differential metabolic routes and rates may make bioaccumulation factors incomparable. For example, the higher BSAF for PCBs than for PAHs in earthworms has been attributable to a greater recalcitrance to biotransformation (Ma et al. 1998).

Given the diversity of organisms, soils, and chemicals in which bioaccumulation potential is necessary to estimate exposure, the most immediate need for research is to broaden the database for uptake among specific contaminant–organism exposure scenarios. These types of data should be collected among several soil types representative of low, medium, and high organic matter content. Use of standardized soils is not necessary for these experiments because sorption equilibria seem highly correlated with organic matter, and this process has been well characterized by empirical models relating K_{ow} to K_{oc}.

Eventually, predictions of bioaccumulation potential will need to rely on deterministic models that are more defined. A greater understanding of metabolism of contaminants is required before these can be developed. To accomplish this, contaminants in ecological indicator species will have to be studied using toxicodynamic approaches. Ultimately, toxicodynamic models will need to be coupled with soil desorption and surface adsorption models. These studies will not be easy for highly hydrophobic, persistent contaminants, owing to the difficulty of carrying out studies long enough to accurately measure toxicokinetic parameters.

In order to better understand fate processes in contaminated soils, mathematical simulation models provide useful tools to test hypotheses. Models are becoming increasingly complex in order to better simulate and understand the complex, interacting processes affecting contaminant fate. On the other hand, these complex, data-intensive models are not necessarily practical or useful at the spatial and temporal scales at which risk assessments occur. Simpler models that capture the most important processes affecting fate of contaminants are greatly needed by risk managers.

References

Alexander M. 1980. Biodegradation of chemicals of environmental concern. *Science* 211:132-138.

Alexander M. 1985. Biodegradation of organic chemicals. *Environ Sci Technol* 18:106-111.

Alexander M. 1994. Biodegradation and bioremediation. San Diego CA, USA: Academic Press. 248 p.

Alexander M. 1995. How toxic are toxic chemicals in soil? *Environ Sci Technol* 29:2713-2717.

Alexander M, Scow KM. 1989. Kinetics of biodegradation in soil. In: Sawhney BL, Brown K, editors. Reactions and movement of organic chemicals in soil. Madison WI, USA: American Society of Agronomy. Soil Science Society of America Special Publication 22. p 243-269.

[ASTM] American Society for Testing and Materials. 1996. Standard guide for remediation by natural attenuation at petroleum release sites. In: Annual book of standards. Philadelphia PA USA: ASTM. p 1024-1065.

Bartha R, Pramer D. 1965. Features of a flask and method for measuring the persistence and biological effects of pesticides in soil. *Soil Sci* 100:68-70.

Baumhardt GR, Welch LF. 1972. Lead uptake and corn growth with soil-applied lead. *J Environ Qual* 1:92-94.

Bekins BA, Warren E, Godsy EM. 1998. A comparison of zero-order, first-order, and Monod biotransformation models. *Ground Water* 36:261-268.

Belfroid A, Van den Berg M, Seinen W, Hermens J, Van Gestel CAM. 1995. Uptake, bioavailability and elimination of hydrophobic compounds in earthworms (*Eisenia andrei*) in field-contaminated soil. *Environ Toxicol Chem* 14:605-612.

Boersma L, McFarlane C, Lindstrom FT. 1991. Mathematical model of plant uptake and translocation of organic chemicals: Application to experiments. *J Environ Qual* 20:137-146.

Borden RC, Gomez CA, Becker MT. 1995. Geochemical indicators of intrinsic bioremediation. *Ground Water* 33:180-189.

Brandes LJ, den Hollender H, van de Meent D. 1996. SimpleBox 2.0: A nested multimedia fate model for evaluating the environmental fate of chemicals. Bilthoven, NL: RIVM. Report Nr 719101029. 156 p.

Briggs GG, Bromilow RH, Evans AA. 1982. Relationships between lipophilicity and root uptake and translocation of non-ionized chemicals by barley. *Pestic Sci* 13:495-504.

Briggs GG, Rigitano RLO, Bromilow RH. 1987. Physicochemical factors affecting uptake by roots and translocation to shoot of weak acids in barley. *Pestic Sci* 19:101-112.

Bromilow RH, Chamberlain K. 1995. Principles governing uptake and transport of chemicals. In: Trapp S, McFarlane JC, editors. Plant contamination. Modeling and simulation of organic chemical processes. Boca Raton FL, USA: Lewis Publishers. p 37-68.

Bromilow RH, Chamberlain K, Evans AA. 1990. Physico-chemical aspects of phloem translocation of herbicides. *Weed Sci* 38:305-314.

Burnside OC. 1974. Prevention and detoxification of pesticide residues in soils. In: Guenzi WD, editor. Pesticides in soil and water. Madison WI, USA: Soil Science Society of America. p 387-412.

Chapman PM, Allen HE, Godtfredsen K, Z'Graggen MN. 1996. Evaluation of bioaccumulation factors in regulating metals. *Environ Sci Technol* 30:448A-452A.

Chiou CT. 1989. Partition and adsorption on soil and mobility of organic pollutants and pesticides. In: Gerstl Z, Chen Y, Mingelgrin U, Yaron B, editors. Toxic organic chemicals in porous media. Ecological studies Volume 73. New York NY, USA: Springer Verlag. p 163-175.

Cowen CE, Mackay D, Fiejtel TJC, van de Meent D, DiGuardo A, Davis J, Mackay N. 1995. The multi-media fate model: A vital tool for predicting the fate of chemicals. Society of Environmental Toxicology and Chemistry (SETAC) International Task Force on the Application of Multi-Media Fate Models to Regulatory Decision Making; 1994 Apr 4-16; Leuven, BE; 1994 Nov 4-5; Denver, CO, USA. Pensacola FL, USA: SETAC. 78 p.

Cox WJ, Rains DW. 1972. Effect of lime on lead uptake by five plant species. *J Environ Qual* 1:167-169.

Criddle CS. 1993. The kinetics of cometabolism. *Biotechnol Bioeng* 41:1048-1056.

Dauterman WC, Hodgson E. 1978. Detoxification mechanisms in insects. In: Rockstein M, editor. Biochemistry of insects. New York NY, USA: Academic Press. p 541-577.

Dean JD, Huyakorn PS, Donigian AS, Voos KA, Schanz RW. 1989. Risk of unsaturated/saturated transport and transformation of chemical concentrations (RUSTIC), Volume 2, user's guide. Athens GA, USA: USEPA, Environmental Research Laboratory. EPA-600-3-89-048B. 357 p.

Devine M, Duke SO, Fedtke C. 1993. Physiology of herbicide action. Englewood Cliffs NJ, USA: PTR Prentice Hall. 56 p.

Di Toro DM, Zarba CS, Hansen DJ, Berry WJ, Swartz RC, Cowan CE, Pavlou SP, Allen HE, Thomas NA, Paquin PR. 1991. Technical basis for establishing sediment quality criteria for nonionic organic chemicals using equilibrium partitioning. *Environ Toxicol Chem* 10:1541-1583.

[ECETOC] European Centre for Ecotoxicology and Toxicology of Chemicals. 1994. HAZCHEM: A mathematical model for use in risk assessment of substances. Brussels, BE: ECETOC. Special Report Nr 8. 108 p.

Edwards CA, Beck SD, Lichtenstein EP. 1957. Bioassay of aldrin and lindane in soil. *J Econ Entomol* 50:622-626.

Eijsackers H. 1994. Ecotoxicology of soil organisms: Seeking the way in a pitch-dark labyrinth. In: Donker MH, Eijsackers H, Heimbach F, editors. Ecotoxicology of soil organisms. Boca Raton FL, USA: Lewis Publishers. p 3-32.

Felsot AS, Lew A. 1989. Factors affecting bioactivity of soil insecticides: Relationships among uptake, desorption, and toxicity of carbofuran and terbufos. *J Econ Entomol* 82:389-395.

Ferguson CF, Darmendrail D, Freier K, Jensen BK, Jensen J, Kasamar H, Urzelai A, Vegter J. 1998. Risk assessment for contaminated sites in Europe. Volume 1. Scientific bases. Nottingham, UK: LQM Pr. 165 p.

Fielden MR, Chen I, Chittim B, Safe SH, Zacharewski TR. 1997. Examination of the estrogenicity of 2,4,6,2',6'-Pentachlorobiphenyl (PCB 104), its hydroxylated metabolite 2,4,6,2',6'-Pentachloro-4-Biphenylol (HO-PCB 104), and a further chlorinated derivative, 2,4,6,2',4',6'-Hexachlorobiphenyl (PCB 155). *Environ Health Perspect* 105:1238-1248.

Fries GF, Marrow GS. 1981. Chlorobiphenyl movement from soil to soybean plants. *J Agric Food Chem* 29:757-759.

Getzin LW, Chapman RK. 1959. Effect of soils upon the uptake of systemic insecticides by plants. *J Econ Entomol* 52:1160-1165.

Glass BL. 1972. Relation between the degradation of DDT and the iron redox system in soils. *J Agric Food Chem* 20:324-327.

Guo L, Bicki TJ, Felsot AS, Hinesly TD. 1991. Phytotoxicity of atrazine and alachlor in soil amended with sludge, manure and activated carbon. *J Environ Sci Health* B26:513-527.

Guo L, Bicki TJ, Felsot AS, Hinesly TD. 1992. Sorption and movement of alachlor in soil modified by carbon-rich wastes. *J Environ Qual* 22:186-194.

Harris CI. 1966. Adsorption, movement, and phytotoxicity of monuron and s-triazine herbicides in soil. *Weeds* 14:6-10.

Hartenstein R, Neuhauser EF, Collier J. 1980. Accumulation of heavy metals in the earthworm *Eisenia fetida*. *J Environ Qual* 9:23-26.

Harvey SD, Fellows RJ, Cataldo DA, Bean RM. 1991. Fate of the explosive hexahydro-1,3,5-trinitro-1,3,5-triazine (RDX) in soil and bioaccumulation in bush bean hydroponic plants. *Environ Toxicol Chem* 10:845-855.

Helmke PA, Robare WP, Korotev RL, Schomberg PJ. 1979. Effects of soil-applied sewage sludge on concentrations of elements in earthworms. *J Environ Qual* 8:322-327.

Hendriks AJ. 1995. Modelling and monitoring organochlorine and heavy metal accumulation soils, earthworms, and shrews in Rhine-Delta floodplains. *Arch Environ Contam Toxicol* 29:115-127.

Hermanson HP, Forbes C. 1966. Soil properties affecting dieldrin toxicity to *Drosophila melanogaster*. *Soil Sci Soc Am Proc* 30:748-752.

Howard PH, Boethling RS, Jarvis WF, Meylan WM, Michalchenko EM. 1991. Handbook of environmental degradation rates. Chelsea MI, USA: Lewis Publishers. 725 p.

Huang JW, Chen J, Cunningham SD. 1997. Phytoextraction of lead from contaminated soils. In: Kruger EL, Anderson TA, Coats JR, editors. Phytoremediation of soil and water contaminants; Aug 25-29; Orlando FL, USA. Washington DC, USA: American Chemical Society. Symposium Series 664. p 283-298.

Huylster A, Muller JF, Marschner H. 1994. Soil-plant transfer of polychlorinated dibenzo-*p*-dioxins and dibenzofurans to vegetables of the cucumber family (Cucurbitaceae). *Environ Sci Technol* 28:1110-1115.

John MK. 1972. Lead availability related to soil properties and extractable lead. *J Environ Qual* 1:295-298.

John MK, Van Laerhoven C. 1972. Lead uptake by lettuce and oats as affected by lime, nitrogen, and sources of lead. *J Environ Qual* 1:169-171.

Jones KC, Sewart AP. 1997. Dioxins and furans in sewage sludges: A review of their occurrence and sources in sludge and of their environmental fate, behavior, and significance in sludge-amended agricultural systems. *Crit Rev Environ Sci Technol* 27:1-85.

Jungk AO. 1996. Dynamics of nutrient movement at the soil-root interface. In: Waisel Y, Eshel A, Kafkafi U, editors. Plant roots. The hidden half. New York NY, USA: Marcel Dekker. p 529-556.

Khan SU. 1982. Bound pesticide residues in soil and plants. *Residue Rev* 84:1-25.

Kieatiwong S, Nguyen LV, Hebert VR, Hackett M, Miller GC, Miille MJ, Mitzel R. 1990. Photolysis of chlorinated dioxins in organic solvents and on soils. *Environ Sci Technol* 24:1575-1580.

Kinerson RK, Leonard SK, Travis CC, Hetrick DM. 1989. Qualitative validation of pollutant transport components of an unsaturated soil zone model (SESOIL). Oak Ridge TN, USA: U.S. Department of Energy, Oak Ridge National Laboratory. Report Nr. ORNL/TM-10672, DE89-008965. 39 p.

Knisel WG, Davis FM, Leonard RA. 1994. GLEAMS: Groundwater loading effects of agricultural management systems. Part III: User manual, version 2.0. Tifton GA, USA: USDA, Agriculture Research Service, Coastal Plain Experiment Station, Southeast Watershed Research Laboratory. 57 p.

Kramer PJ. 1969. Plant and soil water relationships. A modern synthesis. New York NY, USA: McGraw-Hill Book Company. 482 p.

Larsen RA, Weber EJ. 1994. Reaction mechanisms in environmental organic chemistry. Boca Raton FL, USA: CRC Press. p 103-167.

Larson RJ. 1979. Role of biodegradation kinetics in predicting environmental fate. In: Make AW, editor. Biotransformation and fate of chemicals in the aquatic environment. Washington DC, USA: American Society of Microbiology. p 67-86.

Larson RJ, Cowan CE. 1995. Quantitative application of biodegradation data to environmental risk and exposure assessments. *Environ Toxicol Chem* 14:1433-1442.

Lindstrom FT, Boersma L, McFarlane C. 1991. Mathematical model of plant uptake and translocation of organic chemicals: Development of the model. *J Environ Qual* 20:129-136.

Linz DG, Nakles DV, editors. 1997. Environmentally acceptable endpoints in soil: Risk-based approach to contaminated site management based on availability of chemicals in soil. Annapolis MD, USA: American Academy of Environmental Engineers. 630 p.

Loehr RC, Webster MT. 1996. Behavior of fresh vs aged chemicls in soil. *J Soil Contam* 5:361:383.

Ma WC, Van Kleunen A, Immerzeel J, De Maagd PGJ. 1998. Bioaccumulation of polycyclic aromatic hydrocarbons by earthworms: Assessment of equilibrium partitioning theory in in situ studies and water experiments. *Environ Toxicol Chem* 17:1730-1737.

Mackay D. 1991. Multimedia environmental models: The fugacity approach. Chelsea MI, USA: Lewis. 272 p.

Mackay D, Shiu WY. 1992. Estimating the multimedia partitioning of hydrocarbons: The effective solubility approach. In: Calabrese EJ, Kostecki PT, editors. Volume 2. Hydrocarbon contaminated soils and groundwater. Chelsea MI, USA: Lewis Publishers. p 137-154.

MacLeod M, Mackay D. 1999. An assessment of the environmental fate and exposure of benzene and chlorobenzenes in Canada. *Chemosphere* 38:1777-1796.

Madsen EL. 1991. Determining in situ biodegradation rates: Facts and challenges. *Environ Sci Technol* 25:1603-1673.

Marinussen MPJC, van der Zee SEATM, de Haan FAM, Bouwman LM, Hefting MM. 1997. Heavy metal (copper, lead, and zinc) accumulation and excretion by the earthworm, *Dendrobaena veneta*. *J Environ Qual* 26:278-284.

Marschner H, Romheld V. 1996. Root-induced changes in the availability of micronutrients in the rhizosphere. In: Waisel Y, Eshel A, Kafkafi U, editors. Plant roots. The hidden half. New York NY, USA: Marcel Dekker. p 557-579.

Maybey W, Mill T. 1978. Critical review of hydrolysis of organic compounds in water under environmental conditions. *J Phys Chem Ref Data* 7:383-415.

McBride M. 1994. Oxidation-reduction Reactions. In: Environmental chemistry of soils. New York NY, USA: Oxford University Press. p 240-271.

McFarlane JC. 1995. Anatomy and physiology of plant conductive systems. In: Trapp S, McFarlane JC, editors. Plant contamination. Modeling and simulation of organic chemical processes. Boca Raton FL, USA: Lewis Publishers. p 13-34.

McFarlane JC, Pfleeger T, Fletcher J. 1987. Transpiration effect on the uptake and distribution of bromacil, nitrobenzene, and phenol in soybean plants. *J Environ Qual* 16:372-376.

McKone TE. 1992. CalTOX: A multimedia total exposure model for hazardous waste sites. Part 2. The dynamic multimedia transport and transformation model. Washington DC, USA: U.S. Department of Energy, Lawrence Livermore National Laboratory. Report Nr. UCRL-CR-111456-PT2. 97 p.

McKone TE. 1994. CalTOX: A multimedia total exposure model for hazardous waste site. Spreadsheet User's Guide, Version 1.5. Sacramento CA, USA: Department of Toxic Substances Control, Office of Scientific Affairs, California Environmental Protection Agency. 45 p.

Miller RM, Singer GM, Rosen JD, Bartha R. 1988. Sequential degradation of chlorophenols by photolytic and microbial treatment. *Environ Sci Technol* 22:1215-1219.

Moreshet S, Huang B, Huck MG. 1996. Water permeability of roots. In: Waisel Y, Eshel A, Kafkafi U, editors. Plant roots. The hidden half. New York NY, USA: Marcel Dekker. p 659-678.

Moza P, Scheunert I, Klein W, Korte F. 1979. Studies with 2,4',5-trichlorobiphenyl-14C and 2,2',4,4',6-pentachlorobiphenyl-14C in carrots, sugar beets, and soil. *J Agric Food Chem* 27:1120-1124.

Mullins JA, Carsel RF, Scarborough JE, Ivery AM. 1993. PRZM-2: A model for predicting pesticide fate in crop root and unsaturated soil zones, user's manual for release 2.0. Athens GA, USA: USEPA, Environmental Research Laboratory. EPA-600-R-93-046. 400 p.

Nicklow CW, Comas-Haezebrouck PH, Feder WA. 1983. Influence of varying soil lead levels on lead uptake of leafy and root vegetables. *J Am Soc Hort Sci* 108:193-195.

Nofziger DL, Hornsby AG. 1986. A microcomputer-based management tool for chemical movement in soil. *Appl Agric Res* 1:50-56.

Nofziger DL, Rajender K, Sivaram KN, Su P-U. 1989. CHEMFLO: One-dimensional water and chemical movement in unsaturated soils, User's manual, version 1.3. Ada OK, USA: USEPA, Robert S. Kerr Environmental Research Laboratory. 131 p.

Nofziger DL, Williams JR, Short TE. 1988. Interactive simulation of the fate of hazardous chemicals during land treatment of oily wastes: RITZ user's guide. Ada OK, USA: USEPA, Robert S. Kerr Environmental Research Laboratory. 81 p.

Odencrantz JE, Farr JM, Robinson CE. 1992. Transport model sensitivity for soil cleanup level determination using SESOIL and AT123D in the context of the California leaking underground fuel tank manual. In: Kostecki PT, Calabrese EJ, Bonazountas M, editors. Hydrocarbon contaminated soils. Volume II. Proceeding of the 5th Annual Conference on Hydrocarbon Contaminated Soils; 1990 Dec 24-27; Amherst, MA, USA. Chelsea MI, USA: Lewis. p 319-341.

Oertli JJ. 1996. Transport of water in the rhizosphere and in roots. In: Waisel Y, Eshel A, Kafkafi U, editors. Plant roots. The hidden half. New York NY, USA: Marcel Dekker. p 607-633.

Olson WP, O'Brien RD. 1963. The relation between physical properties and penetration of solutes into the cockroach cuticle. *J Insect Physiol* 9:777-786.

Pine SH, Hendrickson JB, Cram DJ, Hammond GS. 1980. Organic chemistry. 4th ed. New York NY, USA: McGraw-Hill. p 583-590; 967-985.

Ravi V, Johnson JA. 1994. PESTAN: Pesticide analytical model. Version 4.0. Ada OK, USA: USEPA, Robert S. Kerr Environmental Research Laboratory. 54 p.

Ravi V, Johnson JA. 1997. VLEACH: A one-dimensional finite difference vadose zone leaching model. Version 2.2. Ada OK, USA: USEPA, Robert S. Kerr Environmental Research Laboratory. 72 p.

Robinson JA. 1985. Determining microbial kinetics parameters using nonlinear analysis: Advantages and limitations in microbial ecology. *Adv Microb Ecol* 8:61-114.

Schmidt SK, Simkins S, Alexander M. 1985. Models for the kinetics of biodegradation of organic compounds not supporting growth. *Appl Environ Microbiol* 50:323-331.

Schroeder PR, Lloyd CM, Zappi PA, Aziz NM. 1994. Hydrologic evaluation of landfill performance (HELP) model, User's guide for version 3.0. Vicksburg MS, USA: U.S. Army, Waterways Experiment Station. 103 p.

Schwarzenbach RP, Gschwend PM, Imboden DM. 1993. Environmental organic chemistry. New York NY, USA: Wiley Interscience. 681 p.

Silvia DM, Fuhrmann JJ, Hartel PG, Zuberer DA. 1998. Principles and applications of soil microbiology. Saddleback NJ, USA: Prentice Hall. 550 p.

Simkins S, Alexander M. 1984. Models for mineralization kinetics with the variables of substrate concentration and population density. *Appl Environ Microbiol* 47:1299-1306.

Sims JT, Kline JS. 1991. Chemical fractionation and plant uptake of heavy metals in soils amended with co-composted sewage sludge. *J Environ Qual* 20:387-395.

Simunek J, Vogel T, van Genuchten MTh. 1994. The SWMS-2D code for simulating water flow and solute transport in two-dimensional variably saturated media, version 1.21. Research Report Nr. 132. Riverside CA, USA: USDA, Agriculture Research Service, U.S. Salinity Laboratory. 197 p.

Snoeyink VL, Jenkins D. 1980. Water chemistry. New York NY, USA: Wiley and Sons. 463 p.

Spencer WF, Adams JD, Shoup TD, Spear RC. 1980. Conversion of parathion to paraoxon on soil dusts and clay minerals as affected by ozone and UV light. *J Agric Food Chem* 28:1295-1300.

Sterritt RM, Lester JN. 1980. The value of sewage sludge to agriculture and effects of the agricultural use of sludges contaminated with toxic elements: A review. *Sci Total Environ* 16:55-90.

Stumm W, Morgan JJ 1981. Aquatic chemistry. 2nd ed. New York NY, USA: Wiley and Sons. p 418-496.

Trapp S, Matthies M, Scheunert I, Topp EM. 1990. Modelling the bioconcentration of organic chemicals in plants. *Environ Sci Technol* 24:1246-1252.

Trapp S, McFarlane C, Matthies M. 1994. Model for uptake of xenobiotics into plants: Validation with bromacil experiments. *Environ Toxicol Chem* 13:413-422.

Travis CC, Arms AD. 1988. Bioconcentration of organics in beef, milk, and vegetation. *Environ Sci Technol* 22:271-274.

[USEPA] U.S. Environmental Protection Agency. 1994. Identification and compilation of unsaturated/vadose zone models. Ada OK, USA: USEPA, Robert S. Kerr Environmental Research Laboratory. EPA-600-R-94-028. 137 p.

Voss CI. 1984. A finite-element simulation model for saturated-unsaturated, fluid-density-dependent ground-water flow with energy transport or chemically-reactive single-species solute transport. Report 84-4369. Reston VA, USA: USGS, Water Resources Investigations. 409 p.

Yeh G-T, Cheng J-R, Short TE. 1997. 2DFATMIC: Two-dimensional flow, fate and transport of microbes and chemicals model. User's Manual, Version 1.0. Washington DC, USA: USEPA, Office of Research and Development. EPA-600-R-97-052. 132 p.

Zepp RG, Schlotzhauer PF, Sink, RM 1985. Photosensitized transformations involving electronic energy transfer in natural waters: Role of humic substances. *Environ Sci Technol* 19:74-81.

Zimdahl RL, Skogerboe RK. 1977. Behavior of lead in soil. *Environ Sci Technol* 11:1202-1207.

CHAPTER 6

Transport of Contaminants in Soil

Randall J. Charbeneau, David G. Linz, Charles J. Newell, William G. Rixey, James A. Ryan

Introduction

This chapter will review major mechanisms by which organic compounds and metals move from the location of their initial deposit in the soil, and how resistant desorption and sorption processes may influence contaminant transport. Approaches for incorporating realistic partitioning, including slow desorption, into transport models are reviewed. Discussion of how the extent and rate of release of contaminants enters into transport calculations and equations is presented.

The primary focus of the chapter is on presentation of a framework for representing the relationship between the mass transport and the source term that is used in risk assessments. Mass transport from the contaminated soil source region decreases the source term mass, and the other processes occurring in the source region determine the mass that is available for transport from the source. Thus, the behavior of mass transport and the source term are coupled. One objective of this chapter is to present a simple, physicochemical characteristic-based source term model that includes sequestration of contaminant mass and slow desorption of mass from soil. The transport mechanisms that are included in the source term model are leaching, volatilization, and erosion or soil resuspension. The fate mechanisms include sequestration, biphasic desorption, various transformations, and bioaccumulation of contaminant mass within the soil biotic community.

This discussion assumes that a mobile nonaqueous-phase liquid (NAPL) phase is not present. If present, a mobile NAPL phase can dominate the release and transport of organic chemicals. These sites are typically managed on a case-by-case basis. The present discussion relates to the release of chemical from a soil–waste matrix, which can then be available for further transformation and/or migration. The presence of an immobile, residually trapped NAPL phase is included in this chapter.

The source term represented in this chapter provides the contaminant mass source term for release to environmental pathways for risk assessment. In addition, because many remediation methods involve the passage of fluids through the

Contaminated Soils: From Soil–Chemical Interactions to Ecosystem Management. Roman P. Lanno, editor.

contaminated soil mass, this representation provides the framework for estimating efficacy of remediation efforts.

This chapter is based on 3 significant findings and conclusions:

1) The slow release mechanisms that operate in soil and subsurface environments have the greatest effect directly within the source itself. Although resistant sorption and desorption processes may occur in the downgradient zone, they are relatively less important to environmental transport and risk compared to the release from the source. As such, it is assumed that conventional transport methods outside of the source are adequate, and this chapter will focus exclusively on interactions of chemicals in the source.
2) The presence of residually trapped NAPL in the soil-waste significantly affects the release of chemicals and should be considered in risk and site management determinations.
3) While the detailed mechanisms for chemical interactions in the soil–waste matrix are not well understood, computational and experimental methods can be applied today that provide improved estimates of release from field soils and that are environmentally protective and useful in a site risk management context.

Context of Contaminant Transport in Risk Characterization

The transport of contaminants through soil depends on physicochemical characteristics of the contaminants and on site-specific characteristics such as soil type, soil heterogeneity, geochemical environment, and moisture. Within the analysis of risk characterization, transport analysis provides the link between contaminant source and potential exposure. From the subsurface environment, common transport pathways include volatilization (diffusion through soil air), leaching (advection, diffusion, dispersion in soil water), and for some contaminant releases, transport through NAPL flow. Transport pathways that are not commonly considered include facilitated transport (e.g., colloidal or particulate transport) and transport through the food chain.

Discussions of contaminant transport in soil are intimately tied to those of contaminant fate. At the contaminant source, the mechanisms that determine the availability of contaminants for transport also may be important in determining contaminant bioavailability to organisms, and these may be the same mechanisms involved in contaminant fate. Uncertainty in quantifying contaminant fate (release) mechanisms adds directly to uncertainty in quantifying contaminant transport and ultimately to uncertainty in quantifying potential exposure. These uncertainties are compounded by limitations on predicting transport through heterogeneous media and by transport through preferential flow paths. Furthermore, there also is considerable uncertainty in identifying significant transport pathways. For example, contaminants can move out of the soils and into the food chain in a variety of ways. Metals tend to move into a portion of the soil invertebrate food web and are subsequently taken up into the aboveground system by insectivorous wildlife species. Other contaminants accumulate in plants that, in turn, are ingested by herbivores.

Conventional Risk Assessment Models

The assessment of risks from contaminated sites relies on 3 pieces of information, as shown in Figure 6-1. First, one must know certain characteristics of the contaminant source zone. These include the chemicals that are present, along with their concentrations within the various phases (water, air, soil, NAPL, biomass), and the significant mass transport mechanisms, along with estimates of important parameters that characterize the environment. The second piece of information that is required concerns the potential transport pathways from the point of contaminant release to the point of exposure. For ecological risks, the point of exposure may coincide with the location of the release, and environmental transport is not involved. However, for other potential human or ecological exposures, contaminant migration through environmental pathways is necessary. The mechanisms for transport through environmental pathways include volatilization, leaching, and suspension or erosion of surface soil particles. The third requirement for assessing risks is to estimate the uptake and dose at the point of exposure. One of the primary objectives of transport modeling in risk assessment is to relate the contaminant concentration at the source to the exposure concentration (Charbeneau 2000).

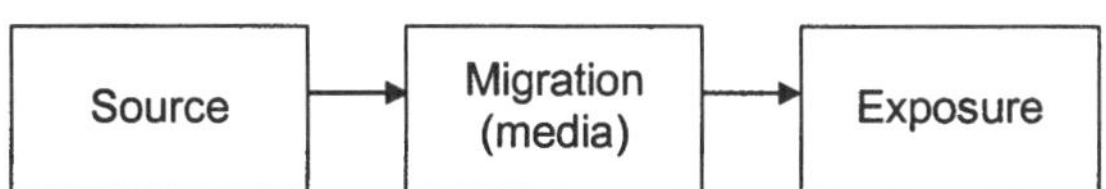

Figure 6-1 Assessment of risks from contamination

There are 3 basic mechanisms by which miscible and immiscible contaminants are transported in the subsurface environment: advection, diffusion, and mechanical dispersion. For transport within the vadose zone, the primary mechanisms are leaching, which involves advective and dispersive transport in the aqueous phase, and volatilization, which involves diffusive transport through the soil air.

"Leaching" refers to the soluble species being carried along with the flow of subsurface water. In representing the groundwater source term from contaminated soil, it is usually assumed that the leachate release rate is equal to the product of the net infiltration rate and the average porewater concentration. This assumes that water moves uniformly through the contaminated soil and that the soil water concentration is uniform. Neither of these assumptions is true. Because of soil heterogeneities and the presence of preferential flow paths, contaminant transport occurs only through a fraction of the soil mass. This is especially true in well structured soils. In addition, the contaminant mass is not evenly distributed.

Volatilization is the mass transfer of chemical substances from soil to the atmosphere. Contaminants may leave the soil by vaporizing into the soil air and then exiting to the atmosphere by diffusion in soil air to the ground surface or by being carried in soil air as it is forced from the soil by vacuum extraction in wells. The rate of volatilization is affected by many factors, such as soil properties, chemical properties, and environmental conditions. It is ultimately limited by the chemical vapor concentration that is maintained at the soil surface and by the rate at which this vapor is carried away from the soil surface to the atmosphere. The factors that control the rate of volatilization have been studied mostly for pesticides, but there is little that distinguishes pesticides from other organic chemicals in this context.

Mass transport models may be used to relate contaminant concentrations at the source to exposure concentrations at a receptor. Groundwater pathway models usually include the processes of advection, transverse and vertical dispersion, retardation, and decay. It is also usually assumed that the source is infinite and that steady-state conditions exist. Under these assumptions, the results of the transport model may be expressed using dilution-attenuation factors (DAFs), or the ratio of the porewater concentration at the source to the groundwater concentration at the receptor. Generic default DAF values have been suggested by the U.S. Environmental Protection Agency (USEPA 1994) and site-specific DAF values have been calculated for transient conditions with a finite source and residual NAPL in contaminated soil (Johnson et al. 1998).

Similar models have been developed for atmospheric transport and indoor air. Atmospheric dispersion models are used for air pathways, and screening models are available that relate the pore air concentration at the source to the air concentration at the receptor (ASTM 1998).

Incorporation of Abiotic and Biotic Processes in Transport Models

Chemical processes (e.g., adsorption, precipitation, complexation, hydrolysis, oxidation, or reduction) and bacterial processes (e.g., bacterial degradation) that occur in the soil system and alter the solubility of the contaminant in the soil system need to be considered as input variables in mass transfer processes. Additionally, loss mechanisms (e.g., plant uptake, animal ingestion of soil) need to be accounted for in mass transfer processes. Discussion of these topics and their importance can be found in Chapter 5.

From the perspective of contaminant transport, these processes can be considered as simple linear constants or as mathematical expressions requiring independent solutions. Bioaccumulation (of metals) in plants can present an additional storage reservoir for contaminant mass and, if plants are growing and are harvested, additional loss terms. Both representations can be incorporated in the sequestered source model.

Representing Sorption and Desorption and Nonaqueous-Phase Liquids in Transport Models

The conventional approach for modeling solute partitioning in multimedia environments has been through application of a linear partition model that states that the concentration in one phase (water, air, soil, or NAPL) is proportional to the concentration in another phase. For risk characterization, the solute concentration in soil water is significant for estimating sources to environmental transport pathways. For soil–water interactions, air–water interactions, and NAPL–water interactions, respectively, the linear partition models are

$$q = K_d\, C_w \qquad (6\text{-}1),$$

$$C_a = K_H\, C_w \qquad (6\text{-}2),$$

$$C_o = K_o\, C_w \qquad (6\text{-}3),$$

where q is the sorbed concentration (mg/kg), the parameter K_d is called the "soil-to-water distribution coefficient" (L/kg), C_w is the soil porewater concentration (mg/L), C_a is the soil–pore–air concentration (mg/L), K_H is the dimensionless Henry's Law constant, C_o is the solute concentration (mg/L) within the NAPL phase (oil, organic immiscible liquid), and K_o is the dimensionless NAPL-to-water partition coefficient. If a contaminated soil sample is taken and the entire chemical mass is extracted, then the total concentration, C_T, is measured, which is the total contaminant mass per total (bulk) volume. The total concentration is related to the concentration within the various phases through

$$C_T = \varphi\left(S_w\, C_w + S_a\, C_a + S_o\, C_o\right) + \rho_b\, q \qquad (6\text{-}4),$$

where ϕ is the soil porosity, S_w is the water saturation, S_a is the air saturation, S_o is the NAPL saturation, and ρ_b is the bulk density of the soil (kg/L). When combined with Equations 6-1 through 6-3, Equation 6-4 may be written as

$$C_T = [\varphi(S_w + S_a K_H + S_o K_o) + \rho_b K_d] C_w \equiv B_w C_w \quad (6\text{-}5).$$

Equation 6-5 defines the bulk water partition coefficient, B_w, which is dimensionless. Equivalent coefficients may be defined for each phase in a multiphase system (Charbeneau 2000). Equation 6-5 shows that the relationship between the porewater concentration for a solute and its bulk (or total) concentration in soil is dependent on soil porosity and bulk density, saturation of the 3 phases that might be present, and linear partition coefficients for the solute between the phases and water. The equation assumes that the solute concentrations within the various phases are in equilibrium with one another. In the application of Equation 6-5, it is assumed that the same distribution coefficient (K_d) applies for both the sorption and desorption processes. However, this assumption must be used with caution, because the distribution coefficient for desorption may be quite different from the distribution coefficient for sorption (Chapter 7). If the total soil concentration is analytically determined for a soil sample, and if the other parameters are estimated, then the corresponding aqueous concentration is calculated from

$$C_w = \frac{C_T}{B_w} \quad (6\text{-}6).$$

In risk assessments, Equation 6-5 may be used to back-calculate the allowable concentration that may be left in soil. If C_{WE} is the allowable screening level concentration in water at a potential receptor or exposure location, and (DAF) is the dilution attenuation factor that accounts for environmental transport between the contaminant source and the exposure location, then the allowable porewater concentration at the source is $C_w = (\text{DAF}) \times C_{WE}$. The allowable concentration that may be left in soil is then calculated from

$$C_T = B_w (\text{DAF}) C_{wE} \quad (6\text{-}7).$$

If the presence of NAPL is not accounted for so that the $S_o K_o$ term is neglected in B_w, then application of Equation 6-7 is overly restrictive because the B_w value would be underestimated.

Recent research suggests that application of Equation 6-1 can lead to overestimation of aqueous solute concentrations when applied to a source region where the contaminant has been present for sufficient time to become sequestered (aged contaminants in soils). If only a fraction, *F*, of the sorbed mass was readily available for desorption, then the corresponding porewater concentration would be

$$C_{\mathrm{w}} = \frac{F_{\mathrm{q}}}{K_{\mathrm{d}}} \qquad (6\text{-}8).$$

Neglecting the sequestered mass results in overestimation of the source porewater concentration and subsequently the resulting potential exposures in risk assessments and remediation performance using techniques such as pump-and-treat, soil vapor extraction, and soil flushing. Sequestered chemicals are not readily released to soil water, and the kinetics of desorption should be considered for source term models.

Empirical evidence suggests that sorption sites may be classified into 2 types: those that are readily available for desorption (or allow rapid though kinetically controlled desorption) and those that only allow desorption at a slow rate. Relatively simple models may be developed to produce the observed behavior, and there are cases where use of such models may be important in risk assessment, just as consideration of the presence of residual NAPLs may be important. The following subsections describe such models and their applications and limitations.

Development of multiprocess sequestered source models

Development of new, contaminant source zone models that include sequestered contaminant mass may be significant for risk calculations and for assessing the significance of some soil remediation efforts. In the following subsections, a multiprocess source zone model is presented that specifically includes sorbed mass that is sequestered and slowly released and NAPLs that may be present at residual saturation. Assessment of mobile NAPL source zones is not considered because these require more detailed assessments of NAPL migration. A schematic view of the resulting model is shown in Figure 6-2. The figure shows a region of contaminated soil with water, air, soil, and NAPL phases depicted. Within the soil phase, contaminant mass is shown in 3 different locations that are designated by accessibility to interact with soil pore water. Part of the soil mass is readily available to interact with pore water, and this mass is assumed to be in equilibrium with the porewater concentration. This corresponds to a fraction *F* of the sorbed mass. Another part of the sorbed mass is designated as "sequestered." This mass is available for desorption, but its desorption rate is kinetically limited, and it is released at a slower rate. Finally, some of the soil mass is designated as "immobile." This designation applies to mass that cannot be released to the soil solution. It does not imply that the soil particle cannot be transported through suspension or erosion or through ingestion by mobile soil invertebrates.

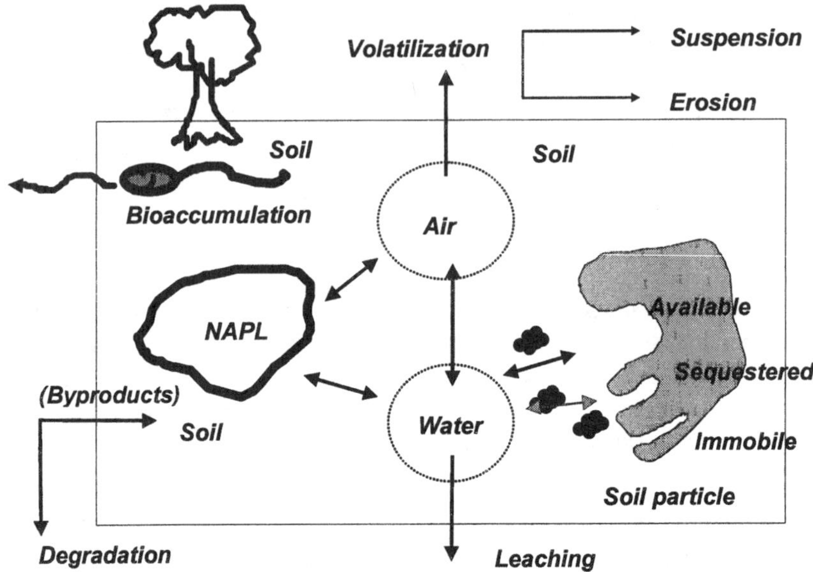

Figure 6-2 Schematic view of the sequestered source model

Figure 6-2 also shows 3 different mass transport mechanisms. Contaminant mass may be leached from the source zone by water moving through the soil. Contaminant mass may partition into soil air and diffuse to the soil surface and volatilize into the atmosphere. Mass that is sorbed to soil particles may be transported in bulk if the particles are moved through suspension or soil erosion.

Three general fate mechanisms are also shown in Figure 6-2. For organic contaminants, the contaminant may biodegrade, resulting in a loss of mass. However, degradation of a parent compound may generate products also considered to be soil contaminants, so the degradation fate of one constituent may result in a source term for byproduct of organic chemical degradation. The soil contaminants may also bioaccumulate, which may result in a loss term as far as migration potential is concerned. Finally, through biotic or abiotic processes, the soil contaminant may be immobilized.

The following discussion outlines how these transport and fate processes may be represented in a sequestered source model (SSM).

Methods for representing sorption and desorption in transport models

In this section, a simple 2-site partitioning model will be presented that will describe the release of contaminant from the soil in the presence of other processes that occur in the source zone. As described previously we will limit our discussion to characterizing the processes that are important to transport in the source zone, since it is in the source region where physicochemical unavailability will impact transport to receptors relative to the case when the contaminant is entirely available.

Treating the source zone as a well mixed cell in a region of the vadose zone, a mass balance accounting for both equilibrium– and mass transfer–limited release from soil with simultaneous losses due to volatilization, degradation, and uptake into organisms can be written as

$$A\,L\frac{\mathrm{d}C_{\mathrm{T}}}{\mathrm{d}t}=-\Lambda\,C_{\mathrm{w}}\,A\,L \qquad (6\text{-}9).$$

In Equation 6-9, A is the cross-sectional area of the source zone (m^2), L is the vertical thickness of the source zone (m), C_T is the total or bulk constituent concentration (mg/L). On the right side of Equation 6-9, Λ is the first-order loss rate coefficient that includes losses due to leaching, degradation, volatilization, and biological uptake (day^{-1}), and C_w is the aqueous concentration (mg/L). Equation 6-9 is a mass conservation equation that states that the rate of change of constituent mass within the source zone volume (AL) is equal to the constituent loss rate due to leaching, degradation, volatilization, and biological uptake (loss). However, because the source zone volume is constant in size (as assumed by this model), Equation 6-9 may be written in the simpler form

$$\frac{\mathrm{d}C_{\mathrm{T}}}{\mathrm{d}t}=-\Lambda\,C_{\mathrm{w}} \qquad (6\text{-}10).$$

Equation 6-10 is the conventional first-order decay (loss) equation, except that it relates the total mass loss to the concentration within an individual phase of a multiphase system.

Mass storage capacity of the source zone

In Equation 6-10, C_T is the total concentration of the constituent within the source zone and includes both the available sorbed mass and the sequestered or slowly released mass. It may be expressed

$$C_{\mathrm{T}}=\rho_{\mathrm{b}}\left(q_1+q_2\right)+\varphi\left(S_{\mathrm{w}}\,C_{\mathrm{w}}+S_{\mathrm{a}}\,C_{\mathrm{a}}+S_{\mathrm{o}}\,C_{\mathrm{o}}\right)+\rho_{\mathrm{B}}\,q_{\mathrm{B}} \qquad (6\text{-}11),$$

where q_1 is the available mass in soil (mg/kg soil), q_2 is the slowly released amount in soil (mg/kg soil), ρ_B is the biological mass per unit volume of soil within which the constituent may accumulate, including plants, soil invertebrates, etc. (kg/L soil), and q_B is the constituent concentration in the biological matter (e.g., plants; mg/kg biomass).

The readily available concentration, q_1, will be assumed to be in equilibrium with the porewater concentration and will be characterized by a linear equilibrium partitioning coefficient:

$$q_1 = K_d \; F \; C_w \qquad (6\text{-}12).$$

The less available contaminant concentration, q_2, will be characterized by a linear driving force, mass transfer rate-limited process as follows:

$$\frac{dq_2}{dt} = k_2 \left[K_d \left(1 - F\right) C_w - q_2 \right] \qquad (6\text{-}13),$$

where k_2 is the rate constant for slow release (day^{-1}).

The relationship between the mass of constituent per unit plant biomass and the soil concentration is defined through the plant-to-soil concentration ratio (Peterson 1983)

$$q_B = K_B \; q_1 \qquad (6\text{-}14).$$

In the definition of the plant-to-soil concentration ratio, K_B (dimensionless), given in Equation 6-14 is assumed, though there are some questions of whether it is consistent with reported data. The mass of the soil is usually expressed on a dry weight basis but results based on both dry and wet (fresh) weights for crops have been reported. Further, the soil constituent mass likely includes both the aqueous and sorbed concentrations in the soil sample.

With these results, Equation 6-11 may be written

$$C_T = \rho_b \, q_2 + \left[\left(\rho_b + \rho_B \, K_B \right) F \, K_d + \varphi \left(S_w + S_a \, K_H + S_o \, K_o \right) \right] C_w \qquad (6\text{-}15),$$

where Equation 6-13 is used to model the changing concentration of the slowly released constituent.

Mass loss rate from the source zone

The right side of Equations 6-9 and 6-10 represents the mass loss rate from the source zone from leaching, volatilization, degradation, and biological uptake. These are written in the form of first-order rate expressions. Explicitly, the right side of Equation 6-9 is written

$$-\Lambda C_w A L = -(\lambda_L + \lambda_V + \lambda_D + \lambda_B) C_w A L \quad (6\text{-}16).$$

Each of the loss rate coefficients is based on a physicochemical description of the loss process.

The leachate loss-rate coefficient is defined by

$$\lambda_L = \frac{u}{L} \quad (6\text{-}17),$$

where u is the water volumetric flux (Darcy velocity) through the source zone (m/d). With this formulation, the advection flux from the source zone is written as follows:

$$\lambda_L C_w A L = u C_w A \quad (6\text{-}18).$$

The volatilization first-order constant can be expressed in terms of the effective vapor diffusivity as follows:

$$\lambda_V = \frac{D_e K_H}{L L_v} \quad (6\text{-}19),$$

where D_e is the effective vapor diffusivity through soil (m^2/day), and L_V is the diffusion length for vapor diffusion (m). With this formulation,

$$\lambda_V C_w A L = \frac{D_e C_a}{L_v} A \quad (6\text{-}20),$$

which is Fick's first law for vapor diffusion through soil with effective diffusion coefficient D_e, source vapor concentration C_a, concentration zero at the ground surface, and effective diffusion length L_V. The length for vapor diffusion will be time dependent. However, for our development here, we will assume an average constant value of L_V.

The first-order decay coefficient λ_D is related to the constituent degradation half-life $T_{1/2}$ through

$$\lambda_D = \frac{\ln(2)}{T_{1/2}} \tag{6-21}$$

Finally, the biomass uptake loss rate coefficient is

$$\lambda_B = \frac{\dot{m} K_B K_d F}{L} \tag{6-22}$$

where $\dot{m}$ is the biomass harvest rate (kg biomass per m^2 per day). With this formulation, the biomass loss rate is expressed by the following

$$\lambda_B C_w A L = \dot{m} A q_B \tag{6-23}$$

Sequestered soil model

Equations 6-10 and 6-15 may be combined to give

$$\rho_b \frac{dq_2}{dt} + [(\rho_b + \rho_B K_B) F K_d + \varphi(S_w + S_a K_H + S_o K_o)] \frac{dC_w}{dt} = -\Lambda C_w \tag{6-24}$$

Equations 6-24 and 6-13 can be solved simultaneously, and the concentrations in the soil phase can be related to the concentrations in the pore water for any time t. The solution to these equations for the soil and porewater concentrations as a function of time for the various parameters that appear in these equations is presented in the appendix to this chapter (p 248).

Figure 6-3 shows how the slow release of a fraction of the sorbed contaminant can impact the porewater concentrations. The equations from the appendix to this chapter were used to calculate the curve for $k_2 = 0.1$ yr^{-1}. For Figure 6-3, it was assumed that the fraction of contaminant available was originally 50% of the total sorbed amount. It was assumed that the contaminant, naphthalene for this example, was in contact with the soil (aged) for a sufficient amount of time so that the slowly released fraction is in equilibrium with the pore water. As weathering of the contaminant in the soil begins to occur through volatilization, degradation, leaching, and other processes, the concentrations in the soil and pore water will decrease with time. As the soil concentrations decrease, the corresponding porewater concentrations will be affected by the slow release of the contaminant from the sequestered fraction as shown by the 3 curves in Figure 6-3.

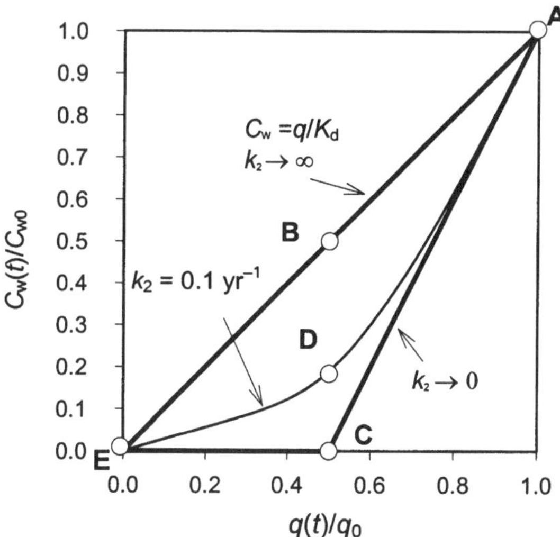

Figure 6-3 Effect of slow release on the porewater concentrations as a naphthalene-contaminated soil is weathered. The initial soil and porewater concentrations are $q_0 = 12.3$ mg/kg and $C_{w0} = 1.9$ mg/L, respectively, and $K_d = 6.5$ L/kg. Equations and parameters for determining curve A–D–E are given in the appendix (p 248).

If the release rate of the slow fraction is fast relative to the weathering processes within the source, then the porewater concentrations will always be related to the soil concentrations by a constant value of K_d and will follow the path A–B–E shown in Figure 6-3. When the release rate for the sequestered fraction is slow, however, the porewater concentrations will be less for a given soil level as the weathering processes proceed. If the slow release rate constant is zero (i.e., the desorption process is irreversible), then for a soil that originally had 50% of the naphthalene sequestered, the porewater concentration would follow a path given by A–C in Figure 6-3. At point C, the porewater concentrations would be 0. In other words, the effective K_d for the sorbed amount remaining at point C would be infinite.

For many soils, the release rate constant will be finite. The effect of a finite release rate on porewater concentrations for a given soil concentration will be dependent upon how this rate of release compares with the rate of loss of contaminant from weathering. For a slow rate constant that is small relative to the weathering processes, the relationship between the porewater and soil concentrations will follow a path like that given by A–D–E in Figure 6-3. Curve A–D–E was determined using the equations presented in the appendix (p 248) for porewater and soil concentrations. Calculations were carried out with a rate constant 0.1 yr^{-1}. (Values for the various parameters for these equations that would be expected to characterize typical rates of volatilization, degradation, and leaching for naphthalene are also presented in the appendix). A comparison of Points B and D shows the effect that a slow desorption rate constant of 0.1 yr^{-1} would have on porewater concentrations when the majority of the available contaminant has been removed from the soil, and all that remains is the slowly released fraction.

From Figure 6-3, it should also be clear that contaminated soil would have values of F that can range from 0 to 1. The actual value of F will depend upon how much weathering has occurred and how much of the contaminant was originally sequestered in the soil.

It is instructive to decouple the slow and fast release processes because a very simple equation (simpler than those presented in the appendix to this chapter) describes the relationship between soil and porewater concentrations for slow release. The point on the curve will be described when the release from the soil becomes rate limiting, that is, the readily available fraction has been removed.

We can describe the porewater concentration by neglecting the rate of change in the aqueous concentration (see Equation 6-24). The following relationship between the concentration in the soil and the porewater concentration is obtained:

$$\frac{q_2}{C_w} = K_d(1-F)\left[1+\frac{\lambda_L+\lambda_D+\lambda_V+\lambda_B}{K_d(1-F)\rho_b k_2}\right] \quad (6\text{-}25).$$

Alternatively, the porewater concentration can be measured for a given soil concentration by rearranging Equation 6-25 as follows:

$$C_w = \frac{q_2}{K_d(1-F)\left[1+\frac{\lambda_L+\lambda_D+\lambda_V+\lambda_B}{K_d(1-F)\rho_b k_2}\right]} \quad (6\text{-}26).$$

Equation 6-26 shows more simply the relationship between porewater and soil concentrations for a given value of the slow rate constant k_2. If the rate of release is slow (characterized by a long half-life for the sequestered fraction) relative to other processes such as leaching, degradation, volatilization, or other processes that might occur in the source region (e.g., plant or invertebrate uptake), then the second term (in the brackets) becomes large relative to the first and enhances the "apparent" distribution coefficient between soil and pore water. Note that if the rate constant is large (short half-life) relative to the corresponding constants for leaching, degradation, etc., then the partition coefficient approaches the normal equilibrium partition coefficient that characterizes the partitioning of the available contaminant.

Equation 6-26 can be used to estimate the range of slow release constants that will be significant with respect to the exposure of a chemical to various environmental receptors.

To illustrate when k_2 becomes important, consider the release of some representative compounds that reflect reasonable ranges of volatility and sorption potential.

Slow release relative to leaching and volatilization

Assume the following values for source region parameters:

$$L = 1 \text{ m}$$

$$u = 0.2 \text{ m/y (sandy loam)}$$

$$S_w = 0.40$$

$$\phi = 0.35$$

$$L_v = 0.3 \text{ m}$$

Some physicochemical properties of selected organic compounds are given in Table 6-1.

Table 6-1 Physicochemical properties of selected organic compounds[a]

Compound	K_{oc} L/kg	D_{air} cm^2/s	D_{water} cm^2/s	K_H dimensionless
o-Xylene	3.63E+02	8.70E–02	1.00E–05	2.13E–01
Naphthalene	2.00E+03	5.90E–02	7.50E–06	1.98E–02
Anthracene	2.95E+04	3.24E–02	7.74E–06	2.67E–03
Benzo[*a*]pyrene	1.02E+06	4.30E–02	9.00E–06	4.63E–05

[a]Attachment C, USEPA soil screening-level guidance (1996).

Slow release relative to leaching, volatilization, and degradation

Using the values presented above and in Table 6-1, one can estimate when slow release is important relative to the various loss processes that occur in the source zone. In Table 6-2, half-lives for slow releases correspond to losses for leaching, volatilization, and degradation. These were determined by equating the second term in the brackets in Equation 6-26 to 1 and translating the rate constant, k_2, to a half-life ($T_{1/2} = \ln(2)/k_2$).

Table 6-2 Half-lives (years) necessary for slow release to be important relative to various loss processes in the vadose zone

Compound	Leaching	Volatilization	Degradation
o-Xylene	5	0.1	2.5
Naphthalene	3(10)	10	1.5
Anthracene	$4(10^2)$	$2(10^3)$	6
Benzo[*a*]pyrene	$1.5(10^4)$	$3(10^6)$	16

Table 6-2 shows that slow release of contaminants from soils will impact the porewater concentrations calculated by the conventional K_d approach when the rate is slow enough relative to those loss processes. For example, if the only loss process occurring in the vadose zone was loss from leaching due to infiltration, then the half-life for slow release for a compound with K_d values similar to xylene would have to be of the order of 5 to 10 years to have an impact on porewater concentrations. For more strongly partitioning compounds, required half-lives would be longer in proportion to the partition coefficient of the compound. However, when degradation and/or volatilization losses are considered, there will be cases when smaller half-lives (larger values of k_2) for the slow release of contaminants from soil yield lower porewater concentrations for a given soil concentration. To put these values in perspective, half-lives for slow release reported in the literature range from days to several months (Ball and Roberts 1991; Brusseau et al. 1991; Grathwohl and Reinhard 1993; Farrell and Reinhard 1994; Kan and Tomson 1994; Pignatello and Xing 1995; Williamson et al. 1998; Rixey et al. 1999). It should be made clear, however, that the lowest values of the rate constants reported in the literature have been limited by the duration of laboratory experiments. Note that, for the calculations given in Table 6-2, it was assumed that the source of contamination is close to the surface, and for contamination further down in the vadose zone, the losses due to volatilization will be considerably reduced relative to those represented in Table 6-2.

Incorporation of residual nonaqueous-phase liquid into the source term

In general, NAPLs almost always are present in soils with organic chemical concentrations greater than 10,000 mg/kg, and often are present in soils containing room temperature organics exceeding concentrations on the order of 100 mg/kg. Feenstra et al. (1991) provide a method to estimate if NAPLs are present in soils based on equilibrium partitioning relationships, and a variety of field techniques are available to help determine if NAPLs are present in soil samples. In practice, NAPLs are probably present in contaminated soils at many or most petroleum hydrocarbon and chlorinated solvent sites.

The presence of NAPLs is important to the source term in risk assessments for several reasons:

- partitioning between the NAPL phase and the aqueous and vapor phases is controlled by different relationships than partitioning between the aqueous, sorbed, and vapor phases in soils that do not contain NAPLs;
- because of the relatively large mass associated with NAPLs, chemicals in soils containing NAPLs can have a longer half-life than chemicals in soils without NAPLs;
- many NAPLs contain relatively inert organic compounds (e.g., long-chained alkanes in gasoline) that act as a relatively large reservoir of organic carbon that will slowly release soluble contaminants from the NAPL to the aqueous phase.

Nonaqueous-phase liquid–water partitioning

When infiltrating water comes in contact with residual or free-phase NAPL in the soil, the aqueous-phase concentration can be estimated from the concentration in the NAPL phase with K_0, the NAPL (or "oil") partition coefficient. The partition coefficient K_0 can be based on actual measurements of NAPL-containing soils, it can be approximated using K_{ow}, or it can be calculated more rigorously from the following equation (Cline et al. 1991; Lane et al. 1992):

$$K_0 = \frac{C_0}{C_L} = \frac{1}{\gamma_i^0} \frac{MW_i \times 10^6}{MW_0 \, S_i^{aq}} \qquad (6\text{-}27),$$

where γ_i^0 is the activity coefficient of the chemical in the oil phase (as in Raoult's Law; unit less; typically assumed to be ρ_0 is the oil density (g/mL), S_i^{aq} is the subcooled liquid solubility of chemical compound in water (mg/L), MW_i is the molecular weight of chemical compound, and MW_0 is the molecular weight of oil phase. Note that the aqueous-phase concentration (C_L) is often approximated by multiplying the pure-phase solubility by the mole fraction of the contaminant in the NAPL.

Banerjee (1984) and Broholm and Feenstra (1995) have used experimental evidence to conclude that the effective solubility relationship produces reasonable approximations of effective solubilities for mixtures of structurally similar compounds and that the relationship works best for binary mixtures of similar compounds. In general, the method is appropriate for many environmental studies for which there are many other uncertainties (Pankow and Cherry 1996).

For complex mixtures (e.g., multiple identified and unidentified solvents, or mixed fuels and solvents), it is necessary to estimate the weight percent and an average molecular weight of the unidentified fraction of the NAPL before the calculation can be completed. For example, a molecular weight of 90 to 100 is typically used for gasoline. Pankow and Cherry (1996) provide an example of these calculations for a mixture of chlorinated and nonchlorinated compounds.

With this partition coefficient, the total concentration of the contaminant dissolved in the leachate remains constant, even if the total concentration of the NAPL in the soil increases. Aqueous-phase concentrations will increase together with soil concentrations only while the soil contaminants are sorbed (no NAPL). With NAPL present, the leachate contaminant reaches a maximum concentration determined by the mole fraction of the contaminant in the NAPL and the aqueous solubility of the contaminant. This is shown in Figure 6-4.

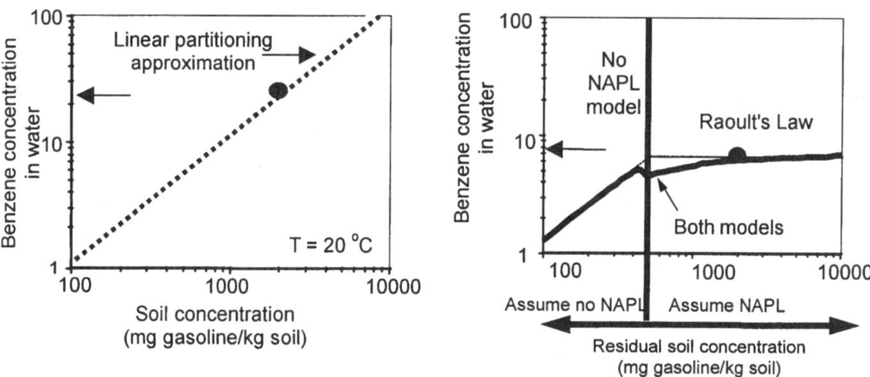

Figure 6-4 Conventional linear partitioning model compared to combined NAPL model

For example, soils with residual gasoline concentrations of 2000 mg/kg would yield a predicted aqueous-phase benzene concentration of 25 mg/L using the conventional linear partition approach, as shown in Figure 6-4. Using the NAPL partition relationship, the predicted aqueous-phase benzene concentration would be about 6 mg/L. Therefore, accounting for the presence of NAPL would result in a 4-fold reduction in the predicted aqueous-phase concentration. The error would increase for soils with higher concentration, while no error would result in use of linear partitioning alone for soils with concentrations below the level where NAPL is expected (typically total organic chemical concentrations on the order of a few hundreds of mg/kg).

Note that application of the effective solubility relationship to field sites requires the assumption of homogeneous soils and homogeneous NAPL distribution. However, at many field sites, the distribution of NAPL is heterogeneous, and some flow lines do not come into contact with NAPL, and therefore dilute the bulk leachate. Therefore, at actual field sites, the average leachate concentration reflects the contribution of high-concentration flow lines that came into contact with NAPL and non-NAPL flow lines at much lower concentrations.

Impact of nonaqueous-phase liquids on source mass

The impact of NAPLs on the distribution of contaminants in soils can be seen in Figure 6-5, where 100 mg of benzene in 1 L of soil has been assumed under 2 different conditions: first, a soil with no NAPL, and second, a soil with NAPL. For this calculation, the soil is also assumed to have a natural organic C fraction of 0.005 g C/g soil. Using Equation 6-5 for the bulk water coefficient that describes the relationship between the total contaminant concentration in the soil and the pore water, the distribution of benzene in the water, soil (sorbed to the organic C on the soil), and NAPL was determined. Note that the calculation assumes that the NAPL is a mixture of benzene and other chemicals, with the other compounds having an average molecular weight of 150 g/mole. The NAPL–water partition coefficient was determined from Raoult's Law using Equation 6-27 with $\gamma_i^o = 1$.

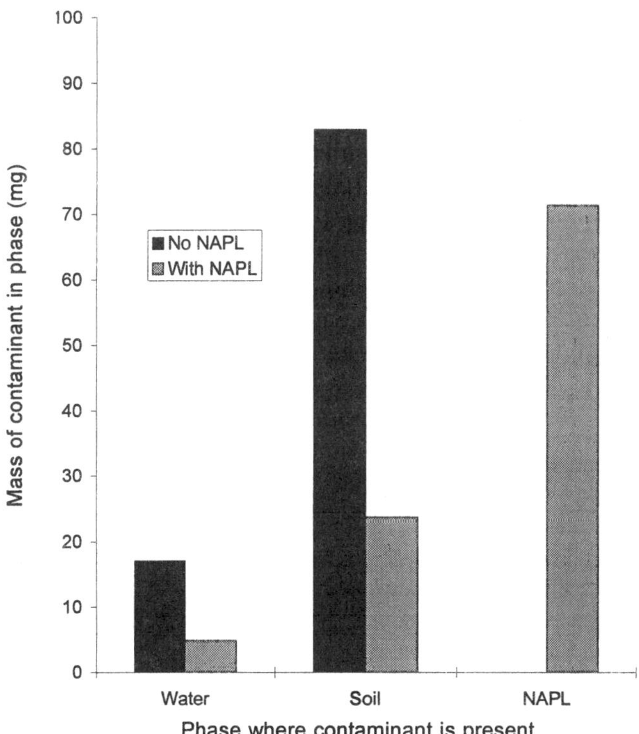

Figure 6-5 Distribution of benzene mass among various phases for soils with and without residual NAPL (Basis: Total concentration of benzene = 100 mg/L of media.)

As can be seen in Figure 6-5 for the "no NAPL" case, most of the benzene (a total of 82 mg) is in the soil (sorbed) phase, with only 18 mg in the water phase. When NAPL is present, however, the majority of the benzene is in the NAPL (71 mg) with lesser amounts (24 mg) in the soil phase and only a small amount in the water phase (5 mg). In other words, for a fixed amount of contamination in soil, the presence of a NAPL changes the relative distribution of the contaminant among the different phases (water, sorbed, and NAPL) so that water-phase concentrations are lower when NAPL is present.

This type of calculation was extended for 5 different contaminants (benzene, toluene, *p*-xylene, naphthalene, and phenanthrene) in Table 6-3, using the same assumptions (100 mg contaminant/L soil, fraction natural organic C on soil = 0.005). The table shows that for each of the 5 contaminants, the water-phase concentration is lower when NAPL is present, compared to the no-NAPL case. For a fixed amount of contamination in soil, the presence of NAPL will reduce the resulting concentration of contaminants in the water phase, that is, the contaminant is less available to the aqueous phase.

Table 6-3 The effect of NAPL on water-phase concentrations (Basis: Total concentration of contaminant = 100 mg/L soil; fraction of natural organic C in soil = 0.005)

	Molecular weight	K_{ow} (L_w/L_o)	Solubility (mg/L)	Concentration in water phase (mg/L)	
				Soil without NAPL	Soil with NAPL
Benzene	8.1	135[6]	1780	38.3	34.9
Toluene	92.1	490[6]	515	38.3	9.1
p-Xylene	106.2	1413[7]	198	13.8	3.1
Naphthalene	128.2	2290[6]	31	8.6	1.5
Phenanthrene	178.2	37,150[6]	1.1	0.5	0.1

Impact of nonaqueous-phase liquids on contaminant release

While NAPLs in soils greatly increase the mass of contaminants in soil, they also restrict the release of these contaminants to other phases. Both the vapor-phase and aqueous-phase partitioning are based on the mole fraction of the contaminant in the NAPL, and as contaminant is released, the mole fraction is reduced. Therefore, the mass flux of contaminants leaving the source zone over time (either via vaporization or dissolution) will be reduced over time.

In other words, the NAPL phase acts as if the organic C content of the soil was greatly increased. If a soil with a natural organic C content of 0.001 g C/g soil (0.1%) is contaminated with NAPL so that only 5% of the pore space is filled with

NAPL (5% residual saturation), then the total organic C content of the soil is increased by 0.009 g C/g soil. The total of 0.010 g C/g soil is 10 times the original C content of the soil. Furthermore, the effective partitioning of an organic-based NAPL is larger than an equivalent amount of natural organic C, as it may be assumed that only a fraction, approximately 60%, of the natural organic C is available for sorption (Olsen and Davis [1990]) note that $K_{oc} = 1.724\ K_{om}$, where K_{om} is the partition coefficient for organic matter).

The key partitioning relationships described above are based on the amount of organic C in the soil where many organic contaminants will be sorbed (in the case of natural organic C) or dissolved (in the case of NAPL). When NAPL invades the soil, the total mass of the contaminants is increased, but extremely rapid release rates do not occur as the NAPL–aqueous-phase partitioning moderates the release of the soluble contaminants to the aqueous phase. In summary, the presence of NAPL changes both the magnitude of the contaminated soil problem (greatly increases the mass to be managed) and the nature of contaminant release to the environment by moderating release compared to classic linear desorption.

Parameter Estimation

If it is assumed that there is not a mobile (or potentially mobile) NAPL phase in the contaminated source area, and that facilitated transport mechanisms are unimportant in subsurface environments, then the most significant considerations for determination of release from a source are the partitioning behavior, especially relative to desorption, and the release rate of chemical from the soil matrix. Partition coefficients are typically estimated from 1) a default or measured soil organic carbon fraction and the organic carbon–water partition coefficient for the chemical of interest or 2) a measurement based on the standard American Society for Testing and Materials (ASTM) sorption method. As discussed in the previous sections, the calculation of release from field soils based on either a calculated or measured sorption partition coefficient is likely to overpredict, perhaps significantly, chemical release from the soil–waste matrix. In addition, the chemical release process can be kinetically limited, further affecting the accuracy of prediction of release.

In this section, methods will be suggested to obtain the parameters needed for improved prediction of release from the source. These methods are based on the empirical observation that release from field soils is not instantaneous but tends to occur over long periods of time. Although the mechanisms for this behavior are complex and not well-defined, it is useful to depict the release behavior as occurring from a readily releasable compartment (releasing the readily releasable fraction F in a relatively short time) and from a slowly releasable compartment (releasing the slowly releasable fraction over a relatively long period of time at a slow release rate k_2). Though mechanistically simplistic and inaccurate, this depiction is useful for developing improved estimates of kinetically constrained release while still being conservative and environmentally protective. It also allows for a reliable measurement of a partition coefficient to be made based on a desorption experiment in a reasonable period of time (1 or 2 days). Specifically, methods suggested for improved partition coefficient estimates are

- measurement of field soil organic C content,
- estimation of partition coefficients for soils containing residually trapped NAPL using Raoult's Law,
- desorption method for measurement of site-specific partition coefficient,
- estimation and measurement of F, and
- estimation and measurement of k_2.

Improved partition coefficients

The first step that can be taken to improve a partition coefficient estimate is typically to use a laboratory measurement of soil organic C. Standard laboratory methods for soil organic C are adequate.

A further refinement of the estimation is appropriate for soils containing residually trapped NAPL (total petroleum hydrocarbons or oil and grease concentrations on the order of 0.1% to 0.2%). A better K_d value is calculated by adding the $K_{oc} \times f_{oc}$ value to the $K_o \times f_o$ value (Rutherford et al. 1997). The K_o value is obtained by Raoult's Law knowing the subcooled liquid solubility of the chemical of interest and the average molecular weight of the NAPL phase (Equation 6-27). The molecular weight can be estimated from composition data or boiling point data or can be measured using osmometry.

A further possible refinement is to measure an experimental partition coefficient for a field sample of the source. The standard ASTM procedure for measuring a partition coefficient involves sorption of a chemical from a solution onto a soil sample. The values determined for field soils by this method may be significantly less than values measured for desorption from field soils, especially if the field soils contain sequestered and weathered residues. This leads to an overprediction of partitioning from the soil to water. Because of slow release from these soils, an unreasonably long desorption experiment may be required to derive reliable partition coefficients. If a relatively short desorption experiment of 1 or 2 days is conducted, it is possible that only a fraction of the chemical is available to be released. If this is not accounted for in the calculation of a partition coefficient, the result will be in error.

To address this situation, an alternate procedure is suggested, in which the partition coefficient for the readily available fraction is measured in a reasonable time period such as 1 or 2 days. The assumption in using this approach is that there is a readily available fraction of chemical in equilibrium with the soil that determines the porewater concentration, and that the partition coefficient for the readily available fraction is relatively constant with concentration (linear desorption isotherm). If it can be determined what the readily available fraction is in a 1- or 2-day time period, then by a mass balance, an estimate of the partition coefficient for the sample, can be obtained. A second experiment is needed to determine the readily available fraction in order to develop a partition coefficient estimate normalized to the total soil mass.

The following is a brief description of 2 possible approaches for which procedures need to be developed and validated. In the first approach, 2 water desorption experiments on a split sample are conducted, one with a resin sorbent to remove the chemical from the water phase. The partition coefficient normalized to the entire soil mass can be calculated from a material balance of the 2 experiments, assuming the readily available fraction of chemical in the soil is in equilibrium with the water phase.

Alternatively, the second approach uses multiple batch desorption experiments of split samples at different soil-to-water leachate ratios. A material balance of these experiments at different soil-to-water leachate ratios will allow an estimate of the equilibrium porewater concentration, which is generally the number of interest. This estimate of equilibrium porewater concentration can be obtained without knowing the starting total concentration of chemical in the soil sample. If the starting soil concentration is known, an estimate of the partition coefficient can be obtained as well.

It must be noted that these methods assume that the chemical in the sorbed and water phases is in equilibrium in the field. If a soil is highly weathered and biodegraded in the environment, then these methods may not yield the maximum possible porewater concentration. Preliminary studies show that these methods may underestimate the maximum porewater concentration slightly, but in no case by more than a factor of 2. These methods are an improvement over methods that are typically used to obtain desorption partition coefficient measurements. However, their usefulness may be limited for very low solubility chemicals because of achievable detection levels in the water phase.

Estimation and measurement of k_2

The first approach involves estimation of mass diffusion through nanopores (Wu and Gschwend 1986; Carroll et al. 1994). This approach requires approximations of soil parameters such as pore size and tortuosity. It may be possible to estimate k_2 within 1 or 2 orders of magnitude using this approach.

The preferred approach would be to experimentally measure the slow release rate. It must be recognized that this may require unreasonably long time periods because of the slow desorption mechanisms discussed throughout this publication. One experimental approach involves the measurement of leaching from a fixed bed. The k_2 value can be derived from the data by using the standard mathematical formulation for leaching from a porous media bed, modified to incorporate a 2-compartment mathematical model for the release term in the mass transfer balance (Garg 1998). This second approach involves the use of multiple batch aqueous and resin extractions carried out for different lengths of time. Each extraction provides a measurement of the mass released (available fraction) for the time period. A plot of this release data against time is then fit to a first-order rate expression to yield the k_2.

Both of these methods may require several months to measure slow rates as low as 0.0001 day^{-1}. It may be possible to reduce the time by using supercritical fluid extraction, co-solvents, or elevated temperatures. More work is needed to correlate these accelerated methods to actual field leaching behavior before they can be used for determining k_2.

Practical Utility of Transport Models in Risk Characterization

Application of simple models

Transport models can be used in risk characterization and establishment of soil-quality limits at several levels or contexts. At the lowest level, simple equilibrium assumptions using linear partitioning are used along with conservative assumptions related to source size and location to relate source concentrations to some concentration at a receptor. An example of this is the development of risk-based screening levels (RBSLs in risk-based corrective action [RBCA] Tier 1) or soil-screening levels (USEPA Superfund SSLs) for soil. RBSLs and SSLs are developed based on a risk-based concentration limit for groundwater or air, a default partition coefficient, and a default DAF. This is essentially a trivial transport application as the concentration in the transport media (e.g., groundwater or air) is given by equilibrium partitioning from an assumed infinite source. This may be useful for site management screening purposes for many sites, but it can result in unrealistic prediction of air or groundwater concentrations, or soil-quality limits. The information in this chapter provides straightforward approaches for refining this lowest-level estimate. In this case, the improved transport information is in the form of an improved estimate or measurement of partition coefficient. It is also possible to develop improved site-specific DAF values based on simplified transport methods and nomographs (Johnson et al. 1998).

A second level of use of the information in this chapter is improved representation of the release from a source that takes into account mass transfer and other kinetic limitations in field soils. This can be done in the context of development of site-specific soil-quality criteria, or the development of a site-specific risk assessment. The methods outlined in this chapter provide guidance on implementing these approaches.

Biphasic release

This chapter suggests a method to model release from source areas that accounts for the biphasic (or multiphasic) slow release behavior exhibited for many field soils. It is concluded that this available or unavailable fraction approach provides a more accurate and useful method to define interactions and release in source areas. It is suggested that this method be formalized and incorporated into routine site management practice.

Incorporation of nonaqueous-phase liquids into risk characterization

Currently some soil models and risk assessment protocols (e.g., API 1994; Connor et al. 1997; ASTM 1998) do not incorporate NAPLs in the source-leachate calculations. Instead, these approaches make the assumption that soil contaminants exist in only 3 phases (vapor, aqueous, or sorbed), with no contaminants in the fourth phase (NAPLs). In general, this assumption of instantaneous equilibrium partitioning will tend to overestimate the contaminant mass transferred from the affected soil zone to infiltrating rainwater. Connor et al. (1997) showed that use of a 3-phase rather than a 4-phase partitioning relationship resulted in a conservative overestimation of leachate concentrations by a factor of 1.3 to 4.2 times, corresponding to residual NAPL saturations of 0.1% and 1%, respectively. Because most of the models and risk assessment protocols assume either an infinite source or perform a mass balance on the source using soil concentrations, the large mass associated with the NAPL phase is either directly or indirectly incorporated into these models.

In summary, 3-phase equilibrium partitioning is a convenient and common modeling assumption that simplifies transport modeling of contaminants from soil source zones. Use of slightly more complex modeling approaches that incorporate NAPL-leachate modeling will make improvements in leachate concentration estimates on the order of 5- to 10-fold. This can be significant for many site situations that contain mixtures of contaminants, for example, petroleum hydrocarbons.

Steady-state source assumption

One assumption routinely used for determination of soil-quality criteria and risk assessment transport and fate boundary conditions is that of an infinite steady-state source. This assumption typically leads to a gross overprediction of transport away from the source. More realistic estimates can be made using methods that account for transients and source depletion (Johnson et al. 1998). The slow release method suggested in this chapter provides a significantly more accurate estimate of porewater concentrations than does the conventional infinite source assumption approach. It should be noted that if time-dependent release approaches are to be useful, the exposure duration for completed pathways must be known (or assumed).

Incorporating erosion or particulate loss terms in source transport

At some sites, mechanical surface processes that move contaminated soils may be important drivers for evaluating ecological risk. Two key processes are surface erosion and particulate emission.

Soil erosion can be estimated using empirical relationships, such as the universal soil loss equation (USLE), that correlate soil type, rainfall, and other factors to erosion rate. In practice, the erosion rate estimates reflect any soil that is eroded, and does not account for subsequent deposition processes. In other words, if the erosion rate from the USLE is extrapolated to an entire watershed, the amount of suspended solids actually reaching the stream will be greatly overpredicted. Sediment delivery ratios are empirical relationships that show the fraction of eroded material that will be transported to receiving streams for watersheds of different area (Shen and Julien 1993).

A second general approach is to use runoff data from nonpoint source studies, such as the Nationwide Urban Runoff Program (NURP) (USEPA 1983). Field studies provide estimates of suspended solids concentrations and annual solids loadings measured in runoff from small study areas (typically 1 to 100 acres). The data, expressed as either concentration or loading per acre, can be used to estimate the amount of solids that are transported from soil source zones by water-driven processes.

The same rate-limited processes discussed previously (i.e., the biphasic release methodology) control the partitioning between contaminated soils and the aqueous phase (in this case, the runoff). While contaminants in suspended soils in runoff are available for transport on a macroscale (i.e., moving in the runoff), some are unavailable for immediate cross-media transport to the aqueous phase because of the hysteretic effects of sorption. The contaminants that are unavailable for cross-media transport to the aqueous phase may or may not be bioavailable, depending on how receptors are exposed to the contaminants (i.e., ingestion of water containing the contaminated sediments versus ingestion of aqueous-phase materials only).

Particulate emission is a wind-driven process that is currently incorporated into some risk assessment protocols (e.g., ASTM RBCA standard; ASTM 1998). For example, the ASTM methodology prescribes a generic value of 6.9×10^{-14} g/cm^2 ×

sec (about 20 $mg/m^2 \times year^{-1}$) for a particulate emission rate. A site-specific method is outlined in USEPA's Soil Screening Guidance (USEPA 1996).

Special situations

It must be noted that the suggested methods may not be applicable in certain site situations such as fractured rock, mobile NAPL phases that can transport much greater levels of contamination than can be transported by solute transport, and fractured clay systems. More detail on the third situation is provided below.

Recently, a new conceptual model for free-phase organic releases to fractured clay systems has been introduced (Parker et al. 1994; Pankow and Cherry 1996). With this conceptual model, relatively rapid rates of transport from the NAPL phase to the aqueous phase are proposed for systems where NAPL first penetrates vertical fractures in clays and then is trapped in the fractures. The high aqueous-phase concentration adjacent to the trapped NAPLs and the low concentrations within the trapped water in clays cause a large driving force for diffusion of the soluble contaminants into clays. Based on examples provided by Parker et al. (1994), large masses of soluble contaminants can be transferred from NAPLs to fractured clays (typical fracture spacing on the order of feet) in relatively short time periods (a few years). While this transport mechanism is reversible, the kinetics of the diffusion out of clay are very slow, because the driving force is typically much smaller. Therefore, fractured clays that have been exposed to vertical NAPL transport will serve as long-term sources of contaminants for long time periods and may be immune to most remediation technologies.

Model validation

Finally, it should be noted it is always prudent to provide validation or ground-truthing for model predictions. It is noted that for some site situations, a few strategic samples (e.g., porewater analyses) may provide more reliable, more useful site management decision data more quickly and cost effectively.

Transport and bioavailability

There are a number of issues that arise in developing models to describe the transport of contaminant mass from source regions, especially when these source term models include mass sequestration. One issue concerns the question of whether contaminant availability for transport always coincides with its bioavailability. While in many cases it does, the answer in general is negative. For example, if Pb is complexed with EDTA, then it is available for transport. It may even be consumed by soil organisms, but it is not bioavailable. Because it is strongly complexed, it will pass through organisms and not be absorbed or metabolized. A second example concerns chemicals that are surface complexed on soils and are not available for transport but can be bioavailable through secretion of enzymes that result in release of the chemical. Similarly, sequestered chemicals in soil can be consumed by soil invertebrates and partially metabolized. NAPLs present a fourth example. Long-chain residual NAPLs are not available for transport, but there is evidence that soil microbes can go directly to the interface and promote the degradation of these molecules. Thus, while chemical availability for transport and chemical bioavailability include many similar characteristics, generally they should be considered separately on a case-specific basis.

Preferential flow

Another factor that complicates the description of transport of contaminants from source regions is preferential flow. "Preferential flow" refers to the rapid movement of solutes through fractures, root holes, worm burrows, and other heterogeneities, at rates much greater than expected from consideration of the porous medium as a whole. Dyes and other chemical tracers have been used to study contaminant movement through the vadose zone. Kung (1990) notes that funneling due to heterogeneities may result in flow occurring through less than 1% of the soil. Unstable flow and fingering occurs in layered soils and may cause lateral spreading of contaminants. Factors such as these make flow and transport difficult to predict and measure; point samples can easily miss narrow fingers of solute. Unfortunately, there are no proven predictive models for vadose zone transport with preferential flow.

Summary

This chapter discusses the major mechanisms that control the release of organic compounds and metals from their source zone in contaminated soils. The focus has been on how resistant desorption processes may influence contaminant transport and the assessment of risks from contaminated soils. It is noted that conventional methods of characterizing chemical partitioning may result in prediction of exposure concentrations that are too large, and when these methods are used to establish soil cleanup levels, the resulting requirements may be overly conservative (and thus overly expensive).

The chapter develops and presents a multiprocess sequestered source model for contaminated soils. This model incorporates biphasic desorption and residual NAPL, in addition to the processes of volatilization, leaching, and first-order biodegradation. The behavior of the model is shown through example, and various limiting conditions are discussed. Criteria show when application of the biphasic desorption model leads to significantly different predictions of aqueous-phase concentrations (Table 6-2) than use of conventional equilibrium partitioning models. However, a more complete characterization of the significance of biphasic desorption and residual NAPL will require coupling of the sequestered source model with a contaminant transport model to evaluate the effects on predicted exposure concentrations available for interaction with organisms. Finally, methods for estimating parameters are described, and use and limitations of the models are discussed.

References

[API] American Petroleum Institute. 1994. Decision support system for exposure and risk assessment [computer program]. Version 3.0.

[ASTM] American Society for Testing and Materials. 1998. Standard guide for risk-based corrective action. Philadelphia PA, USA: ASTM. p 104-98.

Ball WP, Roberts PV. 1991. Long-term sorption of halogenated organic chemicals by aquifer material. 2. Intraparticle diffusion. *Environ Sci Technol* 25:1223-1236.

Banerjee S. 1984. Solubility of organic mixtures in water. *Environ Sci Technol* 18:587-591.

Broholm K, Feenstra S. 1995. Laboratory measurements of the aqueous solubility of mixtures of chlorinated solvents. *Environ Toxicol Chem* 14:9-15.

Brusseau ML, Jessup RE, Rao SC. 1991. Nonequilibrium sorption of organic chemicals: Elucidation of rate-limiting processes. *Environ Sci Technol* 25:134-142.

Carroll KM, Harkness MR, Bracco AA, Balcarcel RR. 1994. Application of a permeant/polymer diffusional model to the desorption of polychlorinated biphenyls from Hudson River sediments. *Environ Sci Technol* 28:253-258.

Charbeneau RJ. 2000. Groundwater hydraulics and pollutant transport. Upper Saddle River NJ, USA: Prentice Hall. 593 p.

Cline PV, Delfino JJ, Rao PSC. 1991. Partitioning of aromatic constituents into water from gasoline and other complex solvent mixtures. *Environ Sci Technol* 25:914.

Connor JA, Bowers RL, Paquette SM, Newell CJ. 1997. Soil attenuation model (SAM) for derivation of risk-based soil remediation standards. In: Proceedings of the Petroleum Hydrocarbon and Organic Chemicals in Ground Water Conference; 14-17 Nov 1997; Houston, TX, USA. Westerville OH, USA: Ground Water Publishing Company. p 380-395.

Farrell J, Reinhard M. 1994. Desorption of halogenated organics from model solids, sediments, and soil under unsaturated conditions. 2. Kinetics. *Environ Sci Technol* 28:63-72.

Feenstra S, Mackay DM, Cherry JA. 1991. A method for assessing residual NAPL based on organic chemical concentrations in soil samples. *Ground Water Monit Rev* 11:128-136.

Garg S. 1998. Dissolution and desorption characteristics of aromatic compounds from media contaminated with multicomponent petroleum hydrocarbons [DPhil Dissertation]. Houston TX, USA: University of Houston, Department of Civil and Environmental Engineering. 170 p.

Grathwohl P, Reinhard M. 1993. Desorption of trichloroethylene in aquifer material: Rate limitation at the grain scale. *Environ Sci Technol* 27:2360-2366.

Johnson PC, Abranovic D, Charbeneau RJ, Hemstreet T. 1998. Graphical approach for determining site-specific dilution-attenuation factors (DAFs). Washington DC, USA: Health and Environmental Sciences Department, American Petroleum Institute. API Publication Nr 4659.

Kan AT, Fu G, Tomson MB. 1994. Adsorption/desorption hysteresis in organic pollutant and soil/sediment interaction. *Environ Sci Technol* 28:859-867.

Kung KJS. 1990. Preferential flow in a sandy vadose zone: 1. Field observations. *Geoderma* 46:51-58.

Lane WF, Loehr RC. 1992. Estimating the equilibrium aqueous concentrations of polynuclear aromatic hydrocarbons in complex mixtures. *Environ Sci Technol* 26:983.

Olsen RL, Davis A. 1990. Predicting the fate and transport of organic compounds in groundwater: Part 1. *Haz Mat Control* 3:38-64.

Pankow JF, Cherry JA. 1996. Dense chlorinated solvents and other DNAPLs in groundwater. Waterloo ON, Canada: Waterloo Press. 522 p.

Parker BL, Gillham RW, Cherry JA. 1994. Diffusive disappearance of dense, immiscible phase organic liquids in fractured geologic media. *Ground Water* 32:805-820.

Peterson Jr HT. 1983. Terrestrial and aquatic food chain pathways. In: Till JE, Meyer HR, editors. Radiological assessment: A textbook on environmental dose analysis. Washington DC, USA: U.S. Nuclear Regulatory Commission, NUREG/CR -3332.

Pignatello JP, Xing B. 1995. Mechanisms of slow sorption of organic chemicals to natural particles. *Environ Sci Technol* 30:1-11.

Rutherford PM, Gray MR, Dudas MJ. 1997. Desorption of 14C-naphthalene from bioremediated and non-bioremediated soils contaminated with creosote compounds. *Environ Sci Technol* 31:2515-2519.

Shen HW, Julien P. 1993. Erosion and sediment transport. In: Maidment DR, editor. Handbook of hydrology. New York NY, USA: McGraw-Hill. p 1-61.

[USEPA] U.S. Environmental Protection Agency. 1983. Final report of the nationwide urban runoff program. Washington DC, USA: USEPA, Water Planning Division. NTIS:PB84-185552.

[USEPA] U.S. Environmental Protection Agency. 1994. Technical background document for soil screening guidance (Review Draft). EPA-540-R-94-106.

[USEPA] U.S. Environmental Protection Agency. 1996. Soil screening guidance: Technical background document. Washington DC, USA: USEPA, Office of Solid Waste and Emergency Response. EPA-540-R-95-128, NTIS: PB96-963502.

Williamson DG, Loehr RC, Kimura Y. 1998. Release of chemicals from contaminated soils. *J Soil Contam* 7:543-558.

Wu SC, Gschwend PM. 1986. Sorption kinetics of hydrophobic organic compounds to natural sediments and soils. *Environ Sci Technol* 20:717-725.

Appendix to Chapter 6

Continuity Equation for Contaminated Soil

$$A\,L\left(B_{\mathrm{w}}\,\frac{\mathrm{d}C_{\mathrm{w}}}{\mathrm{d}t}+\rho_{\mathrm{b}}\,\frac{\mathrm{d}q_2}{\mathrm{d}t}\right)=-\left(u\,A\,C_{\mathrm{w}}+\frac{D_{\mathrm{e}}\,A\,K_{\mathrm{H}}\,C_{\mathrm{w}}}{L_{\mathrm{v}}}+\lambda_{\mathrm{D}}\,A\,L\,C_{\mathrm{w}}\right) \tag{6-28}$$

where $B_{\mathrm{w}}=\varphi\left(\Sigma_{\mathrm{w}}+S_{\mathrm{a}}\,K_{\mathrm{H}}+S_{\mathrm{o}}\,K_{\mathrm{o}}\right)+\rho_{\mathrm{b}}\,F\,K_{\mathrm{d}}$. Equation 6-28 may be simplified by writing

$$B_{\mathrm{w}}\,\frac{\mathrm{d}C_{\mathrm{w}}}{\mathrm{d}t}+\rho_{\mathrm{b}}\,\frac{\mathrm{d}q_2}{\mathrm{d}t}=-\Lambda\,C_{\mathrm{w}} \tag{6-29}$$

where $\Lambda=\frac{u}{L}+\frac{\mathrm{D}_{\mathrm{e}}\,K_{\mathrm{H}}}{L\,L_{\mathrm{v}}}+\lambda_{\mathrm{D}}$. The slow release of mass satisfies the kinetic equation

$$\frac{\mathrm{d}q_2}{\mathrm{d}t}=k_2\left[K_{\mathrm{d}}\,(1-F)\,C_{\mathrm{w}}-q_2\right] \tag{6-30}$$

Substitution of Equation 6-30 into Equation 6-29 gives

$$B_{\mathrm{w}}\,\frac{\mathrm{d}C_{\mathrm{w}}}{\mathrm{d}t}=\rho_{\mathrm{b}}\,k_2\,q_2-\left[\Lambda+\rho_{\mathrm{b}}\,k_2\,K_{\mathrm{d}}\,(1-F)\right]C_{\mathrm{w}} \tag{6-31}$$

Equations 6-30 and 6-31 provide 2 coupled ODEs to find the porewater concentration and the sequestered soil concentration. To find the initial conditions, assume that the initial total concentration is C_{T}. Then the initial water concentration is

$$C_{\mathrm{w}}(0)=\frac{C_{\mathrm{T}}}{B_{\mathrm{w}}+\rho_{\mathrm{b}}\,K_{\mathrm{d}}\,(1-F)} \tag{6-32}$$

and the initial sequestered soil concentration is

$$q_2(0) = K_d\ (1-F)\, C_w(0) \tag{6-33}$$

To simplify notation, write Equations 6-31 and 6-30 as

$$\frac{dC}{dt} = \alpha\, q - \beta\, C \tag{6-34}$$

$$\frac{dq}{dt} = \xi\, C - \eta\, q$$

The solution to Equation 6-34 is

$$C(t) = \frac{1}{2\,\psi}\left\{2\,\psi\, C_o + \left[(\psi - \beta + \eta) C_o + 2\,\alpha\, q_o\right]\left(e^{\psi t} - 1\right)\right\} e^{-0.5(\beta+\eta+\psi)t} \tag{6-35}$$

where $\psi = \sqrt{(\beta - \eta)^2 + 4\,\alpha\,\xi}$.

The corresponding solution for the sequestered soil concentration is

$$q(t) = \frac{1}{2\,\psi}\left\{\left[2\,\xi\, C_o + (\beta - \eta + \psi)\, q_o\right] e^{-0.5(\beta+\eta-\psi)t} - \left[2\,\xi\, C_o + (\beta - \eta - \psi)\, q_o\right] e^{-0.5(\beta+\eta+\psi)t}\right\} \tag{6-36}$$

Calculations for curve A–D–E in Figure 6-1 are developed using Equations 6-35 and 6-36.

CHAPTER 7

Contaminant–Soil Interactions

Mason B. Tomson, Herbert E. Allen, Carol W. English, Warren J. Lyman, Joseph J. Pignatello

Introduction

The manner in which contaminants interact with soil is central to understanding their fate, transport, and bioavailability. Most contaminants of environmental concern interact with the organic or the inorganic portions of soil, a process called "sorption." The term "sorption" may be defined as the transfer of a chemical from one phase, such as water, to another phase, such as soil organic matter. It implies a general process of association with a stationary phase, where the precise physicochemical mechanism of interaction (e.g., adsorption, absorption, ion exchange, precipitation) either is not specified or is not known. The stationary phase can be either mineral, organic matter, or some nonaqueous-phase liquid (NAPL). Bioavailability and toxicity are generally associated with aqueous dissolved species, although there have been reports of direct uptake from the solid phase.

Studies show that the total concentration of contaminants in soil is not a good predictor of long-term bioavailability or toxicity. Chemical and biological test results provide valuable information on existing conditions in the soil, but only on a limited temporal scale. Chemical measurements usually determine chemical concentrations and availability at the time of sample procurement, while bioassays integrate chemical exposure over the duration of the toxicity test or the life span of the organism. These tests need to be coupled with models that can predict longer-term interactions between chemicals and soil components, providing methods for estimating the potential for chemical bioavailability to soil biota over extended periods of time (e.g., years).

Abiotic interactions of neutral organic compounds and inorganic ions with soils are described by sorption-desorption and precipitation-ion exchange models. Within the scope of this body of work, the goal of using these models is to help provide an estimate of the concentrations of these chemicals in the soil water that enable a risk assessment to be conducted to determine cleanup standards. As a better understanding of the mechanisms involved in the sorption or desorption process is achieved, the estimate of risk provided by current knowledge, models, etc., will come closer to the actual risk.

Contaminated Soils: From Soil–Chemical Interaction to Ecosystem Management. Roman P. Lanno, editor.
 ISBN 1-880611-31-7

Until recently, models to predict the sorption and desorption of organic and inorganic compounds involved standard kinetic and equilibrium solubility and transport relationships. These models adequately predict the interactions of the bulk of the contamination, often greater than 80%. However, it is the last 15% to 20% of the residual contamination that will have the greatest impact on risk assessment and therefore on the development of cleanup levels required. While this phenomenon is known to exist, there is neither a mechanistic rate nor an equilibrium model currently available for use in predicting the long-term release of this fraction into the soil water. This is due to the limited understanding and inability to identify the mechanisms that influence the release of this fraction under the variety of conditions present in contaminated soils.

While there is currently no model that can predict the rate or extent of release of this residual fraction, the knowledge that it exists has greatly expanded our understanding and evaluation of desorption effects. Standard practice has been to use a linear sorption isotherm to determine desorption, thereby overestimating the concentrations that could be released to the soil water. Models are needed that can provide a mechanistic interpretation of these processes and predict the long-term concentrations of components in the soil water and in the soil particle or soil organic matter.

Sorption and desorption have been modeled using a single linear isotherm, and numerous methods have been developed to estimate the numerical values of the linear partition coefficient constants (Lyman et al. 1990). Several researchers have reported a sorption–desorption hysteresis effect for both neutral organic and inorganic contaminants, wherein the desorption path is different from that of the sorption path, apparently regardless of equilibration time (Vaccari and Kaouris 1988; Fu et al. 1994; Kan et al. 1994, 1997, 1998; Yin et al. 1996; Huang and Weber 1997; Yin, Allen, Huang, Sanders 1997; Yin, Allen, Huang, Sparks, Sanders 1997; Chen et al. 1999). Hysteretic sorption and desorption has also been observed with nonsoil systems (Zawadzki et al. 1987; Hunter et al. 1995, 1996; Hunter 1996). In the past few years, it has been recognized that this hysteresis, or resistant residual, is probably not a simple complication of the linear model, but is "different" and is a consequence of some other interaction mechanism (Everett and Whitton 1952; Adamson 1990; Hunter et al. 1996; Linz and Nakles 1997; Chiou and Kile 1998; Graber and Borisover 1998; Schlebaum et al. 1998).

It is surprising that the existence of such a desorption-resistant residual phase, as a separate process, had not been recognized earlier. The failure to do so probably lies with both laboratory and field testing and their respective interpretations. In the laboratory, numerous researchers have reproduced similar linear sorption isotherms of organic chemicals to soil, and when these isotherms are normalized for organic C content of the soil, the partition coefficients are typically constant, within expected experimental error (Hamaker and Thompson 1972; Schwarzenbach et al. 1993). Therefore, there were few reasons to suspect any other phenomena, such as the existence of additional sorption mechanisms. Also, laboratory experiments were typically conducted by spiking a soil sample with a compound, measuring the amount that sorbed, and then extracting the solid to recover the contaminant for mass balance. First, these spiking experiments were generally done within a day or 2, and it is now recognized that this is too short of a time to simulate field exposures and "develop" the resistant fraction. Secondly, few desorption experiments were performed; instead, it was tacitly assumed that sorption and desorption were reversible, that is, opposite of each other. Finally, failure of pump-and-treat and field modeling was generally ascribed to various uncertainties associated with soil heterogeneities, variations in flow, and biodegradation, to mention a few. In hindsight, there were numerous pieces of evidence that might have been interpreted as implying the existence of additional sorption and desorption mechanisms.

The practical importance of understanding the mechanism of physicochemical sorption and desorption to and from soils cannot be overstated. Once the fundamental processes responsible for resistant desorption are understood, it should be possible to use results from 1 or 2 simple tests, or assays, to predict contaminant release and bioavailability. A research goal should be to elucidate the mechanisms responsible for resistant sorption and desorption in terms of physicochemical or molecular parameters of the soil and of the contaminants.

Soil Properties

The universe of soils

For the purposes of this chapter, soils include any unconsolidated surface material, substantially of natural mineral origin, other than aquatic sediments. The exclusion of aquatic sediments is only for focus in this chapter and is not meant to imply that the scientific information presented does not apply to sediments; to a large extent it does. We specifically mean to include wetland (hydric) soils even though they may spend a substantial portion of the year submerged under water.

Assessments of contaminated soils, and related assessments of contaminant uptake or release, biological effects, and risks, will commonly involve soils that are not virgin, undisturbed topsoils. What will often be involved is a mixture of topsoils, deeper soils (from borrow pits or mines), and anthropogenic materials that have been commingled in conjunction with some purposeful activity such as urban or industrial development, waste disposal and/or reuse, and agriculture. Specific examples of such "soils" would include

- native topsoils (undisturbed or disturbed),
- agricultural soils (with or without soil amendments; likely having pesticide residues),
- residential fill (i.e., fill or topsoil imported for use at a residential plot),
- urban fill (present in essentially all cities for developed land),
- industrial fill (used at most manufacturing and industrial sites),
- dredged aquatic sediments (which may have spent years in an upland disposal area or a near-shore confined disposal area, but then were selected for beneficial reuse as fill or cover material),
- mine tailings, and
- "manufactured" soils (i.e., blends of natural and anthropogenic materials specifically designed for use as topsoils or agricultural soils).

"Fill" can encompass a variety of materials from both virgin and secondary sites. The virgin sites would include "borrow" areas such as sand and gravel pits. The secondary sites might include any of the above listed soils except the native soils, as well as soils from sites undergoing demolition, abandonment, or redevelopment. In the latter case (i.e., secondary sites), there is a significant chance that waste materials and/or chemical contaminants have been purposefully or otherwise added to the soils. As would be expected, the fill material is commonly chosen on the basis of its cost, geotechnical properties (e.g., weight-bearing capacity), and/or drainage properties, not necessarily its cleanliness. Even when cleanliness is a concern, measures to protect against use of contaminated soils in fill operations may have been inadequate.

"Soil amendments," such as might be used for agricultural soils or manufactured soils, might include a variety of materials including manure, composted sewage sludge, composted or ground vegetation, peat moss, vermiculite, and chemicals intended to adjust the soil properties to enhance plant growth or stabilize contaminants (e.g., lime).

Soil constituents or phases

A native soil will consist of the following basic constituents or phases:

- inorganic minerals (in a wide variety of grain sizes),
- organic material (ranging from fresh plant and animal remains to highly weathered humic material),
- living plants and animals (including microbes),
- water (with dissolved minerals and organic C), and
- air.

However, the soils for which contaminant bioavailability assessments can be of interest may commonly include a variety of other materials and phases such as

- construction debris (wood, brick, cement, metal, and glass fragments),
- asphalt,
- coal,
- ash and soot,
- mine tailings,
- NAPLs, and
- tars.

Much of this foreign material may be present as very small particles, in addition to large fragments. Each of these anthropogenic materials may inherently contain, or act as sorbents for, chemical contaminants. Examples of NAPL materials commonly encountered are petroleum fuels (gasoline, diesel, heating oil, jet fuel), industrial oils and lubricants, transformer oils (including some with polychlorinated biphenyls [PCBPs]), and chlorinated solvents (e.g., trichloroethylene and perchloroethylene {Zawadzki et al. 1987). Highly weathered NAPL materials may exist as a thin film on the soil particles, evident only as a dark stain (with or without odor) on the soils. This NAPL material is generally considered to take up and release organic contaminants via the processes of dissolution, that is, a liquid–liquid partitioning (assuming water is present in the soil) rather than a solid–liquid partitioning.

It is important when conducting soil sorption and desorption studies for selected contaminants to have a good understanding of the different materials and phases present in the soil. Most of the characterization that is needed often can be accomplished by an experienced geologist during the excavation of test pits or soil borings at the site. Identification is made visually based on color and texture (and odor for some NAPLs). More definitive identifications may require specialized chemical or microscopic analyses of the soil. In unusual cases, electron microscopes (microprobes) can be used to chemically characterize (via elemental analysis) the different phases.

Important physicochemical properties of soils

There are a number of physicochemical properties of soils that are known to be important in understanding—and modeling—the nature and extent of chemical sorption and desorption to or from the soil. Table 7-1 provides a list of these parameters, segregating those dealing with the sorption of organic and inorganic chemicals. Each list is further subdivided into "primary importance" and "secondary importance" groups. These are the parameters that may need to be measured in soils from the site of interest in order to carry out general assessments of the extent of sorption and/or the desorption modeling described later in this chapter.

Table 7-1 Soil properties important for understanding contaminant–soil interactions

Parameter	Importance
Organic chemicals	
Primary importance	
Organic carbon content (f_{oc})	Controls the extent of sorption (especially if $f_{oc} \geq 0.1\%$ by weight)
Clay content or surface area	May control the extent of sorption (especially if $f_{oc} <0.1\%$)
Temperature	Affects rate and extent of sorption and desorption
Moisture content	Controls mechanism and extent of vapor sorption
Secondary importance	
Characteristics of organic C	Characteristics include aromatic or aliphatic character, polarity, content of natural carbon black or soot (from forest fires), will affect extent of sorption
Aqueous phase cosolutes, dissolved organic matter or nonsettling particles	Presence may diminish extent of sorption to soil
Presence of biota	May affect the extent of sorption or apparent desorption and potential reactivity
Inorganic chemicals	
Primary importance	
Soil pH	Controls many speciation and precipitation reactions
Redox potential	Controls oxidation-reduction reactions
Cation exchange capacity	Controls the extent of sorption by ion exchange
Temperature	Affects rate and extent of sorption and desorption
Presence of key soil phases: organic matter, iron oxides, and manganese oxides	Controls nature and extent of sorption
Solution chemistry (e.g., presence of complexing ligands such as carbonate, sulfide, and phosphate)	Controls extent of complexation and precipitation
Secondary importance	
Presence of biota	May affect the extent of sorption

Soil heterogeneity

Given an understanding of the various origins and phases of "soils" as defined above, it should not be surprising that the concentrations of the various phases, and the concentrations of the chemical contaminants of interest, will be highly variable over both horizontal and vertical dimensions. Also, there will be significant variability in the particle size distribution, and associated soil porosity and moisture retention characteristics. Extreme variability may even be seen in samples as small as 1 gram taken adjacent to one another at a given site. Because many chemical analyses use only about 1-gram aliquots for analysis, it is clear that soil characterization must either include carefully composited or homogenized samples or a large number of samples. Geostatistics can provide guidance on selection of the appropriate sample size needed to obtain specified confidence levels on the means. The use of homogenization (e.g., grinding, sieving) that involves a change in the natural grain size of the soil particles may lead to altered results of any subsequent sorption or bioavailability tests, and thus should be undertaken with due consideration.

Sorption and Desorption of Organic Contaminants

Introductory remarks

The main emphasis in this section will be on neutral organic compounds because they are the most common; however, many of the same principles apply as well to ionic and ionizable compounds. "Sorption" is defined in the broad sense as the association of contaminant molecules with nonfluid, or stationary, phases. "Fluid phase" refers to the soil solution or the soil atmosphere. "Stationary phase" refers to both solids and NAPLs adhering to or filling pores of the solids. Thus, sorption includes adsorption on surfaces, partitioning into natural organic matter, condensation (liquefaction or crystallization) in the pores of particles, and partitioning into NAPLs associated with the solid phase.

The principal factor governing contaminant bioavailability at any instant is its concentration—more accurately, its chemical potential—in the fluid phase in contact with the receptor membrane. The fluid-phase concentration is governed by thermodynamic and kinetic parameters specific to the contaminant–soil–fluid–receptor system and may or may not vary during exposure. In general, the rate of change in the fluid-phase contaminant concentration is a function of the forward and reverse rates of transport across the receptor–membrane interface (uptake and depuration) and the forward and reverse rates of transport across the particle–fluid interface (sorption and desorption). Assimilation by the receptor may be rate limited by receptor uptake from the fluid phase or by desorption from particles, and the rate-limiting step may change during the exposure. At present, the rates of sorption and desorption from soil particles are poorly predictable on the basis of easily obtainable information, such as total contaminant concentration and typical soil properties.

"Available" and "sequestered" states

As is often the case, the contaminant is not rapidly equilibrated between the stationary and fluid phases on time scales appropriate to the exposure period (Pignatello and Xing 1996). Hence, with respect to a given receptor, there exist "available" and "sequestered" states of sorbed molecules (Alexander 1995). No sharp distinction exists between the 2—rather there appears to be a continuum of sorbed states, from instantaneously in equilibrium to very slowly in equilibrium with the fluid phase. Some fraction may even be considered to be completely immobilized in the stationary phase. An understanding of the factors that govern this complex distribution has evolved into a major branch of environmental science. It must be clearly understood that the sequestered fraction is highly context specific—that is, it depends on the contaminant, soil, receptor, mode of uptake, and duration of uptake. What is considered sequestered for one organism may be wholly or partially available for another. It is this specificity that results in difficulty in the development of predictive models.

In cases where the fluid-phase concentration is not altered significantly by uptake, such as when uptake is very slow, or when the receptor moves rapidly through the contaminated medium, bioavailability may be controlled mainly by the status-quo fluid-phase concentration. In all other cases, kinetics (mass transfer laws) describing the flux of contaminant through the particle, across the particle–bulk fluid interface, and across the fluid–membrane interface will be important. When the organism perturbs the fluid-phase concentration by uptake during the exposure, the concentration gradient between the particle and fluid and between the receptor and fluid are both altered. Because sorption and desorption rates from particles are governed primarily by molecular diffusion, and because diffusion depends on the concentration gradient, such biologically induced changes in concentration gradient may influence contaminant flux out of the particle. Therefore, an accurate bioavailability model will require linkage of biological uptake or depuration kinetics with sorption or desorption kinetics. Such coupled models are practically absent from the literature.

Mechanisms of sorption

The molecular scale mechanisms of sorption in ordinary natural soils are still incompletely understood and controversial despite decades of investigation. Characterization of sorption of typical contaminants to anthropogenic materials lags even further behind that of natural soils. It is important to note that intermolecular interactions of most organic compounds with soil components involve only physical forces (dispersion and dipolar interactions) and not chemical forces (covalent bonding). These physical interactions are generally weak (a few kJ/mole) and practically instantaneous in the absence of steric constraints.

Sorption to organic materials

The dominant component of natural soils influencing the sorption of neutral organic compounds is the soil organic matter (SOM) fraction. SOM consists primarily of humic substances originating from the decomposition of plants and microorganisms (Hayes et al. 1989). The modern paradigm of SOM is a random 3-dimensional network of humic macromolecules. Sorption to SOM occurs by absorption into its matrix, analogous to the absorption of small molecules by synthetic polymers. A major driving force for absorption is hydrophobic expulsion from water—the so-called "hydrophobic effect" (Schwarzenbach et al. 1993). However, direct functional group interactions with SOM macromolecules obviously take place in some cases (e.g., hydrogen bonding, charge–transfer interactions) and may contribute to the overall driving force for absorption.

Recent studies (Pignatello 1998; Xing and Pignatello 1998) indicate that humic substances are composed of rubberylike phases that have an expanded, flexible, highly solvated structure, and glassylike phases that have a condensed, rigid, and less solvated structure. More than likely there is a continuum between these 2 extremes. Sorption has been interpreted in terms of a polymer model based on this rubbery-glassy concept. Interaction with the rubbery phase occurs by solid-phase dissolution, analogous to liquid-phase dissolution. Interaction with the glassy phase occurs by a dual mode mechanism, in which both solid-phase dissolution and "hole-filling" processes occur. The holes are postulated to be closed nm-size pores in which the guest molecules undergo an adsorption-like interaction with the pore walls.

Organocations, typically quaternary ammonium ions, sorb by ion exchange at charged sites both on mineral surfaces and within the SOM phase. Sorption may be facilitated by the hydrophobic effect of the apolar (nonpolar or weakly polar) parts of the molecule. Organoanions, typically carboxylates and phenolates, sorb primarily to SOM. Sorption of an organoanion is considerably weaker than its neutral (protonated) form due to greater solvation in water and charge repulsion with the negatively charged groups on SOM.

Soils may contain other forms of C not usually classified as SOM. These include ancient carbonaceous materials like coal, kerogen, and shale, as well as "soot" carbon, which refers to incompletely combusted organic material from natural or anthropogenic sources. Such materials have a high affinity for organic compounds and may be widely distributed in the environment (McGroddy et al. 1996; Gustafsson et al. 1997). Sorption to these carbonaceous materials may occur by adsorption or by condensation in fixed micropores, whose surfaces are hydrophobic.

Sorption to inorganic materials

Unlike SOM, minerals are impenetrable by organic molecules or ions. Sorption therefore occurs at the solid–fluid interface. At high relative humidity, the sorption of apolar compounds to mineral oxide and external clay surfaces is small on a mass basis compared to sorption to SOM or carbonaceous materials (Mader et al. 1997). This is because mineral surfaces, being polar, are coated with strongly bound water molecules. Thus, sorption of apolar compounds to polar minerals is important only at low moisture levels, or, under wet conditions, only in very low C materials such as aquifer sediments.

Sorption to mineral surfaces from solution is proportional to surface area and is solvent motivated—that is, it is driven mainly by the hydrophobic effect (Schwarzenbach et al. 1993). The nature of the association is best described as a concentration enhancement in near-surface (vicinal) water layers, rather than a direct interaction with the surface atoms. Sorption may be greater for compounds that are capable of hydrogen bonding with surface oxygens or coordinated water molecules and inner sphere coordination with surface metal ions (e.g., carboxylates). In addition, sorption of both apolar and polar compounds may be greater in clay interlayers than on exposed surfaces. At low relative humidity (approximately 50%), 2 additional mechanisms become increasingly important: direct interaction with the bare mineral surface and condensation to a liquidlike or crystallinelike state in small pores not completely filled with water.

Sorption to nonaqueous-phase liquids

The partitioning of contaminants between fluid phases and tars, petroleum oils, and other NAPLs may be assumed to occur by a simple liquid-phase dissolution process, analogous to that which occurs between water and organic solvents like hexane or octanol. Partitioning in such systems is therefore governed by Raoult's Law (water–NAPL) or Henry's Law (vapor–NAPL); that is, the fluid-phase concentration is proportional to the mole fraction of contaminant in the NAPL times the solubility (or vapor pressure) of a pure reference state (Schwarzenbach et al. 1993).

Molecular modeling

Recent attempts to understand sorption have employed molecular modeling, and no doubt this will become more common in the future. Lasaga (1990) has used theoretical molecular methods to characterize the energetics of the silica surface in contact with water and then to better understand the mechanisms of weathering. Similar ideas might be applied to sorption of metals. Shulten (1995) has used molecular modeling to study the interactions of pollutant molecules with hypothetical humic acid macromolecules in the presence and absence of silica minerals. Several commercial software products are available to do these calculations. At present, these molecular calculations are limited by the speed and size of computers, but these limitations are decreasing rapidly.

Thermodynamic relationships

A number of isotherms have been used to describe solid-solution sorption, including Freundlich and Langmuir isotherms and modifications thereof (Carter et al. 1995). Historically, the most commonly used isotherm is the Freundlich equation

$$q = K_d C^n \tag{7-1}$$

where q is the sorbed concentration [moles/kg], C the aqueous concentration (moles/L), K_d the sorption distribution coefficient [(moles/kg)(moles/L)$^{-n}$], and n an exponent. The Freundlich model is often simplified to the linear form ($n = 1$) on the assumption that SOM is the predominant sorbent and that sorption to SOM is dominated by solid-phase dissolution, which, in theory, should be linear. The K_d is commonly normalized to the fraction organic C ($K_{oc} = K_d/f_{oc}$). The thermodynamic relationships governing vapor–particle sorption are complex because they involve other processes including pore condensation, adsorption on nonhydrated surfaces, dissolution in liquid water films, and adsorption on water film surfaces. Hence, they will not be discussed here.

The linear isotherm model has been used to develop the proposed Equilibrium Partition Model that is under consideration by the U.S. Environmental Protection Agency (USEPA) for setting sediment-quality criteria for aquatic sediments (Di Toro et al. 1991). The assumption of this model is that bioavailability, and eventually biological effects, can be predicted from the equivalent porewater concentration. The porewater concentration is calculated from the total concentration present in the solids (after exhaustive extraction) using experimentally determined K_{oc} values or K_{oc} values estimated from established linear free-energy relationships (LFERs), such as those with octanol–water partition coefficients (K_{ow}) or solubility (Schwarzenbach et al. 1993). Although there is an extensive database of K_{oc} values and LFERs, most of these data are based on isotherms constructed over short equilibration times (<48 h). Therefore, their relevance to highly aged contaminated systems is questionable. In a number of cases, the apparent K_{oc} for historically contaminated soils and sediments has been as much as 2 orders of magnitude greater than laboratory-determined values with clean soils that are freshly spiked with the contaminant (Pignatello and Xing 1996). In addition, the Equilibrium Partition Model is inapplicable when nonequilibrium conditions prevail during exposure.

Dual-mode model

Sorption isotherms in soils are frequently observed to be nonlinear. Moreover, competitive effects occur between contaminants in multicomponent systems (e.g., chlorinated hydrocarbons, polycyclic aromatic hydrocarbons [PAHs]) and between contaminants and naturally occurring compounds (e.g., aromatic acids). Cosolute competitive effects have been shown to increase the bioavailability of the principal solute (White, Hunter, Nam et al. 1999).

The dual-mode model (Equation 7-2) places nonlinearity and competition in a mechanistic framework for SOM. The model relates total sorption to the sum of solid-phase dissolution, a linear term, and hole filling, described by a summation of Langmuir terms representing each unique type of site:

$$q = K_{\mathrm{D}}C + \sum_{i} \frac{b_{\mathrm{H},i} Q_{\mathrm{H},i} C}{1 + b_{\mathrm{H},i} C} \qquad (7\text{-}2),$$

where K_D is the dissolution domain coefficient, and where $Q_{H,i}$ and $b_{H,i}$ are the maximum sorbed capacity and affinity constant, respectively, for the *i*th hole of the hole-filling domain. The value of K_D can be obtained from the competitive effect, which blocks the holes and thereby eliminates the Langmuir term (Pignatello 1998). Individual $Q_{H,i}$ and $b_{H,i}$ cannot be determined because we do not know the number of unique sites. However, Langmuir parameters for a hypothetical composite site may be obtained, once K_D is known, by fitting the experimental isotherm using 2-parameter nonlinear regression.

Hysteresis

A fundamental motivation for reexamining the question of "environmentally acceptable endpoints" is the observation that desorption is not simply the opposite of sorption, as has been previously thought. It is often observed that a fraction of the sorbed compound is not removable, as would be expected based upon sorption isotherms such as Equation 7-1. The fraction of sorbed compound that is not readily removed is observed to increase with aging times, from days to weeks and longer. When equilibrium desorption is vastly different from sorption, the process is termed "hysteresis." Hysteresis implies the existence of a time-independent process, that is, the desorption points do not approach the sorption isotherm curve even after long equilibration times, yet the original compound can be recovered quantitatively. A necessary thermodynamic requirement for sorption hysteresis is that there must be a rearrangement after sorption, or simply that desorption takes place from a different molecular environment than sorption.

To explain hysteresis, Kan et al. (1997) have proposed that sorption includes an irreversible fraction. The term "irreversible" does not imply permanent immobilization, but only that molecules come off a site by a different microscopic pathway than they go on. They interpret the sorption and desorption as 2-step processes. They propose that the bulk of sorption takes place via normal hydrophobic linear mechanisms (Equation 7-1, with $n = 1$). Then, in a second step, the SOM rearranges and entraps a portion of the sorbed compound. This rearrangement may be triggered by the very process of hydrophobic sorption, as occurs with various enzymes. Such rearrangement is supported by molecular modeling studies of the interaction of pollutant molecules with hypothetical humic acid macromolecules (Schulten 1995) and by direct membrane studies with atrazine (Devitt and Wiesner 1998). Desorption takes place from both the first and the second fractions. Desorption from the first fraction is described by linear partition coefficients (Equation 7-3). For the second fraction, it has been found empirically that for each soil–contaminant combination there is a fixed maximum capacity, q_{max}^{irr} (mg/kg); the value of q_{max}^{irr} is generally found to be about 1 to 10 mg/kg. The maximum soil capacity, q_{max}^{irr}, can be filled in one or multiple steps; the fraction of q_{max}^{irr} filled is termed "f." Generally, about one-half to one-third of the total amount of contaminant sorbed is entrapped in this second fraction until it is filled (i.e., $f = 1$). In models, the value of f is generally set to 1. Desorption from this second fraction is characterized by a single organic C-normalized partition constant for most hydrophobic organic compounds studied to date ($K_{oc}^{irr} = 10^{5.5 \pm 0.5}$ L/kg), which might be expected if the sorption were related to an entrapped material, but much more work is needed to test the general applicability of these ideas. These ideas can be represented with the following isotherm (Fu et al. 1994; Kan et al. 1994, 1998):

$$q(\text{mg/kg}) = K_{oc}\,\text{OC} \times C(\text{mg/L}) + \frac{K_{oc}^{irr} q_{max}^{irr} f \times \text{OC} \times C}{q_{max}^{irr} f + K_{oc}^{irr}\text{OC} \times C} \qquad (7\text{-}3).$$

Only preliminary work has been done on the desorption kinetics from the irreversible compartment.

Sorption and desorption kinetics in relation to bioavailability

Desorption of the fraction of the contaminant that is not instantaneously available to the receptor is subject to mass-transfer rate laws. The factors responsible for retardation of desorption are not completely understood. Because only physical forces are involved in the interaction of most compounds with soil materials, the underlying cause of retardation is not breakage of bonds, but mass transport (molecular diffusion) within the particle. Diffusion is the random movement of molecules under the influence of the gradient in its chemical potential. Stated another way, a molecule migrates from "site" to "site" in the interior of the particle like a drunken sailor until it reaches the surface–bulk-fluid interface where it undergoes a final desorptive step to escape into bulk solution. Thus, diffusion is exquisitely sensitive to the nature and geometry of the diffusive medium. Obviously, natural particles are highly heterogeneous in composition and geometry.

Diffusion in soil particles may be retarded by the following mechanisms:

- chromatographic retention in fixed pores,
- hindered diffusion within SOM owing to its viscous nature, and
- temporary holdup at specific sorption sites within the particle.

Soil particles are porous by nature. Diffusion through the fixed intraparticle pore system is an important retardation process. The void spaces within individual grains (e.g., clays) and between grains that make up particle aggregates include macropores (>50 nm wide), mesopores (2 to 50 nm), and micropores (<2 nm). Pores smaller than about 1 nm have been referred to as "ultramicropores," "submicropores," or "nanopores." Pore size distributions of soil materials, especially in the micropore and nanopore region, are not readily obtainable at present, although progress is being made through innovative techniques using gas adsorption coupled with molecular models. It should be noted that even the smallest cells are larger than approximately 200 nm in diameter, excluding them from most of intraparticle pore system where the majority of the surface area exists.

From studies in mesoporous and microporous fixed-pore reference materials (e.g., zeolites, silica gel, and carbons) (Karger and Ruthven 1992), we know that diffusion may occur in the aqueous fluids filling the pore or along pore surfaces. Diffusion is retarded by 1) tortuosity, which takes into account deviation of pore paths from linearity and dead end pores; 2) sorption to the pore walls, which is most significant when the pore wall is hydrophobic; and 3) steric hindrance by pore walls, which is important only in pores smaller than a few times the molecular diameter (practically speaking, micropores and smaller mesopores).

Diffusion through SOM is another important retardation mechanism. SOM may be regarded as a highly viscous liquid phase. Diffusion of small gas and organic molecules through rubbery and glassy polymers has been well studied (Rogers 1965; Berens 1989). Diffusion through glassy polymers can be many orders of magnitude slower than through rubbery polymers, owing to their more condensed and rigid nature. The existence of nanopores within the glassy phase has important implications for the physical and biological availability of contaminants. Because such pores are of molecular dimensions, it is possible that the entrance and exit of some pores may be sterically hindered but not completely impenetrable by a given molecule. Steric constraints may assume greater importance as molecular size increases. The humic rearrangement hypothesis discussed in relation to hysteresis phenomena represents the extreme case of steric hindrance because, if it were true, a major reorganization of the humic backbone would be required to permit release. It should be appreciated that release from a sterically hindered site is not necessarily directly into solution but instead may be into the surrounding organic matrix where it is then subject to encountering another sterically hindered site.

Rate models

Mechanistic, semiempirical, and stochastic models all have been used to describe sorption and desorption rates. The literature is limited primarily to experiments on samples that have been newly spiked rather than on historically contaminated samples. Common assumptions are that the particles are uniform and spherical, the diffusion coefficient is single-valued, and the contaminant is uniformly distributed at equilibrium. Diffusion models for glassy organic polymers are available (Horas and Nieto 1994). These models require parameters for both the dissolution and hole-filling domains and become exceedingly complex when the holes are energetically inhomogeneous. A combined pore diffusion–organic matter diffusion model has been derived (Yiacoumi and Rao 1996). Additional complexities in the application of classical diffusion models include 1) the possibility that the diffusion coefficient is concentration dependent; 2) the wide particle size distribution ordinarily found in soils; 3) the inhomogeneities of contaminant concentration within the particle; 4) the sorption nonlinearity; 5) the likelihood that local sorption partition coefficients are distributed over a range of values; and 6) the likelihood of nonequilibrium condition at the start of the experiment. It is even unclear what the relevant diffusion length scale is in natural particles; in some cases, it appears to be the nominal particle radius, while in others it is considerably smaller, perhaps on the order of mm or less (Pignatello and Xing 1996).

The most widely used models in the soils literature are the multisite models that assume discrete sites governed by first-order sorption and desorption rate laws. The popular 2-site model includes an instantaneously reversible site and a kinetic site. In recognition of soil heterogeneity, some researchers have employed stochastic models that assume an array of parameters that are distributed according to probability density function (Connaughton et al. 1993; Pedit and Miller 1995). For example, Pedit and Miller (1995) modeled the months-long uptake of the herbicide diuron with a model that treated K_d and desorption rate constant as continuously distributed variables ranging over certain limits. While stochastic models are successful for a given contaminant–soil system, it remains to be seen whether the parameters can translate to other systems.

Correlations of rate with molecular structure, soil properties, and conditions

Two-site model desorption rate constants have been found to depend inversely on the K_d, K_{ow}, or molecular connectivity index, a measure of topological size and degree of chain branching in the molecule (Piatt and Brusseau 1998). Thus, for a given class of compounds, large molecules desorb more slowly than small and branch more slowly than linear. Polar compounds appear to desorb more slowly than apolar compounds having the same K_{ow}, as expected due to the additional drag of functional group interactions (e.g., hydrogen bonding) occurring at each molecular jump through the matrix. It should be noted that these relationships have been developed from sorption experiments occurring over only minutes to hours; although the relative relationships are likely to hold over longer times, it is still not possible to predict desorption rates in aged systems with any certainty.

The rate of desorption depends on the quality of SOM. For example, White, Hunter, Pignatello, Alexander 1999 showed that aged phenanthrene desorbed faster from a peat soil than from the humin fraction of the peat soil, that is, the organic matter left after base extraction of humic acids. This is consistent with the dual-mode model, since the humin fraction is believed to be enriched in glassy SOM. Weber and Huang (1996) and Xing and Pignatello (1996) have observed decreasing isotherm linearity with time and interpreted this result to mean that sorption occurs faster to the "amorphous" (rubbery) SOM than to the "condensed" (glassy) SOM. The rate of desorption is quite sensitive to temperature. For example, highly resistant fractions of 1,2-dibromoethane in an aged field sample gave a desorption activation energy of approximately 60 kJ/mole (Steinberg et al. 1987). Desorption of aged phenanthrene in a sandy loam was accelerated in the presence of a competitive cosolute (pyrene) (White, Hunter, Nam et al. 1999; White, Hunter, Pignatello, Alexander 1999).

Speciation of organics in aqueous solution in relation to bioavailability

Most nonionic organic compounds exist as a single chemical entity; however, in soil solution, the compound may be freely dissolved or be associated with nonsettling particles (NSPs). Ubiquitous in soil systems, NSPs are colloidal size particles made up of humic substances often associated with inorganic materials. Because it is the humic fraction that usually dominates sorption, NSPs are often referred to simply as "dissolved organic matter" (DOM). Sorption of contaminants to NSPs presumably takes place by the same mechanisms as the bulk soil. The fraction of total solution-phase concentration associated with NSPs depends on the hydrophobicity of the compound and becomes important as its K_{ow} exceeds approximately 10^4. Sorption to NSPs may affect the apparent sorption distribution coefficient by giving an artificially high aqueous-phase concentration. Corrections for NSP sorption can be made through, for example, solubility enhancement experiments, which give some measure of the affinity.

The bioavailability of molecules associated with NSPs is unknown. Solutes partitioned into surfactant micelles are less available than freely dissolved, but appear to be somewhat more available than crystalline or sorbed-phase molecules. Due to their small size, rates to or from NSPs might be expected to be faster than the bulk soil particles. However, recent studies suggest that some molecules associated with DOM may become sequestered. For example, desorption of pentachlorobenzene from "dissolved" humic acid gave a fast and a slowly desorbing fraction (Schlebaum et al. 1998). This issue warrants additional study.

Desorption in biologically influenced microenvironments

The presence of an organism may accelerate the flux of contaminant from soil particles in a variety of ways: 1) by increasing the concentration gradient across the particle–fluid interface as a result of depletion of contaminant in the fluid, thereby accelerating particle-to-fluid flux; 2) through organism-caused changes in the composition of the fluid phase that affect the chemical potential of the solute and, as a result, its sorption coefficient; and 3) through organism-caused alteration of soil properties via changes in fluid-phase composition.

Single-cell organisms may attach to soil surfaces by molecular forces or via extracellular exudates. It is inconclusive whether or not attached cells are capable of abstracting sorbed organic molecules directly from the surface. However, bioavailability is probably not enhanced by this mechanism for 3 reasons: 1) most sorbed organic molecules exist within the interstices of SOM and, therefore, are inaccessible to cells at any given instant; 2) molecules adsorbed to surfaces are similarly unavailable because the preponderance of soil surface area is located in mesopores and micropores, which are inaccessible to even the smallest cells; and 3) because the external surface and the solution are in rapid equilibrium at the microscale, surface abstraction, even if possible, can only enhance bioavailability if the rate of abstraction from the surface is significantly greater than the rate of uptake from solution, which seems doubtful.

Multicellular organisms probably are less capable of abstracting organic contaminants directly from the sorbed state than single-cell organisms, but they may affect desorption in other ways. Little work has been done in this area, however. Dermal contact may involve transfer of skin or hair oils to the particle surface, which may facilitate uptake of hydrophobic contaminants. Ingestion of soil particles exposes them suddenly to an aqueous environment that may include acids, biosurfactants, and enzymes. Weston and Mayer (1998) have shown enhanced bioavailability of PAHs in stomach fluid experiments. Acids and enzymes may affect soil structure, while surfactants can affect both soil structure and, through micelle formation, the apparent aqueous solubility of the contaminant. The solution pH affects soil mineralogy, as well as the structure of SOM. Acidification of a soil to below pH approximately 2 was shown to release sequestered fractions of halogenated hydrocarbons (Pignatello 1990), possibly by dissolving Fe or Al oxide cements holding particle aggregates or organic macromolecules together. Soils ingested by birds may be subject to pulverization in the gizzard, thus facilitating desorption by reducing diffusion path lengths. Grinding in a ball mill was shown to release sequestered fractions of halogenated compounds in soil (Steinberg et al. 1987; Pignatello 1990) and aquifer materials (Ball and Roberts 1991). The effect of biological fluids in the lung on desorption of inhaled particles is an open question.

Plants may facilitate uptake by altering the rhizo-microenvironment. Plant exudates could potentially facilitate desorption by a surfactant effect or by a competitive sorption effect. For example, it has been shown that naturally occurring aromatic acids, which are released from living and decomposing plants, facilitate desorption of chlorinated aromatic hydrocarbons, chlorinated phenols, and PAHs by a competitive displacement mechanism (Xing and Pignatello 1998).

Potential release mechanisms for sequestered fractions

An important question for setting soil-quality criteria and establishing environmentally safe remediation endpoints is the potential for release of sequestered fractions through sudden change in environmental conditions at some unanticipated time in the future. The desorption of sequestered fractions is accelerated by abrasion, application of heat, or introduction of solvents and oils. It is possible that competing cosolutes (either natural or anthropogenic) introduced to the soil could displace sequestered fractions, but this has not been clearly established. Natural microbial decomposition of SOM could, in principle, lead to the release of fractions sequestered therein, but it is hard to envision such a process occurring suddenly and without being replaced by newer SOM. However, the younger SOM is likely to have a more rubbery structure.

Soil heterogeneity considerations

Contaminated sites are spatially inhomogeneous with respect to many properties, including contaminant concentration, mineralogy, fraction of organic C, and moisture. Chemical concentrations may vary up to a 1000-fold over a site (Keith 1996). Thermodynamic coefficients and rate parameters may vary correspondingly with these disperse properties and conditions. For example, sorption coefficients and diffusion coefficients are concentration dependent. Diffusion coefficients are highly dependent on soil geometry and other properties. It is quite possible that the "bioavailability factor" of a given compound varies spatially over a site from near zero to near unity. Research is needed to establish the site-scale heterogeneity of thermodynamic and kinetic parameters of desorption if we are to have any real confidence in predicting bioavailability.

Prescription for bioavailability protocols

Sorption and desorption thermodynamic and kinetic models are presently inadequate for predicting bioavailable fractions using only the total contaminant concentration and soil properties typically available. As discussed above, the literature K_{oc} values generally overestimate risk because short-term sorption studies underestimate K_{oc} values of aged contaminants. Therefore, measurements are required on a site-specific basis in order to obtain the minimal thermodynamic and kinetic information necessary to make bioavailability predictions. What is more, the variability in these data within a site is expected to be considerable. It is felt that some simple measurements can be made to help assess bioavailability. These measurements can tell us the following:

- the prevalent porewater concentration, and
- the desorption rate profile for the contaminants that will give the maximum desorbable contaminant mass over a given period.

The prevalent porewater concentration

In cases where NAPLs are known to be present, it may be possible to estimate the prevalent porewater concentration by the Raoult's Law or Henry's Law relationships between the NAPL and water (or air) phases, as discussed above. The reader should be warned that the validity of this method of estimation in the field has not been rigorously tested. In many cases, the prevalent porewater concentration may be measured directly. For reasons mentioned above, the relationship between measured porewater concentration and total solids-based concentration is not likely to be highly predictable based on short-term K_{oc} values. The relationship between biological effect and porewater concentration is still controversial. Provided there is enough moisture in the soil, the pore water can be separated by centrifugation through a filter and the contaminant concentration determined by established methods. Potential pitfalls using this technique include losses by volatilization, adsorption to the apparatus, or sorption to NSPs in the pore water. Techniques for direct measurement of porewater concentration without separation have not been established.

A desorption rate profile

A desorption rate profile can be obtained by suspending the contaminated soil in water and stripping the contaminant from solution as it desorbs. Stripping can be accomplished in the case of volatile compounds by air sparging, or by in situ adsorption with a third-phase polymeric trap, such as Tenax. The amount of polymer must be at least 10-fold excess over the amount of organic C to ensure efficient removal. At all but very short times (tens of minutes), desorption into solution, and not removal from the aqueous phase, is rate limiting. This type of experiment will give a desorption profile over any given time period, and thus will yield the maximum desorbable fraction of contaminant at any point along the time curve. The information gained using this approach should be used cautiously because the soil is placed under conditions where the rate of contaminant desorption is maximized due to the steep concentration gradient created by its removal.

Heavy metals

Metal problems are frequently found at contaminated sites. Forstner (1995) reported that of the 952 National Priority List (Superfund) sites in the U.S. in 1986, 389, or 41%, reported metal problems. Of the sites reporting metals, 71% reported the presence of multiple metal contaminants. The most frequently reported metal contaminant was Pb, followed by Cr, As, Cd, Cu, Zn, Hg, and Ni. Other metals were reported at 10 or fewer of the sites. Of the 8 metals found at more than 10 sites each, all but C and As are predominantly found in the +II oxidation state. C can be present in 2 oxidation states, as Cr(+III) and as the chromate (+VI) oxyanion. As is most commonly present as the (+III) arsenite and (+V) arsenate oxyanions. This section will focus on those metals present in the +II oxidation state, with additional reference to interactions of soils with the arsenite, arsenate, chromate oxyanions, and Cr(+III).

Interaction of heavy metals with soils reduces the toxicity of the metals. Addition of metal salts to soils results in greater bioavailability of the metal than for metal that has been allowed to age in the soil, or for metal that is added together with components with which it can react, such as sewage sludge. Metals added to natural or artificial soils have been used to ascertain dose–response relationships. When the concentrations of chemicals in field samples are compared to these, the level of biological effect is overestimated. If metal concentrations are determined on the basis of aqua regia digests rather than on milder procedures, this effect is exacerbated. The importance of considering bioavailability in setting soil-quality standards has been discussed by Peijnenburg et al. (1997).

Chemical equilibrium

Equilibrium calculations can help estimate the limits of metal–water–soil interactions. In order for these calculations to be useful for most metals of environmental concern, numerous solution-phase, solid-phase, and interface processes must be considered.

Lattice metals

The chemistry of metals in soil is complicated by the fact that metals may be present both in highly inert and nonreactive pools and in pools where there is significant exchange of metal with the soil solution. The inert metal is present in the crystal lattice of the parent material of the soil. Digestion with strong acids, for example, aqua regia, and other strenuous digestion procedures liberate this metal. Aqua regia digestions of soil are employed in many countries for assessing compliance with environmental regulations (Houba et al. 1996). These concentrations of metals cannot be related to the potential for risk because they include metal that will not be accessible by normal environmental processes.

Oxidation–reduction processes

The redox status of a soil can affect the oxidation state of a trace metal directly, and it can affect chemistry of soil components that are responsible for partitioning of trace metals. Of the trace metals of primary concern, both As and C are present in multiple oxidation states in soil systems.

The 2 predominant forms of As, $AsO_3^{3-}$and AsO_4^{3-}, are both anionic species that sorb to iron oxides. As(+III) can be oxidized As(+V) in soil systems at pH greater than 8. Manning and Goldberg (1997) reported that As(+V) sorbs more strongly to soil under most conditions than does As(+III). Both As(+III) and As(+V) are released into solution in anaerobic soils as a consequence of the dissolution of the manganese and iron oxides that are responsible for their sorption.

Carbon can be present in the +III and +VI forms. The toxicity of Cr^{3+}, which forms a very insoluble hydroxide, is very low. Chromate, CrO_4^{2-}, on the other hand, being quite soluble, is very toxic and can be formed by the oxidation of Cr(+III) by MnO_2 in soil. SOM can reduce Cr(+VI) to Cr(+III) (Rai et al. 1989).

Partitioning processes

It is necessary to determine the concentration of metal in the pool that is capable of interchanging with metal in the solution phase. As will be discussed in Chapter 9, there are several extraction methods that have been used for this purpose or for direct correlation with biological responses (Conder and Lanno 2000). The quantity of metal contained in this exchangeable pool constitutes the capacity, and the equilibrium concentration of the dissolved free metal ion in the solution phase constitutes the intensity.

The free metal ion concentration (actually its activity or chemical potential) has been correlated to the availability of metals to organisms in aquatic systems and higher plants in soils (Morel 1983; Pagenkopf 1983; Checkai et al. 1987; Lund 1990; Playle et al. 1992, 1993a, 1993b; Bergman and Dorward-King 1997; Parker and Pedler 1997). Because the partitioning of the metal between the solid and solution phases is dependent on the strength of binding of the metal by the soil, it is likely that with appropriate considerations of changes in the local environment (e.g., gut or dermal conditions), a similar approach can be applied to the prediction of bioavailability to other classes of organisms. On the other hand, for groundwater contamination, it is the total soluble concentration of the metal, irrespective of chemical form, that is of interest.

Metals are retained by the solid phase by 3 processes: precipitation, ion exchange, and adsorption. In addition, the redox status of the system can greatly affect the partitioning of the metal. The extent of removal from the solution to the solid phase is a function of the chemical composition of both the solid and the solution phases.

Precipitation

The concentrations of metals present at hazardous waste sites are commonly several orders of magnitude greater than in undisturbed background soils. At these sites, metals will precipitate to form secondary minerals in the soil. Metal-contaminated soils are often remediated through the addition of chemical reactants to the soil to enhance the secondary mineral formation process. For example, phosphate fertilizer can be added to Pb-contaminated sites to promote the formation of Pb pyromorphite, which has low solubility and bioavailability. The extent of removal of the metal through formation of the mineral, M_aL_b, is described by the solubility product, K_{so}, for the process

$$M_aL_b \leftrightarrow aM^{z+} + bL^{q-};\ K_{so} = \{M^{z+}\}^a \{L^{q-}\}^b \quad (7\text{-}4),$$

where the braces, { }, indicate chemical activity. Common precipitates are hydroxides, oxides, oxyhydroxides, carbonates, hydroxycarbonates, phosphates, and silicates. The extent of metal removal from solution by precipitation is commonly a function of the pH of the solution. For example, for a metal carbonate, MCO_3, the solubility relationship given in Equation 7-4 is valid in terms of the M^{2+} and CO_3^{2-} ions. The total concentrations of metal and carbonate species in the solution phase are functions of the solution pH and will be given by Equations 7-5 and 7-6, respectively:

$$[M]_T = [M^{2+}] + [MOH^+] + [M(OH)_2^0] + [M(OH)_3^-] + [M(OH)_4^{2-}] + [MCO_3^0] \quad (7\text{-}5),$$

$$[CO_3]_T = [H_2CO_3] + [HCO_3^-] + [CO_3^{2-}] + [MCO_3^0] \quad (7\text{-}6),$$

where the brackets, [ ], indicate concentration.

Often, one does not find distinct mineral phases even though the existence of a solid phase, such as $CdCO_3$, is predicted from equilibrium considerations. Mixed solid phases, such as $CdCO_3$ within $CaCO_3$, may be formed through coprecipitation.

Ion exchange

In ion exchange, metal ions in solution react with the negatively charged sites of soil particles. Permanent charge sites occur in layer silicate clays. Relatively weak outer sphere complexes in which the metal ions retain their water of hydration are formed. This exchange is termed "nonspecific adsorption," or more commonly, "ion exchange." The charge available at the constant charge sites, effective in ion exchange, depends on the clay mineralogy and can be far greater than that at the variable charge sites which are responsible for adsorption, which is discussed in the next section. Ion exchange is a particularly important mechanism for the retention of alkali and alkaline earth ions, including Na^+, K^+, Cs^+, Mg^{2+}, Ca^{2+}, as well as for the retention of NH_4^+. Ion exchange is generally not an important process in controlling the solution-phase concentrations of heavy metal ions because the selectivity for exchange of the heavy metals is too low for them to successfully compete with the major ions for the exchange sites.

Adsorption

A third category of partitioning, referred to as "adsorption," occurs at sites that are variably charged, depending on pH and to a lesser degree on ionic strength. The dependency on pH results from the reaction of protons with oxide and hydroxide minerals and with certain functional groups of humic substances (Sposito 1984).

Adsorption onto hydrous metal oxides

A common surface group that reacts with protons is hydroxide. A surface hydroxide, $S\text{-}OH^0$, can undergo 2 protolysis reactions:

$$\equiv S\text{-}OH_2^+ \leftrightarrow \equiv S\text{-}OH^0 + H^+ \qquad (7\text{-}7),$$

$$\equiv S\text{-}OH^0 \leftrightarrow \equiv S\text{-}O^- + H^+ \qquad (7\text{-}8),$$

The corresponding conditional stability constants are

$$K_{\text{cond, a1}} = \frac{[\equiv S\text{-}OH^0]\{H^+\}}{[\equiv S\text{-}OH_2^+]} \qquad (7\text{-}9),$$

$$K_{\text{cond, a2}} = \frac{[\equiv S\text{-}O^-]\{H^+\}}{[\equiv S\text{-}OH^0]} \qquad (7\text{-}10).$$

Because adsorption is a surface phenomenon, concentrations of bound metals and of surface sites are expressed on an area basis, rather than mass basis, to enable one to relate different materials. The surface area is most commonly determined from the N adsorption isotherm at 77 °K. The Brunauer-Emmett-Teller (BET) equation enables the calculation of the area for a complete monolayer surface sorption (Hiemenz 1986).

As the surface undergoes ionization, for instance, during a titration with base, the surface becomes progressively more negatively charged and it becomes more difficult to remove subsequent protons. Thus, the conditional stability constant varies with the charge on the surface. It is necessary to incorporate a Boltzmann, or electrostatic, factor to convert this conditional constant into an intrinsic constant that does not vary with pH (Huang 1981; Stumm and Morgan 1996; Sposito 1984; Hiemenz 1986; Schindler and Stumm 1987). The intrinsic constant, K_{int}, is given by the relationship

$$K_{cond} = K_{int} \exp[-\Psi_0 F/RT] \qquad (7\text{-}11),$$

where Ψ_0 is the electrical potential at the surface, F is the Faraday, R is the gas constant, and T is the absolute temperature. The value of Ψ_0 is also dependent on ionic strength. However, the ionic strength of the soil solution is usually low.

The surface potential is proportional to the surface charge (Hiemenz 1986; Singh and Uehara 1986; Schindler and Stumm 1987; Westall 1987):

$$\sigma_0 = \kappa\Psi_0 \qquad (7\text{-}12).$$

The surface charge is directly determined from the proton or hydroxide consumption by the solid phase in an acid or base titration (Huang 1981).

One of the major difficulties is that sorption constants are expressed using different models of the solution–solid interface. This results in the use of different parameters among the models and an inability to directly compare the parameters that are common among them (Dzombak and Hayes 1992). The electrical double layer theory of Equation 7-11 can be extended to account for the adsorption of ions at planes other than at the surface. The triple layer model of surface complexation requires an additional potential at the Stern layer (James and Parks 1982). The zeta potential, which is the potential at the plane of shear, is subject to easy instrumental measurement and is a good approximation of the Stern potential (Hiemenz 1986). Experimental data fit the simpler model as well as they fit the more sophisticated models (Morel 1981; Westall and Hohl 1980). A widely used model is the diffuse double layer model for which there is a large compilation of internally consistent constants (Dzombak and Morel 1990).

It is now easy to understand the reason that metal sorption is so highly pH dependent. Protons and metal ions compete with each other for available surface binding sites. When adsorption of a divalent metal, such as Cd, onto a solid, such as goethite, is determined as a function of pH, there is little adsorption at low pH and virtually complete adsorption at high pH values. The transition between low and very high adsorption occurs over a region of approximately 2 pH units. For a divalent metal ion, M^{2+}, the adsorption can be written

$$\equiv \text{S-OH}^0 + M^{2+} \leftrightarrow \equiv \text{S-OM}^+ + H^+ \qquad (7\text{-}13),$$

for which the conditional stability constant is

$$K_{\text{cond}} = \frac{[\equiv\text{S-OM}^+]\{H^+\}}{[\equiv\text{S-OH}^0]\{M^{2+}\}} \qquad (7\text{-}14).$$

This conditional constant is related to an intrinsic constant in a similar fashion to that in Equation 7-11. An analogous reaction to Equation 7-13 can be written for the binding of the metal ion to 2 soil surface sites with the concurrent release of 2 protons, rather than 1. The binding of the hydrolyzed metal, $MeOH^+$, to $S\text{-}O^-$ to give S-OMeOH, with no release of protons, is also possible. If one is to be able to predict the adsorption of a metal at any pH, it is essential to have the acid base equilibrium constants of Equations 7-7 and 7-8.

The adsorption of oxyanions, such as chromate, can be described in an analogous fashion to that of metal cations (Parfitt 1978; Mott 1981). Binding is represented by the reaction

$$\equiv \text{S-OH} + (\text{O-MO}_x)^{-n} \leftrightarrow (\equiv \text{S-O-MO}_x)^{-n+1} + \text{OH}^- \qquad (7\text{-}15).$$

Again, the pH dependency of the reaction is predicted and accounted for by the conditional constants

$$K_{\text{cond}} = \frac{[(\equiv\text{S-O-MO}_x)^{-n+1}]\{\text{OH}^-\}}{[\equiv\text{S-OH}]\{(\text{O-MO}_x)^{-n}\}} \qquad (7\text{-}16),$$

which can be converted to an intrinsic constant by Equation 7-11.

As noted in the section on oxidation-reduction processes, reducing conditions may result in the release of trace metals to the soil pore water as a consequence of the dissolution of metal oxide phases.

Adsorption onto humic materials

Binding of metals by the humic materials, fulvic and humic acids, can be viewed in a similar manner as the sorption onto metal oxides. The carboxyl and phenolic functional groups give rise to the variable charge. For example, the protonation and metal complexation reaction of a metal with a carboxyl group on the humic surface can be represented by the reactions

$$\equiv \text{R-COOH} \leftrightarrow \equiv \text{R-COO}^- + \text{H}^+ \qquad (7\text{-}17),$$

$$\equiv \text{R-COOH} + \text{M}^{2+} \leftrightarrow \equiv \text{R-COOM}^+ + \text{H}^+ \qquad (7\text{-}18).$$

Transition metal ions form inner sphere complexes with humic materials, which accounts for the high stability of these complexes relative to those with the more prevalent alkali and alkaline earth elements. Neither humic nor fulvic acid is a discrete chemical entity, rather they are composed of molecules that differ in properties, including functional groups. Because humic substances contain a large number of ligands differing in concentration and stability constants, simple models that consider reactions with only 1 or 2 sites do not reproduce the measured degree of chemical interaction. Titration of humic materials with either protons or metal ions results in nearly featureless titration curves, which do not rise sharply like titration curves of distinct chemical substances. To model these titration curves, it is necessary to recognize that they represent a mixture of many substances. A number of different approaches have been taken to describe the binding of protons and metal ions by humic substances (Perdue and Lytle 1983; Ephraim and Marinsky 1986; Fish 1986; Cabaniss and Shuman 1988; Tipping 1990). These approaches include discrete site models, models with a continuum of binding sites of varying $_{p}K$, and models that incorporate electrostatic interactions.

Sorption onto soils

A large body of literature exists on the adsorption of metals onto soils. Most studies show that adsorption onto soil shares characteristics with the adsorption of metals onto the iron and manganese oxides and humic material coatings on particles. Most adsorption isotherms are L-shaped. Adsorption is a function of the solution pH with a transition from low to high adsorption over a pH range of approximately 2 units.

The dominant factor controlling the partitioning of metals between the soil and solution phases is the pH. For example, Anderson and Christensen (1988) studied the sorption of Cd on 38 soils with samples being taken at several depths. For their 117 sorption measurements made at very low Cd(II) concentrations (0.7 to 12.6 µg/L), regression of log of the Cd(II) sorption K_D on the pH gave a slope of 0.64. The regression had a R^2 value of 0.776, indicating that proton concentration is the principal factor affecting the partitioning process. The scatter of data about the line suggests that the strength of binding of Cd by the different samples was not the same. To improve the level of prediction, the concentrations of the components of the soil responsible for the metal binding and the strength of the sorption by these materials should be included.

Lexmond (1980) studied the sorption of Cu onto a Dutch, slightly loamy, gravelly sand as a function of concentration and pH. They found that the sorption of Cu could be described by an empirical pH-dependent Freundlich equation:

$$Q_S = K_S\{Cu^{2+}\}^a / \{H^+\}^b \quad (7\text{-}19),$$

where Q_S is the amount of Cu bound to the soil solid phase, $\{Cu^{2+}\}$ and $\{H^+\}$ are the activities of the Cu and hydrogen ions, respectively, and a and b are fitting parameters.

Radovanovic and Koelmans (1998) developed a model to predict metal binding to natural solids that considers sorption onto individual sorption phases as well as the effect of pH on the sorption process. The system is modeled as a series of simultaneous equilibria. The applicability of such a model for the prediction of partitioning assumes that the metal in each of the phases considered will equilibrate between the solid and solution phases during the exposure period. This is a function of the time allowed for equilibration and the geometry of the solid phases. For example, if the surface of the particle is covered by only one of the materials, equilibration with fresh solution phase will be with that sorption material only if there is sufficient time allowed for the diffusion of the metal of interest from the deeper-lying phases to the sorption phase on the surface. One example of such a system would be for redox cycling systems that can have iron hydroxide deposited onto the surface. One would need to know if there was sufficient time for the contaminant metal to diffuse into the iron hydroxide to maintain equilibrium with the underlying solid phases. Davis (1984) suggested that most soils have a surface coating of organic matter. If this is the case, in the equilibration of a soil with water, metal is transferred between the aqueous phase and the particulate organic matter coating of the soil.

Lee et al. (1996) studied the adsorption of Cd onto 15 soils as a function of pH. They found a dependence of partitioning on pH that was nearly the same as that reported by Anderson and Christensen (1988). The variation of the data values (log K_D versus pH) about the line reported by Lee et al. (1996) was about a factor of 30. When the value of K_D was normalized to the fraction of organic matter, the variation was reduced to a factor of approximately 3, a 10-fold improvement in the correlation. These results suggest either that the surface is indeed coated by organic matter and that in the 24-hour equilibration period there was not sufficient time for penetration of the added Cd through the organic matter to reach other sorption sites on, for instance, Fe and Mn oxides, or that sorption of the Cd onto the organic matter dominated the sorption relationship.

Yin et al. (1996) studied the adsorption of Hg(II) from dilute solution onto the same 15 soils that had been studied previously by Lee et al. (1996). A much different result was found for the adsorption of Hg onto the soils as a function of pH than is found for most other metals. At the lowest pH studied, approximately 3 pH, 80% to 90% of the Hg was adsorbed. A maximum extent of adsorption, generally >90%, was attained at pH approximately 4, and the extent of adsorption then decreased with increasing pH. By pH 10, the adsorption ranged from 10% to 50%. This decrease in adsorption was found to be caused by the dissolution of particulate organic matter that could then complex the released Hg. The dissolution of organic matter is a strong function of the pH of the soil, increasing with increasing soil pH. The role of pH in sorption processes is therefore many-fold: It controls the protonation of the solid phase, it is important in the speciation of ligands in the solution phase, and it regulates the distribution of ligands, particularly humic substances, between the solid and the solution phases.

Yin, Allen, Huang, Sanders (1997) found that adsorption isotherms on the higher organic matter content soils had an S-shape. The organic matter released to the solution reacted with the Hg. The complex was more poorly adsorbed than were the inorganic forms of Hg, which predominated only after sufficient Hg had been added to complex the dissolved organic matter.

Modeling partitioning

There are a number of computer programs currently available to compute geochemical speciation, including MINEQL+ (Schecher and McAvoy 1992), MINTEQA2 (Allison et al. 1991) and WHAM (Tipping 1994). The most important factor in these models is the quality of the database. WHAM best handles the binding of metals by humic substances, and its database is validated by the large number of independent data sets for which it has been used. It considers humic substance to be a series of discrete ligands whose relative concentration is greatest for ligands having intermediate stability constants. The binding spectrum for reaction with protons forms the basis for its representation and for reaction of metals the binding spectrum is considered to shift in strength. Several questions are of importance. First, is there a small fraction of the ligands that have a very high stability constant? For example, in the case of Hg, which selectively and strongly binds to thiols, there could be a small concentration of ligands that have a very high affinity that is not included in the models of the humic substances. Second, to what extent are metals competitive for the same binding sites? It is assumed that all metals can react with all sites.

Usually organic C is used as a surrogate for the humic substances. If there is a large percentage of the organic matter that is not reactive with metals, an alternative approach is required. A most important input is the total concentration of metal. In general, partitioning of the metal in precipitates and adsorbed phases are the only pools of metal considered. Therefore, an extraction of that pool of metal is required to provide the necessary information.

Partition coefficient

Information on the metal distribution between the solid and the solution phases is typically expressed as a partition coefficient, similar to what is done with organic compounds (Schwarzenbach et al. 1993). The metal in the soil can be present in several solid phases. In the solution phase, the metal will be present in a number of different chemical forms, including association with dissolved organic matter. Inorganic ligands of importance in the speciation of the metal include hydroxide, carbonate, bicarbonate, and chloride. Stability constants for the formation of inorganic complexes are, in general, well known. The chemical species distribution among the inorganic complexes then can be easily computed using any of the computer models previously discussed. Complexation with dissolved organic matter can be included if these constants are available.

Because the partition coefficient is the ratio of the concentration of the metal in the solid and solution phases, its utility is very limited. The value will depend on the metal speciation in each of the 2 phases, and thus it is a sample-specific parameter (van der Kooij et al. 1991). The partition coefficient is solely a representation of measured values without predictive capability. This can be best understood by writing the partition coefficient, K_D, which is

$$K_D = \frac{[\text{Metal in Soil}]}{[\text{Metal in Solution}]} = \frac{[\text{M-POM}]+[\text{M-FeOx}]+[\text{M-MnOx}]+\cdots}{[\text{M}^{2+}]+[\text{MOH}^{+}]+[\text{MCO}_3^0]+[\text{M-DOM}]+\cdots} \quad (7\text{-}20).$$

An alternative formulation of the partitioning relationship may provide a means to predict the distribution of a metal between the solution and solid phases. The partitioning of a metal between the solution and sorption phases in the soil can be rationalized through consideration of conventional chemical equilibrium concepts. For example, the partitioning of the metal with the particulate organic matter (POM) fraction of the soil can be represented as

$$\text{M}^{2+} + \text{POM} \leftrightarrow \text{M-POM} \quad (7\text{-}21).$$

The equilibrium is then represented

$$K_{eq} = \frac{[\text{M-POM}]}{[\text{M}^{2+}][\text{POM}]} \quad (7\text{-}22).$$

This equation can be rewritten to provide a prediction of the concentration of metal ion in solution

$$[\text{M}^{2+}] = K_{eq} \frac{[\text{M-POM}]}{[\text{POM}]} \quad (7\text{-}23).$$

This indicates that the concentration of free metal ion can be predicted from the knowledge of the concentration of the sorbed metal and the concentration of the sorbent. As was discussed, the free metal ion concentration is believed to be directly related to bioavailability. Thus, prediction of the free metal ion concentration would provide a theoretical measure of bioavailability.

Desorption

As for organics, hysteresis is observed in the desorption of metals from soils. For example, Yin, Allen, Huang, Sanders (1997) studied the desorption of Hg that they had adsorbed onto soils. After 4 steps of desorption, about 11% of the adsorbed Hg had been removed from a soil containing 0.12% organic matter, but less than 2% of the Hg that had been adsorbed was removed from a soil containing 5.0% organic matter. Furthermore, the ability to desorb metal decreases as time of aging increases.

Because of the lack of concurrence between adsorption and desorption results, partition results from adsorption experiments cannot be used to infer release of metal from soil. Desorption data from freshly contaminated soils cannot be used to infer the desorption of metal from aged materials. The processes and mechanisms responsible for the hysteresis must be understood to enable an adequate description of chemical equilibria.

A number of potential reasons for hysteresis have been discussed previously in this chapter in the section that deals with the sorption of organic compounds. These processes may be responsible for hysteresis observed in adsorption and desorption of metals too. In the case of metals, hysteresis could occur if adsorbents such as iron and manganese oxides were coated by organic matter. The organic matter surface would equilibrate with the solution. Metals would have to diffuse through the organic matter and then be sorbed by the metal oxide. Desorption would then need to follow a tortuous reverse path.

Kinetics

The adsorption of metal ions to soil is relatively rapid. In typical batch equilibration experiments, the aqueous-phase concentration is rapidly depleted. Most adsorption studies are allowed to proceed for 24 hours. However, the hysteresis observed on desorption, which is discussed above, indicates that the process is more complicated than that of a simple surface adsorption and desorption process.

Gerth and Brummer (1983) and Brummer et al. (1988) showed that there is a slow diffusion of heavy metals into the goethite structure. The diffusion rates are related to the ionic diameters of the ions. They postulated that micropores or defect positions were responsible for this slow diffusion.

Yin, Allen, Huang, Sparks and Sanders. (1997) found that there was a fast and a slow kinetic step for both the uptake and release of Hg(II) by soil. Both uptake and release rate coefficients were related to the organic matter content of the soil. They reported that a portion of the adsorbed Hg was resistant to desorption and that the greater the organic matter content, the greater the fraction that was resistant to desorption. They felt that diffusion of the Hg through intraparticle micropores in the organic matter was responsible for the irreversibility of sorption.

As a consequence of the slow adsorption and desorption process and/or hysteresis, metals added to soils as salts and metals added as a result of environmental contamination become less mobile and available with age. Short-term laboratory equilibration may indicate a higher risk than that present in field conditions.

The Need for Protocols

The most critical output of contaminant–soil interaction studies in support of bioavailability assessments would be reliable information on the porewater concentration of the contaminant of interest. This can come from discrete measurements of porewater concentrations and/or from measurements of chemical-specific and soil-specific parameters that can be used in some models to predict porewater concentrations under a variety of soil and site conditions, as well as over a reasonable time frame. The desired chemical-specific parameters include partition coefficients, rate constants, and diffusion coefficients. The soil-specific parameters would include those listed in Table 7-1 (p 256). This focus on porewater concentration is not meant to detract from the likelihood that other, artificial means might provide a more direct basis for estimating bioavailability in specific circumstances. These other means might involve, for example, specialized leaching with liquids other than natural waters and/or uptake by manufactured items (e.g., plastic beads or semipermeable membrane devices) that mimic biological systems.

For either of the above approaches, direct porewater measurements of concentration or the measurement of model-related parameters, protocols are desperately needed. There are, however, some significant questions that need to be answered before we decide what protocols are needed. For direct measurements, the questions are relatively easy. They would include number of samples needed for characterization, porewater contact time, and porewater recovery technique. For model-related parameters, the initial issue is which model? Should the model accurately describe the mechanisms that are responsible for the sorption and desorption, or is it felt that empirical models, or models based on a simplistic representation of the sorption and desorption processes, will be adequate?

Both mechanistic and empirical models are needed. The ones that will accurately describe the sorption and desorption mechanisms are needed, at least in these research programs, to better understand the underlying processes. They will also allow more realistic (and sophisticated) assessments at high priority sites where such detailed assessments (and the related resource use) are warranted. However, the problems with this "mechanistic" modeling approach are that the mechanisms are not currently understood; there is insufficient consensus among experts on which mechanisms and models are the most representative; they require too many inputs and are, consequently, not easy to use; and there is little likelihood of regulatory near-term acceptance.

Empirical models are needed in order to deal with near-term problem sites, that is, those for which management decisions will be made in the next few years. While these models may be based on strictly empirical correlations, or on simplistic (or even wrong) assumptions regarding the sorption and desorption process, they are all we have today that are relatively easy to use and generally acceptable to the regulatory agencies (at the level that most sites are given regulatory oversight). Properly applied and qualified, the use of such models should provide reasonable estimates of porewater concentrations under many conditions of concern. Perhaps what a protocol could spell out is the conditions under which the simplistic models do and do not provide a reasonably reliable basis for assessment. By conditions, we mean soil and contaminant types, spatial and temporal scales, exposure scenarios and receptors. In many cases, it may be that the uncertainty in the outputs from the simplistic and empirical models is acceptable.

Available protocols

There is very little in the way of standard protocols—issued by regulatory agencies or standard-setting organizations—that can be properly used to measure porewater concentrations or to measure model-related parameters for use in sorption and desorption models. The American Society for Testing and Materials (ASTM) has published a protocol for measuring organic C normalized sorption constants, that is, K_{oc} values (ASTM 1993). However, this protocol stipulates measurement via spiking of a clean soil and suggests the use of a relatively short contact time (4 to 48 hours) for the soil and the chemical in the soil slurry used. This protocol would not provide a reliable sorption constant that would apply to situations where hindered desorption was a factor.

In addition to the standard method for K_{oc} determination, ASTM provides specifications for other sorption tests including, for example, 24-hour batch sorption (D 4646-87 and D 5285-92), single batch extraction of wastes (D 5233-92), sequential batch extraction of wastes (D 4793-93), and leaching in a column apparatus (D 4874-98). Other methods that might provide roughly equivalent information include toxicity characteristic leaching procedure, synthetic precipitation leaching procedure, multiple extraction procedure, and other leaching procedures (Bricka et al. 1991; USEPA 1995). Other leaching tests are described in an USEPA guidance document for testing solidified wastes (USEPA 1989). However, protocols do not exist for measuring most of the parameters that may be required in the more mechanistic models, including rate constants and diffusion coefficients.

Technical issues for a protocol for measuring a desorption constant

There are a number of technical issues that would need to be addressed during protocol development for the measurement of a desorption constant. The desorption constant is a solid-to-liquid concentration ratio, presumably measured under near-equilibrium conditions, by means of contaminant desorption from a historically contaminated soil. Estimating chemical desorption from historically contaminated soils specifically avoids problems associated with traditional approaches for measuring soil sorption constants that involved either spiking a clean soil (in a soil–water slurry) with the chemical and measuring the amount of chemical remaining in the aqueous phase after a relatively short mixing time (hours to several days), or spiking a clean soil (dry or in a slurry) and allowing a long contact time (weeks to months or years) before conducting desorption studies. Many technical issues need to be addressed in the development of a protocol for determining desorption constants. These include

- Soil sampling: A sufficient number of soil samples should be collected from the site to sufficiently represent the range of soil types, habitats (if ecological concerns are involved), contaminant concentration range, and aerial extent and depth of contamination. Geostatistics can help in determining the number of samples required to achieve a defined confidence level in mean values. Initial (on-site) analyses of soil contaminant concentrations may be helpful in assuring sufficient coverage of the range of concentrations.
- Sample preparation: The protocol will need to specify when (if ever) and how to 1) prescreen the soil (to remove oversized particles), 2) composite samples, 3) homogenize samples (e.g., via grinding), and/or 4) sterilize the samples (to prevent unwanted contaminant biodegradation).
- Leachant selection: The protocol might allow the use of natural waters (e.g., rain water, site groundwater and/or tap water) or specify that a more aggressive aqueous solution needs to be prepared, for example, addition of acetic acid for organic contaminants and an inorganic acid for inorganic contaminants.
- Contact mode: The protocol could stipulate batch contact in a soil slurry or a packed column. Alternatively, contact in a flow-through column could be specified.
- Solids-to-liquids ratio: For slurries, the protocol will need to stipulate a solids-to-liquids ratio. This is necessary because the extent of sorption, and thus the derived sorption constant, has been found to be a function of this ratio.
- Mixing method: For slurries, the method of mixing will need to be specified. In most cases, a low-energy mixing method (e.g., slow rolling on a roller table) will be preferred in order to avoid excessive alteration of the particle size configuration in the sample.

- Biological inhibition: Although the initial soil may have been sterilized, the soil–water mix may have to be further treated to prevent unwanted biodegradation of the contaminant. Such treatments may involve the addition of mercuric chloride, acid, azide, or formaldehyde.
- Chemical control: If the desorption test is long, changes in the basic soil chemistry (e.g., pH and redox potential) may take place. If such changes could adversely affect the desorption process, then the protocol will need to allow for adjustment of the chemistry during the test.
- Desorption time: Because a desorption "constant" is being measured, the protocol will have to stipulate a time period for the soil–water contact that will allow a reasonable approach to an equilibrium partitioning state. The protocol may require that an initial time-series test be carried out to determine how long a time is required. Desorption times of days to months may be required.
- Solids–liquids separation: The protocol will need to stipulate what means are acceptable for the separation of the solid and liquid phases at the end of the contact period. The typical choices include fabric filtration and centrifugation. In either case, the protocol may need to require special tests to determine if the separation was adequate, that is, there was no solids carry-over into the aqueous sample. For strongly sorbed chemicals, even a small amount of solids carry-over would ruin the results.
- Analyses: The final analyses of the contaminant in the aqueous phase (i.e., the desorbed chemical) can probably be done by standard methods as long as the detection limits are adequately low. The solid phase may also have to be analyzed if a significant fraction of the chemical was desorbed from the soil by the desorption test.

As mentioned above, this material was provided only as an example. It is intended to illustrate the types of technical issues that would need to be addressed and the types of solutions that might be allowed or stipulated. Putting together a single protocol, even in a limited case, that would achieve the right balance of scientific merit, ease of use, and regulatory acceptance will require time.

Research and Development Needs

Specific research and development needs include

- obtaining a thorough understanding of sorption hysteresis, specifically including an elucidation of the responsible mechanism (where possible, develop methods to correlate and predict data on sorption hysteresis);
- developing models, preferably mechanistic, to describe the kinetics of contaminant desorption from soils and, where appropriate, coupled sorption and desorption-bioaccumulation models;
- developing easy-to-use tools for the routine application of state-of-the-art information or models relating to soil sorption and desorption (tools must include protocols for measuring or obtaining the chemical-specific and soil-specific parameters required by the models);
- evaluating the impacts of soil heterogeneity at different scales (i.e., micro and macro) on contaminant–soil interactions. Particularly, determining the range in kinetic and thermodynamic sorption parameters expected across a site;
- assessing the significance of biofacilitated desorption, both in the environment and in vivo; and
- establishing technical protocols for measuring sorption and desorption parameters.

References

Adamson AW, editor. 1990. Physical chemistry of surfaces. 5th ed. New York NY, USA: John Wiley and Sons, Inc. 777 p.

Alexander M. 1995. How toxic are toxic chemicals in soil? *Environ Sci Technol* 29:2713-2717.

Allison JD, Brown DS, Novo-Gradac KJ. 1991. MINTEQA2/PRODEFA2, a geochemical assessment model for environmental systems [computer program]. Version 3.0.

Anderson PR, Christensen TH. 1988. Distribution coefficients of Cd, Co, Ni, and Zn in soils. *J Soil Sci* 39:15-22.

[ASTM] American Society for Testing and Materials. 1993. Test method E1195-87(1993)el standard test method for determining a sorption constant (K_{oc}) for an organic chemical in soil and sediments. In: Annual book ASTM standards, Volume 11.05. E47.06. Philadelphia PA, USA: ASTM. p 443-450.

Ball WP, Roberts PV. 1991. Long-term sorption of halogenated organic chemicals by aquifer material. 2. Intraparticle diffusion. *Environ Sci Technol* 25:1237-1249.

Berens AR. 1989. Transport of organic vapors and liquids in poly (vinyl chloride). *Macromol Chem Macromol Symp* 29:95-108.

Bergman HL, Dorward-King EJ. 1997. Reassessment of metals criteria for aquatic life protection: Priorities for research and implementation. Pensacola FL, USA: SETAC. 114 p.

Bricka RM, Holmes TT, Cullinane JM. 1991. Comparative Evaluation of Two Extraction Procedures: The TCLP and the EP. Rep. for 1 Oct 86 – 30 Sep 89. USEPA, Cincinnati OH, USA. EPA/600/2-91/049. 194 p.

Brummer G, Gerth J, Tiller KG. 1988. Reaction kinetics of the adsorption and desorption of nickel, zinc, and cadmium by goethite. I. Adsorption and diffusion of metals. *J Soil Sci* 39:37-52.

Cabaniss SE, Shuman MS. 1988. Copper binding by dissolved organic matter: I. Suwannee River fulvic acid. *Geochim Cosmochim Acta* 52:185.

Carter MC, Kilduff JE, Weber Jr WJ. 1995. Site energy distribution analysis of preloaded adsorbents. *Environ Sci Technol* 29:1773-1780.

Checkai RT, Corey RB, Helmke PA. 1987. Effects of ionic and complexed metal concentrations on plant uptake of cadmium and micronutrient cations from solution. *Plant Soil* 99:335-345.

Chen W, Kan AT, Fu G, Vignona LC, Tomson MB. 1999. The adsorption-desorption behavior of hydrophobic organic compounds in sediments of Lake Charles, LA. *Environ Toxicol Chem* 18:1610-1616.

Chiou CT, Kile DE. 1998. Deviations from sorption linearity on soils of polar and nonpolar organic compounds at low relative concentrations. *Environ Sci Technol* 32:338-343.

Conder JM, Lanno RP. 2000. Evaluation of surrogate measures of cadmium, lead, and zinc bioavailability to Eisenia fetida. *Chemosphere* 41: 1659-1668.

Connaughton DF, Stedlinger JR, Lion LW, Shuler ML. 1993. Description of time-varying desorption kinetics: Release of naphthalene from contaminated soils. *Environ Sci Technol* 27:2397-2403.

Davis JA. 1984. Complexation of trace metals by adsorbed natural organic matter. *Geochim Cosmochim Acta* 48:679-691.

Devitt EC, Wiesner MR. 1998. Dialysis investigations of atrazine-organic matter interactions and the role of a divalent metal. *Environ Sci Technol* 32:232-237.

Di Toro DM, Zarba CS, Hansen DJ, Berry WJ, Sqartz RC, Cowan CE, Pavlou SP, Allen HE, Thomas NA, Paquin PR. 1991. Technical basis for establishing sediment quality criteria for nonionic organic chemicals using equilibrium partitioning. *Environ Toxicol Chem* 10:1541-1583.

Dzombak DA, Hayes KF. 1992. Comment on Recalculation, evaluation and prediction of surface complexation constants for metal adsorption on iron and manganese dioxides. *Environ Sci Technol* 26:1251-1253.

Dzombak DA, Morel FMM. 1990. Surface complexation modeling: Hydrous ferric oxide. New York NY, USA: Wiley Interscience. 393 p.

Ephraim J, Marinsky JA. 1986. A unified physicochemical description of the protonation and metal ion complexation equilibria of natural organic acids (humic and fulvic acids). 3. Influence of polyelectrolyte properties and functional heterogeneity on the copper ion binding equilibria in an Armadale horizons Bh fulvic acid sample. *Environ Sci Technol* 20:367-376.

Everett DH, Whitton WI. 1952. A general approach to hysteresis. *Trans Faraday Soc* 48:749-757.

Fish W. 1986. Metal-humate interactions. 1. Application and comparison of models. *Environ Sci Technol* 20:676-683.

Forstner U. 1995. Land contamination by metals: Global scope and magnitude of problem. In: Allen HE, Huang CP, Bailey GW, Bowers AR, editors. Metal speciation and contamination of soil. Boca Raton FL, USA: Lewis Publishers. p 1-33.

Fu G, Kan AT, Tomson MB. 1994. Adsorption and desorption hysteresis of polycyclic aromatic hydrocarbons in surface sediment. *Environ Toxicol Chem* 13:1559-1567.

Gerth J, Brummer G. 1983. Adsorption und Festlegung von Nickel, Zink und Cadmium durch Goethit (a-FeOOH). *Fresenius J Anal Chem* 316:616-620.

Graber ER, Borisover MD. 1998. Evaluation of the glassy/rubbery model for soil organic matter. *Environ Sci Technol* 32:3286-3292.

Gustafsson O, Haghseta F, Chan C, MacFarlane J, Gschwend PM. 1997. Quantification of the dilute sedimentary soot phase: Implications for PAH speciation and bioavailability. *Environ Sci Technol* 31:203-209.

Hamaker JW, Thompson JM. 1972. Adsorption. In: Goring CAI, Hamaker JW, editors. Volume 1, Organic chemicals in the soil environment. New York NY, USA: Marcel Dekker. p 49-143.

Hayes MHB, MacCarthy P, Malcolm RL, Swift RS. 1989. Humic Substances II. London, UK: John Wiley and Sons, Inc. 764 p.

Hiemenz PC. 1986. Principles of colloid and surface chemistry. 2nd ed. New York NY, USA: Marcel Dekker. 815 p.

Horas JA, Nieto F. 1994. A generalization of dual mode transport theory for glassy polymers. *J Polymer Sci Part B-Polymer Physics* 32:1889-1898.

Houba VJG, Lexmond TM, Novozamsky I, van der Lee JJ. 1996. State of the art and future developments in soil analysis for bioavailability assessment. *Sci Total Environ* 178:21-28.

Huang CP. 1981. The surface acidity of hydrous solids. In: Anderson MA, Rubin AJ, editors. Adsorption of inorganics at solid-liquid interfaces. Ann Arbor MI, USA: Ann Arbor Science. p 183-217.

Huang W, Weber Jr WJ. 1997. A distributed reactivity model for sorption by soils and sediments. 10. Relationships between desorption, hysteresis, and the chemical characteristics of organic domains. *Environ Sci Technol* 31:2562-2569.

Hunter MA. 1996. Irreversible adsorption of hydrocarbons to natural and surrogate sediments [DPhil dissertation]. Houston TX, USA: Rice University. 141 p.

Hunter MA, Kan AT, Tomson MB. 1995. In: 209th National Meeting of the American Chemical Society. Volume 35. San Diego CA, USA: ACS. p 598-602.

Hunter MA, Kan AT, Tomson MB. 1996. Development of a surrogate sediment to study the mechanisms responsible for adsorption/desorption hysteresis. *Environ Sci Technol* 30:2278-2285.

James RO, Parks GA. 1982. Characterization of aqueous colloids by their electrical double layer and intrinsic surface chemical properties. In: Matijevic E, editor. Surface and colloid science. New York NY, USA: Plenum Press. p 119-216.

Kan AT, Fu G, Hunter M, Chen W, Ward CH, Tomson MB. 1998. Irreversible adsorption of neutral organic hydrocarbons: Experimental observations and model predictions. *Environ Sci Technol* 32:892-902.

Kan AT, Fu G, Hunter MA, Tomson MB. 1997. Irreversible adsorption of naphthalene and tetrachlorobiphenyl to Lula and surrogate sediments. *Environ Sci Technol* 31:2176-2185.

Kan AT, Fu G, Tomson MB. 1994. Adsorption/desorption hysteresis in organic pollutant and soil/ sediment interaction. *Environ Sci Technol* 28:859-867.

Karger J, Ruthven DM. 1992. Diffusion in zeolites and other microporous solids. New York NY, USA: John Wiley and Sons, Inc. 605 p.

Keith LH. 1996. Principles of environmental sampling. Washington DC, USA: ACS. 848 p.

Lasaga AC. 1990. Atomic treatment of mineral-water surface reactions. In: Hochella JMF, White AF, editors. Mineral-water interface geochemistry. Volume 23. Washington DC, USA: Mineralogical Society of America. p 17-85.

Lee S-Z, Allen HE, Huang CP, Sparks DS, Sanders PF, Peijnenburg WJGM. 1996. Predicting soil-water partition coefficients for cadmium. *Environ Sci Technol* 30:3418-3424.

Lexmond TM. 1980. The effect of soil pH on copper toxicity to forage maize grown under field conditions. *Neth J Agric Sci* 28:164-183.

Linz DG, Nakles DV. 1997. Environmentally acceptable endpoints in soil: Risk-based approach to contaminated site management based on the availability of chemicals. Annapolis MD, USA: American Academy of Environmental Engineers. 630 p.

Lund W. 1990. Speciation analysis: Why and how? *Fresenius J Anal Chem* 337:557-564.

Lyman WJ, Reehl WF, Rosenblatt DH. 1990. Handbook of chemical property estimation methods: Environmental behavior of organic compounds. Washington DC, USA: ACS.

Mader BT, Goss K, Eisenreich SJ. 1997. Sorption of nonionic hydorphobic organic chemicals of mineral surfaces. *Environ Sci Technol* 31:1079-1086.

Manning BA, Goldberg S. 1997. Arsenic (III) and arsenic (V) adsorption on three California soils. *Soil Sci* 162:886-895.

McGroddy SE, Farrington JW, Gschwend PM. 1996. Comparison of the in situ and desorption of sediment-water partitioning of polycyclic aromatic hydrocarbons and polychlorinated biphenyls. *Environ Sci Technol* 30:172-177.

Morel FMM. 1981. Adsorption models: A mathematical analysis in the framework of general equilibrium calculations. In: Anderson MA, Rubin AJ, editors. Adsorption of inorganics at solid-liquid interfaces. Ann Arbor MI, USA: Ann Arbor Science. p 263-295.

Morel FMM. 1983. Principles of aquatic chemistry. New York NY, USA: John Wiley and Sons, Inc. 446 p.

Mott CJB. 1981. Anion and ligand exchange. In: Greenland DJ, Hayes MHB, editors. The chemistry of soil processes. New York NY, USA: John Wiley and Sons, Inc. p 179-219.

Pagenkopf GK. 1983. Gill surface interaction model for trace metal toxicity to fishes: Role of complexation, pH, and water hardness. *Environ Sci Technol* 17:342-347.

Parfitt RL. 1978. Anion adsorption by soils and soil materials. *Adv Agron* 38:231-266.

Parker DR, Pedler JF. 1997. Reevaluating the free-ion activity model of trace metal availability to higher plants. *Plant Soil* 186:223-228.

Pedit JA, Miller CT. 1995. Heterogeneous sorption processes in subsurfaces systems. 2. Diffusion modeling approaches. *Environ Sci Technol* 29:1766-1772.

Peijnenburg WJGM, Posthuma L, Eijsackers HJP, Allen HE. 1997. A conceptual framework for implementation of bioavailability of metals for environmental management purposes. *Ecotoxicol Environ Saf* 37:163-172.

Perdue EM, Lytle CR. 1983. Distribution model for binding of protons and metal ions by humic substances. *Environ Sci Technol* 17:654- 660.

Piatt JJ, Brusseau ML. 1998. Rate-limited sorption of hydrohobic organic compounds by soils with well characterized organic matter. *Environ Sci Technol* 32:1604-1608.

Pignatello JJ. 1990. Slowly reversible sorption of aliphatic halocarbons in soils. II. Mechanistic aspects. *Environ Toxicol Chem* 9:1117-1126.

Pignatello JJ. 1998. Soil organic matter as a nanoporous sorbent of organic pollutants. *Adv Colloid Interf Sci* 76-77:445-467.

Pignatello JJ, Xing B. 1996. Mechanisms of slow sorption of organic chemicals to natural particles. *Environ Sci Technol* 30:1-10.

Playle RC, Dixon DG, Burnison K. 1993a. Copper and cadmium binding to fish gills: Modification by dissolved organic carbon and synthetic ligands. *Can J Fish Aquat Sci* 51:2667-2677.

Playle RC, Dixon DG, Burnison K. 1993b. Copper and cadmium binding to fish gills: Estimates of metal-gill stability constants and modelling of metal accumulation. *Can J Fish Aquat Sci* 51:2678-2687.

Playle RC, Gensemer RW, Dixon DG. 1992. Copper accumulation on gills of fathead minnows: Influence of water hardness, complexation and pH of the gill microenvironment. *Environ Toxicol Chem* 11:381-391.

Radovanovic H, Koelmans AA. 1998. Prediction of in situ trace metal distribution coefficients for suspended solids in natural waters. *Environ Sci Technol* 32:753-759.

Rai D, Eary LE, Zachara JM. 1989. Environmental chemistry of chromium. *Sci Total Environ* 86:15-23.

Rogers CE. 1965. Solubility and diffusivity. In: Fox MMLD, Weissberger A, editors. Volume 2. Physics and chemistry of the organic solid state. New York NY, USA: Interscience Publishers. p 510-601.

Schecher WD, McAvoy DC. 1992. MINEQL+: A software environment for chemical equilibrium modelling. *Comp Environ Urban Syst* 16:65.

Schindler PW, Stumm W. 1987. The surface chemistry of oxides, hydroxides, and oxide minerals. In: Stumm W, editor. Aquatic surface chemistry. New York NY, USA: Wiley-Interscience. p 83-110.

Schlebaum W, Badora A, Schraa G, Riemsdijk WHV. 1998. Interactions between a hydrophobic organic chemical and natural organic matter: Equilibrium and kinetic studies. *Environ Sci Technol* 32:2273-2277.

Schulten HR. 1995. The three-dimensional structure of humic substances and soil organic matter studied by computational analytical chemistry. *Fresenius J Environ Anal Chem* 351:62-73.

Schwarzenbach RP, Gschwend PM, Imboden DM. 1993. Environmental organic chemistry. New York NY, USA: John Wiley and Sons, Inc. 681 p.

Singh U, Uehara G. 1986. Electrochemistry of the double-layer: Principles and applications to soils. In: Sparks DL, editor. Soil physical chemistry. Boca Raton FL, USA: CRC Press. p 1-38.

Sposito G. 1984. The surface chemistry of soils. New York NY, USA: Oxford University Press. 234 p.

Steinberg SM, Pignatello JJ, Sawhney BL. 1987. Persistence of 1,2-dibromoethane in soils: Entrapment in intraparticle micropores. *Environ Sci Technol* 21:1201-1208.

Stumm W, Morgan JJ. 1996. Aquatic chemistry: Chemical equilibria and rates in natural waters. New York NY, USA: John Wiley and Sons, Inc. 1022 p.

Tipping E. 1990. Modeling electrostatic and heterogeneity effects on proton dissociation from humic substances. *Environ Sci Technol* 24:1700.

Tipping E. 1994. WHAM-A chemical equilibrium model and computer code for waters, sediments, and soils incorporating a discrete site/electrostatic model of ion-binding by humic substances. *Comp Geosci* 21:973-1023.

[USEPA] U.S. Environmental Protection Agency. 1995. Guidance for the sampling and analysis of municipal waste combustion ash for the toxicity characteristic. USEPA, Office of Solid Waste and Emergency Response. EPA530-R-95-036. 32 p.

Vaccari DA, Kaouris M. 1988. A model for irreversible adsorption hysteresis. *J Environ Sci Health* A23:797-822.

Van der Kooij LA, Meent VD, Leeuwen CV, Bruggeman WA. 1991. Deriving quality criteria for water and sediment from the results of aquatic toxicity tests and product standards: Application of the equilibrium partitioning method. *Water Res* 25:697-705.

Weber Jr WJ, Huang W. 1996. A distributed reactivity model for sorption by soils and sediments. 4. Intraparticle heterogeneity and phase-distribution relationships under nonequilibrium conditions. *Environ Sci Technol* 30:881- 888.

Westall J, Hohl H. 1980. A comparison of electrostatic models for the oxide/solution interface. *Adv Colloid Interface Sci* 12:265-294.

Westall JC. 1987. Adsorption mechanism in aquatic surface chemistry. In: Stumm W, editor. Aquatic surface chemistry: Chemical processes at the particle/water interface. New York NY, USA: Wiley-Interscience. p 3-32.

Weston DP, Mayer LM. 1998. In vitro digestive fluid extraction as a measure of the bioavailability of sediment-associated polycyclic aromatic hydrocarbons: Sources of variation and implications for partitioning models. *Environ Toxicol Chem* 17:820-829.

White JC, Hunter M, Nam K, Pignatello JJ, Alexander M. 1999. Correlation between biological and physical availabilities of phenanthrene in soils and soil humin in aging experiments. *Environ Toxicol Chem* 18(8):1720-1727.

White JC, Hunter M, Pignatello JJ, Alexander M. 1999. Increase in bioavailability of aged phenanthrene in soils by competitive displacement with pyrene. *Environ Toxicol Chem* 18(8):1728-1732.

Xing B, Pignatello JJ. 1998. Competitive sorption between 1,3-dichlorobenzene or 2,4-dichlorophenol and natural aromatic acids in soil organic matter. *Environ Sci Technol* 32:614-619.

Xing B, Pignatello JJ. 1996. Time-dependent isotherm shape of organic compounds in soil organic matter: Implications for sorption mechanism. *Environ Toxicol Chem* 15:1282-1288.

Yiacoumi S, Rao AV. 1996. Organic solute uptake from aqueous solutions by soil: A new diffusion model. *Water Resour Res* 32:431-440.

Yin Y, Allen HE, Huang CP, Sanders PF. 1997. Adsorption/desorption isotherms of Hg(II) by soil. *Soil Sci* 162:35-45.

Yin Y, Allen HE, Huang CP, Sparks DL, Sanders PF. 1997. Kinetics of mercury(II) adsorption and desorption by soil. *Environ Sci Technol* 31:496-503.

Yin Y, Allen HE, Li Y, Huang CP, Sanders PF. 1996. Adsorption of mercury (II) by soil: Effects of pH, chloride and organic matter. *J Environ Qual* 25:837-844.

Zawadzki ME, Harel Y, Adamson AW. 1987. Irreversible adsorption from solution. 2. Barium dinonylnaphthalenesulfonate on anatase. *Langmuir* 3:363-368.

Section III

Measuring Bioavailability

Chapter 8

Biological Measures of Bioavailability

Sara J. McMillen, Cornelis A.M. Van Gestel, Roman P. Lanno, Greg L. Linder, Stan J. Pauwels, Gladys L. Stephenson

Introduction

Bioavailability is a concept that describes the interaction between a chemical and a biological entity. In this context, "chemical" refers to individual chemicals (e.g., Hg), mixtures of chemicals (e.g., petroleum hydrocarbons, Arochlor 1254, polychlorinated biphenyls [PCBs]), or even chemical properties of a medium, such as pH (H^+ concentration). Bioavailability can be measured as the surficial contact or actual entry of chemicals into ecological or human receptors, but it always involves the interaction of an organism with a chemical. It is specific to the receptor, route of entry, time of exposure, type of chemical, chemical concentration, and matrix containing the chemical. By this definition, bioavailability is not necessarily measured as an adverse effect but can simply be measured as uptake of a chemical by an organism. The continuum of interaction and uptake can proceed to a point where bioavailability is of sufficient magnitude to result in an adverse or toxicological effect.

For example, chemicals may accumulate in plants or soil-dwelling invertebrates with no observable adverse effects to those organisms. However, continued uptake and bioaccumulation of chemicals by such organisms may result in adverse effects and also may be an important component of assessing risk to organisms in higher trophic levels that may feed upon those plants or invertebrates. Measures of bioavailability may be direct, such as quantifying contaminant concentrations in tissues, or indirect, such as toxic responses in soil invertebrate bioassays or alterations in microbial processes. Indirect measures are not necessarily indicative of bioavailability or uptake. For example, certain chemicals such as petroleum hydrocarbons may adversely impact soil structure, wettability, and nutrient availability, resulting in inhibition of plant growth with no actual uptake of petroleum hydrocarbons by the plant (Chaineau et al. 1997). Hydrocarbons may also impair the uptake of water from the soil by arthropods. Nevertheless, indirect test method results may be more easily obtained or less expensive than direct measurements and might be well correlated with bioavailability. Preferably, indirect and direct methods should be combined to improve insight into the potential risk for all trophic levels in an ecosystem.

Contaminated Soils: From Soil–Chemical Interactions to Ecosystem Management. Roman P. Lanno, editor.
 ISBN 1-880611-31-7

Why measure bioavailability?

For ecological risk assessments (ERAs), a chemistry-based approach is usually applied by comparing the results of site-specific chemical analyses with generic soil-quality criteria, if criteria exist. When concentrations of specific contaminants in soil exceed criteria values, human health and terrestrial ecosystems might be at risk. Also, linkages between contaminated soil and surface water and ground water may be evaluated, most often through fate and transport modeling (Chapters 5 and 6). For terrestrial ecosystems, a toxicity-based approach should be applied as part of a Tier 2 or site-specific assessment wherein bioavailability and toxicity are measured (directly or indirectly) to more accurately estimate ecological risk. Bioavailability tests, when combined with chemical analyses to identify and quantify contamination in soil and biota, may show that potentially adverse effects result from exposure to chemicals in soils at field sites. This information, when used in conjunction with biological surveys, can be used to establish a causal link between contaminants and adverse ecological responses. Without field and laboratory data, other potential causes for the observed effects, such as habitat alteration or natural variability, which are not directly related to toxic effects of the contaminants, cannot be eliminated. This chapter focuses on biological test methods that measure directly and/or indirectly the bioavailability of contaminants in site soils.

The rationale for using site-specific bioavailability assessments coupled with toxicity-based approaches include the following:

- Soil-quality criteria, when available for individual chemicals present in site soils, are usually derived from the results of laboratory toxicity tests which do not properly reflect bioavailability of these substances under more natural conditions (e.g., aged field soils).
- Soil-quality criteria are usually developed for single chemicals and do not account for additive, synergistic, or antagonistic interactions among chemicals in a complex mixture.
- Laboratory toxicity tests are usually conducted with single species, some of which may not be ecologically relevant to field site conditions. However, the test species may sometimes be appropriate surrogates for ecologically relevant species.
- Soil-quality criteria rarely reflect the influence of abiotic factors (e.g., weathering, organic matter, sequestration, clay mineral content) that reduce chemical bioavailability.

- Toxicity data should not be evaluated without considering bioavailability, because the observed toxic effect might be due to factors other than contaminant availability.
- Mixtures of contaminants and soil interactions will be site specific, depending primarily on soil properties, contaminant and soil contact time, and the local climate.
- Unknown cocontaminants or physical stressors associated with observed adverse effects may exist at a site.

Based on these considerations, it is now generally believed that a bioavailability and toxicity-based approach is appropriate for estimating potential risks to a soil ecosystem (USEPA 1985; Bergman et al. 1986; Linder et al. 1992; Posthuma et al. 1998; ASTM 1999a). Such an approach may combine direct and indirect measurements of bioavailability.

When should bioavailability be measured?

If Tier 1 soil screening levels are exceeded, then bioavailability should be measured as part of subsequent tiers during an ERA. In some cases, bioavailability may be helpful in determining the efficacy of in situ remediation. Bioavailability tests must be conducted on a site-specific basis because bioavailability will be receptor, chemical, and soil dependent. The basis for the screening or threshold values may not be applicable to a given site because the species used in the laboratory toxicity tests may not be representative of the species present within the site ecosystem.

Differences in sensitivity among species are clearly demonstrated in Figure 8-1. The insecticide dimethoate is toxic to arthropods, only moderately toxic to earthworms and isopods, and virtually nontoxic to nematodes.

As outlined in Chapter 2, ecological risk at contaminated sites is best evaluated using a tiered approach. The preliminary tier reviews regulatory requirements, establishes a problem formulation framework, and determines the need to proceed with an ecological evaluation. Tier 1 consists of comparing representative contaminant concentration values in soil to preestablished, generic, soil-quality guidelines. In many situations, one or more contaminants at a site will exceed these guidelines. This may require a more detailed assessment in Tier 2, when bioavailability measurements would be more appropriate.

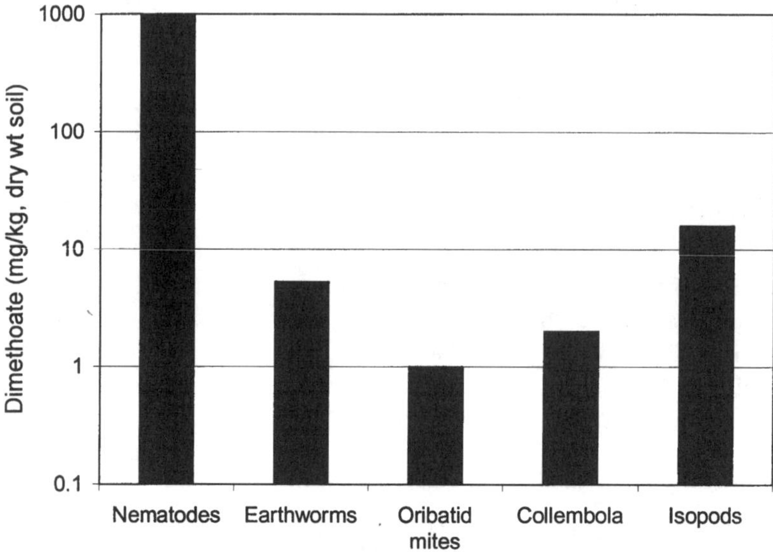

Figure 8-1 EC50 values for the effect of dimethoate on the reproduction of different soil invertebrates (adapted from Løkke and Van Gestel 1998)

Bioavailability and toxicity testing may be conducted in the laboratory using soil samples collected at the site. Many toxicity test results published in the literature are based on chemicals or substances freshly spiked into well-defined synthetic soils or well-characterized field soils. These types of measurements tend to overestimate the toxicity of a chemical at a field site (Alexander 1995; Smit et al. 1997; Smit and Van Gestel 1998). If laboratory tests using chemicals or substances spiked in defined soils were predictive of toxicity to receptors in the field, there would be no need to refine site or ecological risk assessments (ERAs). However, due to the uncertainties of extrapolation of laboratory toxicity test results to field situations, there is a need to examine contaminant bioavailability at field sites. As a consequence, in situ testing may be critical to the assessment process (Stephenson et al. 2002).

Currently, a Tier 2 evaluation could include testing organisms exposed to contaminants in site soils by placing them into containers with the site soil; establishing vegetation plots and monitoring for changes in growth, community structure, and/or biomass; and measuring contaminant residues in tissues of soil invertebrates and plants collected from field sites. Additional approaches are also available to assess in situ bioavailability of contaminants to terrestrial receptors.

The following flow chart is proposed for the site evaluation and assessment of bioavailability:

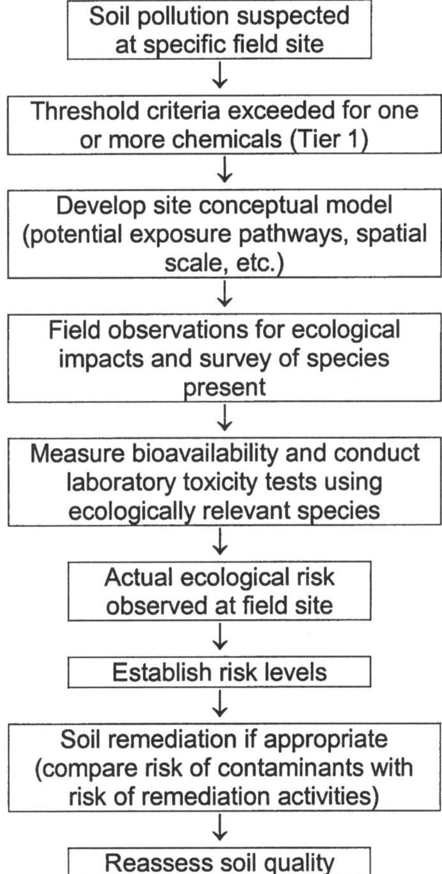

Measurement Techniques to Assess Bioavailability

Several techniques might serve as direct or indirect measures of bioavailability. It is important that techniques quantify the amount of contaminant potentially available to or taken up by an organism exposed to contaminated soil. Tables 8-1 and 8-2 summarize various methods and their application to contaminated soils. A short description of each method is also provided.

Table 8-1 Comparison of test methods to measure bioavailability and toxicity of chemicals in field soils (in situ or laboratory test methods)

Test	Direct measure of bioavailability	Contaminant specificity	Specific for which chemical?	Standard protocol available?[a]	Regulatory use[b]	Validated[c]	Comparative cost of test[d]	Test duration[e]
Microbial processes or assays	No	No	–	Yes	No (?)	Yes (soils)	M[f]	M to H[g]
Decomposition	No	No	–	Yes	Yes (?) But not with contaminated soils	??	L[h] to M	M
Pollution-induced community tolerance (PICT)	No	No	–	No	No	No	M	M
Microtox solid phase	No	No	–	Yes	Yes	Yes (soils)	L	L
Genotoxicity or mutagenicity	No (?)	No	–	Yes	Yes	Yes (?)	L to M	L
Toxicity	No	Both	Depends on how tool is applied	Yes	Yes	Yes, for some soil species	M to H	L to H

[a] Have standard protocols or guidelines been published by regulatory or other organizations (such as USEPA, ASTM, OECD, FIFRA, others)?

[b] Has the test been applied on contaminated soils within a regulatory context (Superfund, RCRA, others)?

[c] Has the test been validated using a "round-robin" type of assessment?

[d] Comparative cost: L = <1K; M = 1 to 5K; H = >5 K.

[e] Test duration: L = <7 days; M = 7 to 30 days; H = >30 days.

[f] M = Medium.

[g] H = High.

[h] L = Low.

Table 8-2 Test methods for the measurement of bioavailability and/or toxicity in organisms (in situ or laboratory test methods)

Test	Direct measure of bioavailability	Contaminant specificity	Specific for which chemical?	Standard protocol available?[a]	Regulatory use[b]	Validated[c]	Comparative cost of test[d]	Test duration[e]
Organism level								
Bioaccumulation	Yes	Yes	Most chemicals	Yes	Yes	Yes (aqueous phase)	H[f]	H
Critical body residue (CBR)	Yes	Both	PAH, metals, halogenated hydrocarbon (HC), nonpolar chemicals (except aliphatic HCs)	No (?)	No	Yes (aqueous phase)	H	H(?)
Suborganism level								
Metabolites	Yes	Yes	PAH, others (?)	??	Yes	No	M[g] to H	L[h] to H
Biochemical biomarkers	Yes	Yes	Metals, PAHs (any P450 inducer), DNA adducts	No	Yes	No	M to H	L to H
Histology	No	No	–	No	No (?)	No	M to H	L to H
Histochemistry	Yes	Yes	Metals (others??)	No	Yes	No	M to H	L to H
Physiological	No	No	–	No (?)	??	No	M to H	L to H
Immunological assays	No	No	–	No	No	No	M to H	L to H

[a] Have standard protocols or guidelines been published by regulatory or other organizations (such as USEPA, ASTM, OECD, FIFRA, others)?

[b] Has the test been applied on contaminated soils within a regulatory context (Superfund, RCRA, others)?

[c] Has the test been validated using a "round-robin" type of assessment?

[d] Comparative cost: L = <1K; M = 1 to 5K; H = >5K The estimated cost includes the cost of exposing the test organisms in the lab; costs may differ if organisms are field collected.

[e] Test duration: L = <7 days; M = 7 to 30 days; H = >30 days. The estimated test duration includes the time necessary to expose the test organisms in the lab; duration may differ if organisms are field collected.

[f] H = High.

[g] M = Medium.

[h] L = Low.

Rationale for excluding "aquatic" toxicity tests

"Aquatic" toxicity tests were excluded from this assessment. Such tests include frog embryo teratogenis assay–*Xenopus* (FETAX) (*Xenopus laevis*–African clawed frog), fishes (e.g., fathead minnow or rainbow trout), aquatic invertebrates (e.g., *Daphnia magna*), algae (e.g., *Selenastrum capricornutum*), or aquatic plants (e.g., *Lemna minor*). Also excluded were sediment-type toxicity tests. The main reason was that such tests do not provide a true measure of contaminant bioavailability in soils. Instead, they primarily assess the availability of chemicals that can be readily extracted using aqueous solvents, and for aquatic organisms, or for bulk sediment tests, they expose those chemicals that may be available in pore water and/or coincidentally ingested during feeding. Soils may be evaluated using aqueous eluates, but for the most part, aqueous extracts only allow performance of acute or short-term tests that generally focus on lethal endpoints. More importantly, aquatic species used in these soil-derived matrices are not ecologically relevant for soils under realistic exposure pathways, and aqueous extracts may give an indication of instantaneous bioavailability, generally expressed as toxicity, but not longer-term bioavailability. While many researchers, regions, and states have used these tests to assess soil, such tests cannot provide realistic data useful to answer questions about the bioavailability of contaminants in soils and ecological effects in soil ecosystems. In some habitats (e.g., wetlands), or under well-defined conditions (e.g., contaminated soils characterized by high erosion potential, periodically inundated hydric soils), applying soil-derived aqueous test solutions or sediments may be useful to evaluate risk.

Soil dilution tests

When a soil is very toxic, tests may be conducted using a "soil dilution" series to assess toxicity and bioavailability. In these tests, contaminated soil is diluted with control or reference soil, or some other solid phase, but not water. This approach is often referred to as "soil amendment." The result of a soil dilution test is expressed as a dilution factor resulting in a toxicity endpoint (e.g., LC50 = 40% contaminated soil). When considering ERA, one may simply state that the ecological risk associated with a particular site soil is high when initial tests with the soil show 100% effect in an undiluted state, as opposed to no effect in an undiluted state. As a consequence, these results may lead into a second tier of risk assessment, and no further initial testing may be required unless other stressors (e.g., soil type) are suspected as the source of adverse effects. In a second tier of testing, a soil dilution series test may be conducted to determine at what dilution toxicity will be altered. A major consideration in soil dilution tests is deciding what constitutes an appropriate control soil or medium for use as a diluent. Control soils currently used in soil toxicity tests can be formulated from soil constituents (OECD 1984; ASTM

1999b), field-collected "clean" reference soils (Stephenson 1998a, 1998b, 1998c), or a substrate with minimum sorption potential (e.g., glass beads, sand). Regardless, a reference soil should have physicochemical characteristics similar to those of the site soil being evaluated, but it should be demonstrably free of the contaminants present in the site soils. Past practice suggests that it is difficult to obtain a soil which satisfies all of these requirements (Van Gestel et al. 1988). Therefore, rather than a dilution series, it may be more practical to collect soil samples from the site along a gradient of pollution and subject these to direct and/or indirect measures of bioavailability. The least polluted site on the gradient then may serve as an experimental control or as the soil used to amend the site soils with elevated pollution (Van Gestel and Hensbergen 1997). However, if samples are taken along a gradient of pollution, care must be taken to ensure that only contaminant concentration varies along the gradient and not other parameters, particularly those that may act as modifying factors of toxicity (e.g., pH or organic C).

Overview of test methods

The tests shown in Table 8-1 can be divided into different categories. The tests at the community level focus on processes performed by organisms indigenous to the contaminated soils, or the health of the indigenous microbial community (e.g., pollution-induced community tolerance (PICT). These tests, as well as those at the population level and toxicity tests with single species, are directly applied to soil samples collected from the contaminated site (Table 8-1). All other test methods (Table 8-2) can be applied to test organisms either during or at the end of toxicity tests or to organisms captured or sampled at the contaminated site.

An ideal measure of bioavailability should 1) provide a direct, accurate, and relevant measure of bioavailability to a receptor; 2) be codified in a standard protocol; 3) be validated and accepted for regulatory use; and 4) be cost-effective in terms of time, equipment, and labor. It may also be advantageous for the measurement to be specific for one chemical or a well-defined group of contaminants. However, none of the tests in Tables 8-1 or 8-2 meet all these criteria. On closer examination, 2 types of tests probably come closest to this ideal: the bioaccumulation test and the metabolite test. Both approaches measure the concentration of contaminants inside an exposed organism, with the former assessing concentrations of the parent compound and the latter measuring concentrations of metabolic transformation or degradation products. Of the two, the bioaccumulation test may be most relevant because parent compounds are measured directly. Ideally, after steady state is reached between the test organism and chemical in the environment, it may be possible to match this "internal dose" with a measure of chemical bioavailability and/or the exposure concentration in whole soils.

Test methods for site soils

Community-level tests

Soil microorganisms and microbial processes

Microbial populations in the rhizosphere are a particularly important soil microbial community. Rhizosphere microbial populations have a unique relationship with the plant root zone they colonize. As a result, nutrients become available for the mutual benefits of plants and rhizosphere microbes (Gerhardson and Clarholm 1986).

Nutrient cycling is one of the most ecologically significant and potentially most sensitive processes within terrestrial ecosystems. Soil processes involving nutrients, especially C, N, P, and S transformations, are important in maintaining functioning ecosystems, and they contribute to soil stability, soil fertility, and plant productivity. The movement of nutrients in an ecosystem includes cycling within and between below- and aboveground compartments. Such processes are performed by an array of microorganisms, including free-living and symbiotic bacteria and fungi, algae, and various protozoans. For a review of nutrient cycling in soils see Parmelee et al. (1998) or Blair et al. (1995). Because the majority of biochemical transformations in soil result from microbial activity (Alexander 1977), there is concern that soil contaminants that affect microbial life also may alter cycling of nutrients in the environment and ultimately affect soil fertility and plant productivity. For example, processes such as nitrification and sulfur oxidation are mediated exclusively by specific groups of microorganisms, and the rates at which these processes occur are indicative of their specific metabolic activity.

Tests based on inhibition of soil respiration, as an indicator of C transformation and overall microbial activity, have long been used for the risk assessment of pesticides. However, these tests do not appear to be the most sensitive for measuring the impact of these toxicants on soil environments. In addition, certain types of soil contamination, such as biodegradable organic compounds, may stimulate rather than inhibit soil respiration. For that reason, soil respiration is not recommended as a method for determining bioavailability.

The transformation of organic N to inorganic forms is an important microbial function contributing to the fertility of soil and is a microbial process that has become significant as an indicator of the effects of toxicants. The major N transformations mediated by soil microorganisms include ammonification, nitrification, denitrification, and N fixation.

The biological oxidation of ammonia to nitrate in soil is facilitated by 2 groups of chemolithotrophic bacteria: ammonium oxidizers and nitrite oxidizers. Inhibition of either of these groups may significantly alter the dynamics of the soil N pool. Nitrification has been shown to be the most sensitive part of the nitrogen cycle when a microbial community is exposed to toxicants. Degradation products of chlorinated compounds may influence nitrification (Corke and Thompson 1970) and, in general, nitrification is inhibited by the action of heavy metals (Giashuddin and Cornfield 1979; Chang and Broadbent 1982; Bewley and Stotzky 1983). For these reasons, nitrification is recommended as a method for determining bioavailability and effects of soil contaminants. See Chapter 4 for recommended test methods.

In addition to assays for microbial processes, changes in soil microbial communities can be assessed using analysis of cell constituents such as phospholipid fatty acid analysis (PLFA) (Ringelberg et al. 1988; Pinkart et al. 1995; Bååth et al. 1998), or nucleic acid probes or fingerprinting techniques (e.g., amplified ribosomal DNA restriction analysis [ARDRA], amplified fragment length polymorphism [AFLP]). Alternatively, more traditional assays such as direct microscopic counts with differential staining techniques or plate counting may be used (Lawlor et al. 1997). However, due to functional redundancy in the soil microbial communities, the numbers, the types, or even the distribution of types of microbes are not necessarily indicative of soil health in terms of soil processes. Therefore, the direct measurement of soil microbial processes is recommended.

Decomposition

Decomposition of organic matter (leaf litter, dead animal material, etc.) is essential to the cycling of nutrients in the terrestrial ecosystem. The microbial community carries out the majority of decomposition and mineralization processes in the soil. The functions are stimulated by the activities of soil animals, especially saprotrophs feeding on decaying organic matter. Microbivores grazing the microflora can directly contribute to elemental cycles (Verhoef and Brussaard 1990). As a consequence, decomposition processes may be impaired when either the microbial or the animal community or both are adversely affected by soil contamination.

Traditionally, litterbags have been used to study the role of soil animals in decomposition or to determine potential effects of contaminants on decomposition processes. Recently, the bait lamina test has been used as a relatively easy and inexpensive measure of decomposition activity in soils (Kratz 1998). This test quantifies the decomposition of a cellulose-based substrate that is enclosed in small holes drilled into PVC strips (length 15 cm, containing 16 holes). A strip is inserted into the test soil, and decomposition can be measured from the number of empty holes. Organisms in the soil community consume the cellulose-based substrate. The test provides a measure of the reduction in decomposition rate and the potential effect of contaminants on the vertical distribution of the organisms contributing to the decomposition process. This test is not only a measure of microbial activity; it also reflects the level of activity by earthworms, enchytraeids, microarthropods, protozoans, fungi, and other microfauna and flora. For further details on this method, refer to Chapter 4.

Pollution-induced community tolerance

The PICT concept was initially developed for assessing aquatic algal communities (Blanck et al. 1988). The concept assumes that organisms exposed to increased levels of contaminants develop tolerance to these contaminants. The degree of tolerance of a community can be indicative of the level of exposure, which in turn is related to the bioavailability of the contaminants. For soil microbial communities, the technique has not yet been standardized, but promising results have been obtained during application to metal-contaminated soils (Rutgers et al. 1998). The method involves collecting soil samples, preferably from a gradient of contamination. The microbial community then is extracted from these soil samples and exposed under laboratory conditions to different concentrations of the predominant contaminant found in the field gradient. For each field sample, this will yield a dose–response relationship, from which a sensitivity value (e.g., EC50) of the microbial community can be derived. By plotting these sensitivity values as a function of the field exposure concentrations, sensitivities can be compared. If the sensitivity of a community to a contaminant has decreased when compared to the control community, it is concluded that PICT has occurred. This response can be considered a strong indication of contaminant bioavailability and effects on the microbial community (Blanck et al. 1988; Van Beelen and Doelman 1997; Rutgers et al. 1998).

Population-level tests

Solid-phase Microtox assays

Bioluminescence is a phenomenon produced by a branch of the electron transport system in specific microbes, and several investigators have described toxicity and bioavailability assays based on inhibition of this system (Bulich 1984, 1986). The first commercial toxicity and bioavailability test using bioluminescent bacteria was developed at Beckman Instruments, Carlsbad, California, USA (Bulich 1979, 1982). The test, now marketed by Microbics Corporation (still under the trade name of Microtox), uses freeze-dried cultures of the marine bacterium *Vibrio fischerii* and is based on the inhibition of bioluminescence by toxicants. A solid-phase test may be used with soils (and sediments) and is completed using a small amount of soil added to a cuvette, then adding a saline solution (or osmotically adjusted liquid), and shaking the cuvette. The soil is subsequently allowed to settle to the bottom of the cuvette and, in practice, is a soil-extract test. As such, the test can be applied to extracts of contaminated soils or soil slurries ("solid-phase" Microtox). The results of several studies with pure compounds and complex chemical mixtures in water have shown that Microtox responses are in general agreement with several standard fish and aquatic invertebrate bioassays (Curtis et al. 1982; Sanchez et al. 1988).

Microtox has similarly been compared with soil bioassays using *Lactuca sativa* (plants) and *Lumbricus terrestris* (earthworms) for soil enzyme activity (Sheppard et al. 1993). Horvath and Gruiz (1995) compared Microtox with tests on *Bacillus subtilus*, *Azotobacter chroococcum* (bacteria), *Scenedesmus* sp. (algae), *Daphnia magna* (crustaceans), and *Sinapis alba* (plants) and found that, in general, bacteria were usually more sensitive.

Interpretation of the solid-phase Microtox test may be hampered by the fact that particle size composition of the soil may affect the response of the test organism (Benton et al. 1995). Also, for general use with soil extracts, the presence of 2% sodium chloride in the assay medium, and the fact that a marine microorganism is used, makes its applicability to contaminated soils questionable. Because of these aspects, tests on indigenous soil microorganisms exposed to contaminants in whole soil are preferred. This has resulted in the development of other bacterial bioluminescence assays based upon the insertion of lux genes into bacteria that are more relevant to soil ecology such as *Pseudomonas fluorescens* and *Rhizobium leguminosarum biovar trifolii* (Paton et al. 1995; Paton, Palmer et al. 1997; Paton Rattray et al. 1997). Bioassays with these genetically modified bacteria provide similar toxicological response based upon the inhibition of luminescence with increased toxicity to general cellular metabolism.

Bacterial genotoxicity and mutagenicity tests

Several bacterial screening tests have been developed and applied to assess the potential of contaminated soil for genotoxicity. These tests include the Ames assay with *Salmonella typhimurium*, *Escherichia coli* WP2, SOS Chromotest, Toxi-Chromotest, and the *Bacillus subtilis* repair assay, among others (Linz and Nakles 1997). One major use of these tests in the field of environmental assessment has been to screen contaminated soils and waste materials before and after cleanup operations, such as bioremediation.

The tests rely largely on organic solvent extraction procedures to remove the contaminants from the solid matrix. The solvents include dichloromethane, methylene chloride, and distilled benzene or methanol. Typically, genotoxicity varies with the type of solvent employed. After the extraction step, the material is resuspended in dimethylsulfoxide (DMSO) before being applied to the bacteria.

Bacterial genotoxicity assays allow for a rapid screening of potentially mutagenic compounds (i.e., the SOS-Chromotest takes 2 days to complete), are relatively well standardized, have been applied in regulatory contexts, and also are relatively cost-effective. One of the major drawbacks or limitations with these tests is that the solvent extraction step eliminates information relevant to assessing the bioavailability of contaminants in soil. Also, the bacterial strains used in these assays have little or no relevance to bacterial populations in the field or to invertebrate and vertebrate receptors. As such, their use as a potential tool to measure the bioavailability of contaminants in soil is considered to be limited. An exception to this limitation is a new test developed by Alexander et al. (1999) using a rifampycin-resistant bacterium, *Pseudomonas putida,* that is indigenous to soils. It is a solid-phase test and therefore produces more relevant data than those tests performed on soil extracts.

Organism-level tests

Toxicity

Toxicity testing with whole soils and ecologically relevant soil organisms is one of the tools that can be used by a site or risk assessor to indirectly measure biological availability of contaminants in soils. Toxicity may indicate that a chemical is bioavailable; however, the absence of adverse effects in a traditional toxicity test does not infer an absence of bioavailability to test or field-collected organisms. Nonetheless, toxicity is often used as an indicator of the bioavailability of a soil contaminant. As such, toxicity may be associated with acute and chronic effects reflected in various endpoints or measures of biological effects.

Terrestrial toxicity tests have been developed to assess the toxicity of soils amended with potentially toxic chemicals (OECD 1984; Stephenson, Solomon et al. 1997, 1998a, 1998b, 1998c; ISO 1998a, 1998b; Løkke and Van Gestel 1998; ASTM 1999b, 1999c, 1999d). The development and standardization of most of these test methods have evolved gradually since the 1970s as a direct result of the regulatory requirements for registration of new products. The need for additional biological test methods to assess contaminated soils has increased (Greene et al. 1989; Keddy et al. 1995; Stephenson 1995, 1998a, 1998b, 1998c) to meet the regulatory requirements for site-specific ecotoxicological characterization of risk associated with site contamination. For soils, another impetus has been the need to answer the question: How clean is clean?

A number of approaches can be taken to evaluate contaminated soils. Depending upon local environmental conditions (e.g., upland or wetland soils) and the site-specific questions being asked in the evaluation process, numerous methods have evolved (Kapustka et al. 1989; La Point and Fairchild 1989; McBee 1989; Parkhurst et al. 1989; Holland 1990; Simenstad et al. 1991; Linder et al. 1992; ASTM 1999a; Stephenson et al. 2002; see also Chapter 4).

Toxicity endpoints are derived from acute and chronic toxicity tests. The tests most frequently conducted are designed to address survival of the test organisms in 100% site sample (water, soil, or sediment) in the laboratory or in situ (see "Soils and Society," p 76), and to describe dose–response relationships for the survival of laboratory test organisms exposed to a range of contaminant concentrations in soil. This approach yields estimates of acute toxicity that can be used to estimate median lethal concentrations (LC50s). Reproduction and growth assays are used less frequently to estimate median effective concentrations (EC50s), or depending upon dose–response relationships, may yield no-observed-effect concentrations (NOECs) and lowest-observed-effect concentrations (LOECs). These tests and in situ exposure data may provide information on the bioavailability of ambient concentrations of chemical contaminants in soils, if the toxic effects are related to chemical exposure.

Test methods for measurements in organisms

Organism-level tests

Bioaccumulation

Bioaccumulation tests directly measure the uptake of chemicals from soil into test organisms. Measurements of chemical uptake by plants and soil macroinvertebrates are preferred for contaminated soil assessments when the potential for trophic transfer of contaminants is a concern. For example, it is generally assumed that most of the contaminant uptake in plants occurs through the roots and that the concentration of the contaminants in leaves is directly related to the concentrations in the soil or sediment. Bioaccumulation tests are widely cited in the literature pertaining to ERA, and have been carried out under both laboratory and field conditions using a variety of species (Marquenie et al. 1987; Morgan and Morgan 1988; Beyer and Miller 1990; Callahan et al. 1991; Corp and Morgan 1991; Menzie et al. 1992; Hopkin et al. 1993; Paine et al. 1993).

Bioaccumulation of chemicals in soils by plants is of concern in hazard and risk evaluation (e.g., upland disposal of dredge materials; Folsom and Price 1990), and a bioaccumulation test currently is being standardized by ASTM. In general, bioaccumulation in plants has been evaluated using a variety of species, including commercial varieties (e.g., lettuce and alfalfa), native species (e.g., sunflowers), and exotics (e.g., grasses and forbs). To evaluate bioaccumulation of chemicals from bulk soils, plants are grown under controlled conditions (most often in greenhouses or environmental chambers) in 100% site soils for species-specific time periods. Commercial species may be exposed for 14 to 30 days, depending on the test design, but some exposures may be extended for weeks, depending upon test specifications regarding attainment of steady-state conditions. Depending upon the experimental design and hypotheses tested (e.g., evaluation of remediation options), plant bioaccumulation tests may incorporate amended soils with fractional exposures of 50%, 25%, 12.5%, and 0% contaminated soil.

Bioaccumulation tests have also been standardized for soil macroinvertebrates, primarily the earthworm (ASTM 1999b), although alternative species have been used (Løkke and Van Gestel 1998). As with plant bioaccumulation, uptake and depuration of contaminants in earthworms must be considered within the limitations of the test. Test duration must be sufficiently long to ensure the establishment of a steady state or to allow for a significant accumulation of chemical so that depuration can be monitored over time. Bioaccumulation tests may be performed with organisms exposed to contaminated soils in the laboratory as part of chronic toxicity bioassays. Bioaccumulation may also be assessed in situ with "clean" organisms exposed to site soils in the field, or measured in field-collected organisms assuming that the organisms are at steady state with respect to chemical accumulation.

Bioaccumulation factors (BAFs) can help to compare bioavailability differences between metals in polluted and decontaminated soils using bioassays with earthworms and plants (Van Gestel et al. 1988, 1993; Linder et al. 1998). BAFs for metals might be dependent upon metal concentrations in the soil and vary with the bioavailable soil metal concentration, with higher BAFs at lower metal concentrations in the soil. For essential metals, internal concentrations in organisms are often regulated over a wide range of exposure levels, making the interpretation of BAFs difficult (Chapman et al. 1996). For organic chemicals, BAFs are affected by several factors, such as the lipophilicity of the chemical (expressed as K_{ow}), the organic matter content of the soil, and the lipid content of the organism. Ideally, BAFs should be independent of K_{ow} and determined mainly by the latter 2 factors (Connell and Markwell 1990). When significant biotransformation or metabolism of a chemical occurs, BAFs are often lower than expected based upon accumulation kinetics. For the purposes of ERA, there appear to be 2 ways to interpret BAFs:

1) comparing of internal body concentrations with critical threshold levels associated with some type of biological response in the assessment of risk to a particular species and
2) considering internal body concentration in an organism in the potential for food-chain transfer to higher trophic levels.

Critical body residues

Critical body residues (CBRs) are tissue, organ, or whole-body residue concentrations for specific chemicals or groups of chemicals that are correlated with adverse effects (e.g., reduced survival or decreased reproductive success). Results of bioaccumulation tests or tissue residues measured in organisms exposed to contaminated soil in toxicity tests or in the field can be compared with the CBR to obtain an indication of bioavailability and potential risk. Although the CBR concept has been well established in aquatic ecotoxicology (McCarty et al. 1992;

Van Straalen 1996; Lanno et al. 1998), only a few CBR values have been derived for terrestrial soil-dwelling organisms. Lanno and McCarty (1997) suggested the use of the CBR concept for earthworms, using pentachlorophenol as an example. Lethal CBRs have also been developed for phenanthrene in earthworms (Wells and Lanno 2001). Crommentuijn et al. (1994) determined lethal body concentrations for Cd in 6 different soil arthropod species. CBRs for Cd and/or Zn on the growth and/or reproduction of plants, earthworms, and Collembola have also been reported (Smit and Van Gestel 1997, 1998; Van Gestel and Hensbergen 1997; Posthuma et al. 1998). For a critical review of the possible use of CBRs in terrestrial ecotoxicology, refer to Van Straalen (1996) and Van Wensem et al. (1994).

Suborganism-level tests

Metabolites

Within an organism are physiological and biochemical processes involved with the uptake, metabolism, and depuration of contaminants. These processes provide analytical targets reflective of concentrations of chemicals that are bioavailable to organisms exposed to chemical mixture in soil. For example, although the induction of cytochrome P450 is generally characterized by lower enzyme activities in soil macroinvertebrates compared to terrestrial vertebrates (Stokke and Stenersen 1993; Van Straalen 1994; Zanger et al. 1997), the chemical characterization of metabolites from parent compounds may be a useful tool for evaluating the bioavailability of a chemical. The evaluation of tissues for the presence of metabolites and the parent compound provides a potentially sensitive tool to qualitatively identify whether chemical constituents in the soil are bioavailable. This approach has been used on chemicals such as PAHs (Stroomberg et al. 1996; Van Brummelen et al. 1996; Ma et al. 1998) and chlorinated hydrocarbons, including chlorophenols (Haimi et al. 1992; Fitzgerald et al. 1996).

Biochemical biomarkers

Several biochemical techniques may be related to exposure and potential adverse effects of chemicals to organisms. The advantage of such techniques is that they can be applied to organisms after exposure in toxicity tests and to field-collected organisms, thereby possibly reflecting field exposure. A major problem with these biomarkers is that although good correlations with chemical exposure have been achieved in the laboratory, their application to chemical exposure under field conditions is not validated. As an example, phytochelatins in plants cannot be used as indicators of metal exposure because their induction is not proportional to metal levels in soil (Knecht et al. 1994). This poor correlation may be due to plant metabolism or to metal exposure expressed as total metal levels rather than as the bioavailable fraction of metals in the soil. Therefore, the presence or absence of biomarker responses can be misleading and needs to be evaluated with caution.

Several biochemical biomarkers are specific to certain chemicals; other biomarker methods, however, are not specific to any chemical and are simply indicative of general stress.

Examples of specific biomarkers are

- metallothionein in animals, specific to metals;
- phytochelatins in plants, specific to metals;
- acetylcholinesterase activity in animals, specific to organophosphate and carbamate insecticides; and
- d-Aminolevulinic acid dehydratase (ALAD) in animals; specific to Pb exposure.

Examples of biomarkers indicative of exposure to general groups of chemicals:

- cytochrome P450—induced by PAHs and other hydrocarbons, organohalogen compounds, other pesticides; and
- DNA adducts—some PAHs.

Examples of nonspecific biomarkers are

- heat shock proteins (HSPs), and
- Glycogen content.

Peakall and Walker (1994) and several others provide additional details about methods and applications of biochemical markers.

Histology

Histology has long been used to assess the effects of environmental contaminants on the tissues of exposed organisms, both in terrestrial and aquatic systems. Significant morphological changes are typically interpreted to mean that chemicals are responsible for the observed effects (McCarthy and Shugart 1990; Huggett et al. 1992; Hamelink et al. 1994). When histological changes are associated with exposure to specific chemicals, histological techniques could serve as indirect measures of bioavailability.

This approach consists of exposing test organisms to known concentrations of contaminants in the laboratory or collecting organisms naturally exposed to contaminants in the field. With smaller species (e.g., earthworms, nematodes, larvae), the entire organism can be prepared for microscopy. Otherwise, the organ of interest (e.g., testes, liver, kidney) is excised and prepared. Either way, after the sample is fixed, it is sectioned into thin slices (<5 µm), mounted on glass slides, and stained to highlight the features of interest. To assess gross pathological responses, it is usually sufficient to prepare tissue samples for light microscopy (100 to 1000 times magnification) (Glaister 1986).

Observable changes are usually assessed qualitatively and described in terms of a visual effect (e.g., pitting, clumping, necrosis, swelling). Such effects can be further ranked as "slight," "moderate," or "severe" when compared to tissues from control organisms. Under certain circumstances, more quantitative descriptions can be provided using the histomorphometric techniques (Jones et al. 1997). For some terrestrial invertebrates, the literature presents well-characterized methods, including histology on slugs (Triebskorn and Köhler 1992) and Collembola (Pawert et al. 1996).

Several inherent limitations with this approach preclude it from being used routinely as a tool to assess contaminant bioavailability in soils. Some of the more important limitations are

- the method is costly and labor intensive;
- the morphological observations are at best semiquantitative and can be rather subjective;
- the appropriate target organ needs to be identified;
- this is not a routine test with recognized protocols;
- it would be difficult to correlate morphological changes with environmental exposures, especially when the organisms have been field collected; and
- a highly skilled histopathologist is required to detect anomalies.

Histochemical methods

To date, the application of cytochemical and histochemical techniques in evaluating contaminant bioavailability in soil biota and in plants has been relatively limited. However, as a qualitative tool to identify whether a chemical is bioavailable in a given soil type, the techniques are relatively inexpensive and are potentially chemical specific. For example, metal-specific dyes and fluorescent antibody techniques may be used in conjunction with routine histological and histochemical techniques (Marigómez et al. 1998) to evaluate whether chemicals, or their derivatives (e.g., cellular inclusion bodies), are present in tissues following exposure. As a tool for evaluating bioavailability, tissues obtained from laboratory-exposed or field-exposed invertebrates and plants may be evaluated for targeted chemicals. For example, metals in tissues can be measured histologically using electron microprobe analysis. However, such an application is highly restricted, owing to the limited availability of the tool. In contrast, standard histochemical techniques are readily available (e.g., from commercial laboratories) and quite suitable tools for analysis (Graumann and Drukker 1991; Presnell et al. 1997).

Physiological markers

Soil contaminants may interfere with the physiology of an organism; a few guidelines have been developed to detect such effects. For example, a method has been applied to detect the effects of soil contaminants on the physiology of earthworms. The method, referred to as the "Neutral Red Retention assay" (Weeks and Svendsen 1996), determines the capacity to retain neutral red as a measure of cell health. The method uses coelomocytes extracted from the earthworm and plated on a microscopic glass. When a neutral red solution is applied, this indicator colorant is taken up by the coelomocyte lysosomes. Because neutral red is cytotoxic, the lysosomal wall will disintegrate after some time and the dye will leak out, coloring the whole cell content red. The time required for leakage to occur may be quantified as an indicator of stress. For example, in healthy earthworms, it may take almost 60 minutes before leaking of the lysosomes occurs, whereas in stressed animals, leaking will occur within a shorter time. This method has shown to be correlated with Cu uptake and toxicity in earthworms (Svendsen and Weeks 1997a, 1997b).

Physiological biomarkers in terrestrial vertebrates have been used to evaluate contaminant bioavailability in soils. Although most of these biomarkers are nonspecific, they can be effectively used to monitor changes in bioavailability of a contaminant over time, provided baseline conditions are well characterized. For example, periodic sampling of terrestrial vertebrates to determine tissue metal residues may be required in "post-remedial action" evaluations. In conjunction with this direct measure of bioavailability, laboratory-exposed or field-collected biota may also be analyzed for physiological biomarkers such as liver function enzymes and stress hormones (Jones et al. 1997). Although an indirect measure of bioavailability, the combination of direct measures such as tissue residues with physiological biomarkers of exposure can help characterize recovery in remediated terrestrial habitats over time. As a result of such post-remediation or compliance monitoring, site management may be adjusted to meet remediation goals.

Immunological markers

While immunotoxicological test methods are more widely described for vertebrates, similar methods have been applied to terrestrial invertebrates under controlled laboratory conditions. For example, methods have been developed to evaluate the immunocompetence of earthworms exposed to various contaminants (Mohrig et al. 1984; Rodriguez-Grau et al. 1989; Eyambe et al. 1990; Chen et al. 1991). In these subacute tests, coelomocytes are harvested from exposed earthworms and are evaluated for their viability, ability to form secretory rosettes, and total and differential cell counts, among other potential subacute measurement endpoints. Through these and similar measurement endpoints, the health of earthworms exposed to soil contaminants can be inferred, and cellular immune function is evaluated as a measurement of effects, which are indicative of exposure to bioavailable fractions of chemicals in soils. Results from such evaluations, if accompanied by field data, would also reflect potential chronic effects associated with diminished disease resistance that may have effects at the population level. Similarly, immunotoxicological test methods using vertebrates may be used to support studies focused on chemical bioavailability. Various biomarkers focused on immune function have been identified and may support weight-of-evidence arguments in ecological effects assessments (Huggett et al. 1992). Several methods have been modified from the biomedical and veterinary sciences to evaluate sublethal immunotoxic effects that may be associated with contaminant exposures in wildlife species. Tests have been modified for terrestrial vertebrates; these methods may be considered within an ecological effects assessment when chemical bioavailability is a concern.

Standard test methods are available to evaluate immune function in vertebrates (Rose and Friedman 1976). These tests typically evaluate humoral and cellular immune function. For example, plaque-forming cell assays to evaluate humoral function (Tizard and Kersey 1996) have been modified for evaluating contaminant effects in vertebrates. Additional methods are also potentially available to measure immunoglobulins (Davis and Ho 1976; Kochwa 1976). Similarly, cellular immune function (Coles 1986) and other immunohematological functions (Jones et al. 1997) are available. Such tests include in vivo and in vitro phytohemagglutinin tests, delayed-type hypersensitivity tests, leukocyte migration inhibition tests, and macrophage activation, function, and phagocytosis.

Most immune function evaluations are insufficiently developed to act as stand-alone assessment methods to evaluate bioavailability. However, when site-specific contaminant histories indicate potential immunological effects, then multiple immunofunction tests (Exon et al. 1986) may be considered in integrated studies that evaluate, for example, subacute effects on avian receptors. In the future, immunotoxicity assessments may help evaluate long-term ecological risks associated with differential bioavailability of chemicals in soils.

Selection of a test battery to assess bioavailability

Chemical measures of contaminant bioavailability are likely to be useful for modifying Tier 1 soil screening levels on a site-specific basis. It is recommended that a test battery comprising ecologically relevant, soil-dwelling organisms, ecologically relevant measurement endpoints, and standard test protocols be used for a Tier 2 assessment of contaminant bioavailability at a contaminated site.

After a decision has been made to conduct site-specific tests on bioavailability and toxicity of contaminants, major questions to be addressed include Which test organisms should be used in the assessment? Which organisms, or groups of organisms, are representative of a suitable terrestrial test battery for evaluating bioavailability? A number of selection criteria should be considered as part of the selection process:

- The organisms should be representative of the ecosystem that is being evaluated. For example, it would be inappropriate to select a test organism such as a bioluminescent marine microorganism to represent terrestrial organisms that typically inhabit soil ecosystems. If a surrogate species must be used to measure bioavailability, it is recommended that this species have a similar life history strategy, perform a similar ecological function in soil ecosystems, be taxonomically related, and have the same route of exposure to contaminant.
- The organisms in the test battery should be representative of the range of species that could potentially inhabit a site. For example, if no lumbricid earthworms exist locally but other soft-bodied, soil invertebrates (e.g., enchytraeids, insect larvae) are present, then it would be prudent to include an organism representative of this group.
- The organisms in the test battery should reflect the diversity that exists for organism-specific sensitivity to contaminants. For example, if a site was contaminated with carryover soil residues of a normally nonpersistent herbicide, it would not be prudent to select as the test organism to measure bioavailability a plant species that selectively excludes herbicide residues. Alternatively, there are biochemical measures of bioavailability (e.g., acetylcholinesterase) that are specific to a particular group of compounds (e.g., organophosphate and carbamate insecticides). See Figure 8-1 for an example of species differences in sensitivity.
- The organisms should be selected in consideration of the current and future land use, the soil type, and the ecoregion in which the site is located. For example, it would not be prudent to select a native prairie grass species from western Canada that commonly grows in clay loam soils, and is tolerant of dry soil conditions but intolerant of wet soils, as a representative plant species to assess a site in a residential area with sandy loam soils, located in Louisiana, that experience relatively high levels of annual rainfall.
- The selected organisms should be able to survive and thrive in the control or reference soils used as experimental controls or as diluents in a test where the organisms are being exposed to sequential or proportional dilutions of the contaminated soils. (See Chapter 2 and glossary for a review of control and reference soil selection).

The test battery recommended for biological measures of bioavailability at a particular site will depend to a large extent on the purpose of the assessment, the site characteristics, and the magnitude and extent of the contamination problem that has been identified. However, at a minimum, it should include tests with plants, soft-bodied soil invertebrates, and soil arthropods, and tests designed to assess decomposition and/or a microbial process such as nitrification.

Selection of measurement techniques

As shown in Tables 8-1 and 8-2, several techniques could be applied to assess bioavailability to organisms exposed to amended soils under controlled conditions in the lab or to organisms collected from a contaminated site. Although choosing the most suitable techniques depends on the objectives of the study, the nature of the receptor and the stressor, and the site-specific conditions and characteristics, a number of criteria have been established to evaluate the applicability of these methods. In addition, the usefulness of each method has been evaluated for different types of organisms and common contaminants, or groups of contaminants, found in soils.

The assays highlighted in Table 8-1 were further examined to identify those that could be applied to the 3 groups of indicator organisms (i.e., plants, earthworms or other soft-bodied organisms, and arthropods) exposed to contaminated soils. The following assays were deemed relevant to the indicator organisms: toxicity, CBR, bioaccumulation, metabolite measurements, biomarkers, histology, histochemistry, physiological markers, and immunological markers.

For each group of organisms, the applicability of the 9 assays was compared for several classes of contaminants. This is probably somewhat unrealistic because soil contaminants are often present as complex mixtures, not distinct categories. It was felt, however, that the selection process for a suitable bioavailability test would be made more difficult if discrete contaminant groupings were not assessed. Readers should therefore be aware that, under some circumstances, the selection process may not be as straightforward as is indicated in this section. A given test was assigned a "yes" value if it had been applied to at least 1 compound within a contaminant class, a "no" value was assigned when a test had not been applied (i.e., no supporting data).

The comparisons in Tables 8-3, 8-4, 8-5, and 8-6 are not the result of an exhaustive literature search. Instead, they are qualitative and based only on the collective "wisdom" and experience of the authors. It is therefore possible that a given test may actually have been applied when in fact the table indicates that it was not, and vice versa. On the whole, however, it is felt that the information captured in these tables reflects the general state-of-the-science in this area.

Tables 8-3 through 8-5 indicate that, as a general rule, toxicity testing, bioaccumulation analyses, and metabolite analyses (plus biomarkers in earthworms) have received the most attention in soil contaminant assessment studies. Biomarkers (with plants and arthropods) and CBR analyses have received less attention. The remaining tests have not been applied to any significant extent.

This breakdown in practice is not surprising. The organisms used in toxicity testing integrate the entire bioavailable dose from the contaminants in soil into a toxic response. This may reflect toxicity from cocontaminants, the parent compound of concern, intermediate breakdown products, or internally generated metabolites. As

Table 8-3 Soil contaminant bioavailability to earthworms (via dermal uptake and ingestion)

	Potential tests to measure contaminant bioavailability in worms								
Contaminant class	Toxicity	CBR	Bioaccumulation	Metabolites	Biomarkers	Histology	Histochemistry	Physiological markers	Immunological markers
Metals	Rec[a]	Rec	Rec	No[b]	Yes[c]	No	Yes	Yes	Yes
PAH	Rec	Rec	Rec	Yes	Yes	No	No	No	Yes
TPH[d]	Rec	?[e]	?	No	Yes	No	No	No	No
Chlorinated organics	Rec	Rec	Rec	Yes	Yes	No	No	No	Yes
Salt	Rec	No	No	No	No	No	No	No	No
Organophosphates	Rec	No	No	No	Rec	No	No	No	No

[a]Rec = recommended.

[b]No = not used.

[c]Yes = used.

[d]TPH = total petroleum hydrocarbon.

[e]? = unknown.

Table 8-4 Soil contaminant bioavailability to plants (via root uptake)

	Potential tests to measure contaminant bioavailability in plants								
Contaminant class	Toxicity	CBR	Bioaccumulation	Metabolites	Biomarkers	Histology	Histochemistry	Physiological markers	Immunological markers
Metals	Rec[a]	Yes[b]	Rec	No[c]	Yes	No	Yes	No	NA[d]
PAH	Rec	No	Yes	No	Yes (?[e])	No	No	No	NA
TPH[f]	Rec	No	?	No	No	No	No	No	NA
Chlorinated organics	Rec	Yes	Yes	No	No	No	No	No	NA
Salt	Rec	No	Yes	No	No	No	No	No	NA
Herbicides	Rec	Yes	Yes	Yes	Yes (?)	Yes	No	Yes	NA

[a]Rec = Recommended.

[b]Yes = used.

[c]No = not used.

[d]NA = not applicable.

[e]? = unknown.

[f]TPH = total petroleum hydrocarbon.

Table 8-5 Soil contaminant bioavailability to arthropods (via ingestion)

	Potential tests to measure contaminant bioavailability in arthropods								
Contaminant class	Toxicity	CBR	Bioaccumulation	Metabolites	Biomarkers	Histology	Histochemistry	Physiological markers	Immunological markers
Metals	Rec[a]	Rec	Rec	No[b]	Yes[c]	No	Yes	No	No
PAH	Rec	No	Yes	Yes	Yes	No	No	No	No
TPH[d]	Rec	No	No	No	No	No	No	No	No
Chlorinated organics	Rec	No	Yes	No	No	No	No	No	No
Salt	Rec	No	No	No	No	No	No	No	No
Organophosphates	Rec	No	No	No	Yes	No	No	No	No

[a]Rec = Recommended.

[b]No = not used.

[c]Yes = used.

[d]TPH = total petroleum hydrocarbon.

Table 8-6 Soil contaminant bioavailability to terrestrial vertebrates

Contaminant Class	Toxicity	CBR[a]	Bioaccumulation	Metabolites	Biomarkers	Histology	Histochemistry	Physiological markers	Immunological markers
Metals	Yes[b]	Yes	Yes	No[c]	Yes	Yes	Yes	Yes	Yes
PAH	Yes	No	Yes	Yes	Yes	Yes	No	Yes	No (?)[d]
TPH[e]	Yes	No	No	No	Yes (?)	Yes (?)	No	Yes	No
Chlorinated organics	Yes	Yes	Yes	Yes	Yes	Yes	No	Yes	Yes
Pesticides	Yes	Yes	Yes	Yes	Yes	Yes	No	Yes	Yes

[a]In most cases, critical tissue residue is measured instead of CBR (for example, Cd residues in kidneys or Pb residues in bone samples).

[b]Yes = used.

[c]No = not used.

[d]? = unknown.

[e]TPH = total petroleum hydrocarbon.

such, toxicity testing can provide an unambiguous answer to the question "Are soil contaminants bioavailable?" There are, however, some significant limitations with this approach:

- For soils containing complex mixtures, the measured toxicity does not determine which of the contaminants are bioavailable or responsible for the observed response.
- Significant contaminant uptake from the soil matrix can take place even if no toxic responses are observed in the exposed organism; this will occur if the total uptake does not exceed a CBR for the duration of the test.
- Bioavailability will be a function of the toxicity endpoints measured. Short-term, acute exposures may not result in a toxic response (such as mortality), whereas a longer-term chronic exposure may result in significant toxicity (such as adverse effects on reproduction, survival, or growth.) Decisions about the bioavailability of contaminants based only on the result of acute tests might be inaccurate.
- Clean control soils collected in the field, or prepared in the laboratory, might not adequately support the test organism. However, this does not reflect an adverse biological effect associated with contaminant bioavailability but rather the result of unsuitable physicochemical characteristics of the reference soil.

The last bullet item above warrants more discussion, given the role that the natural physicochemical characteristics of a soil might have on the results of an experiment and the interpretation of biological effects. Specifically, in any terrestrial habitat, the physical, chemical, and biological properties of the soil are partially responsible for determining its productivity. Methods for evaluating the bioavailability and toxicity of chemicals in soil must separate adverse effects associated with soil physicochemical characteristics from the adverse effects of contaminants in the soil. In general, chemicals spilled in soil primarily cause changes in soil chemical properties, but their release, or activities associated with their release (e.g., increased traffic or physical disruption of habitat), may also influence physical and biological soil properties. Therefore, the physicochemical reference conditions of a soil must be adequately described if deviations in biological parameters of organisms exposed to contaminated soils are to be attributed to the contaminant itself. In addition, a quantitative knowledge of certain soil properties is often necessary for modeling the fate and transport of the contaminant in question. For example, the conversion of contaminant concentration to loading rates on an areal basis requires knowledge of the bulk density and depth of the contaminated soil horizon. An adequate description of soil properties involves both field observations and laboratory measurements. Field observations include variables that represent traditional agronomic site characteristics (e.g., texture, fertility), while laboratory measurements are those commonly associated with toxicity testing (e.g., pH, organic matter content, cation exchange capacity). Both types of measurements are important in defining soil productivity.

Terrestrial wildlife receptors

Terrestrial wildlife receptors at contaminated sites have received much attention with respect to chemical toxicity and bioavailability in soil. This reflects a societal decision on the importance of such organisms as being indicators of a healthy ecosystem. Most studies have focused on smaller mammalian species, such as field mice and shrews, but song birds and raptors have also been examined. Organisms are usually collected (i.e., using live or snap-traps, mist nets, etc.) directly from the field site or in the vicinity of the site. They are either killed for intrusive analyses (histology, critical tissue residues, bioaccumulation in tissues) or released unharmed after sampling for blood, fur, or feathers (metabolites, immunological biomarkers).

Terrestrial wildlife can be exposed to contaminants in soil directly (via accidental or deliberate soil ingestion) or indirectly (via ingestion of contaminated foods such as soil invertebrates or plant materials). Even compounds that traditionally do not bioaccumulate in higher trophic levels can be ingested in high quantities. Within the context of evaluating ecological risks to wildlife, bioavailability tests may help derive dose estimates focused on dietary exposures to soil contaminants. For example, screening-level estimates of risks to birds and mammals frequently centers on a food chain analysis, the derivation of ingested dose, and the comparison of ingested dose to toxicity reference values (TRVs) for wildlife (Chapter 2). In Tier 1 efforts to describe wildlife risks, literature-based estimates are often used to describe relationships between soil contaminant concentrations and concentrations of these chemicals in food items; hence, the estimated ingested dose may be highly conservative. Risk estimates, such as hazard quotients, may subsequently be refined using bioavailability tests (e.g., bioaccumulation tests), which yield site-specific estimates of tissue concentrations in food items. Bioaccumulation tests may also be used to describe differences between absolute and relative bioavailability.

Case studies of risk assessments incorporating bioavailability measurements

Within the context of ERA, adverse effects associated with exposures to chemically contaminated soil frequently focus on terrestrial vertebrates, most often birds and mammals (i.e., "charismatic megafauna"). The concept of using indicator species to assess terrestrial habitats impacted by chemicals has been interpreted in many different ways and has received some support from regulatory agencies. Although birds and mammals are often used as indicator species, there are many other organisms that also may be used as indicator species, although their use instead of charismatic megafauna often requires some justification. The case studies presented below provide examples of different types of organisms that can be used in measuring chemical bioavailability.

Case Study 1: Availability of hydrocarbons to plants, invertebrates, and small mammals

This case study involved an ERA of an agricultural site where a pipeline rupture resulted in a spill of 22,000 barrels (3,497,350 L) of crude oil onto agricultural fields that were used to grow wheat (*Triticum aestivum*), barley (*Hordeum vulgare*), and canola (*Brassica rapa* or *napus*) in rotation. Immediate response to the spill entailed building a soil berm around the spill site, removing the surface oil by pumping and hauling, and burning the surface residual. The goal of the ERA was to determine the acceptable level of oil contamination that would allow the land to return to its previous level of crop productivity, and to determine the risk to relevant ecological receptors. The ecological receptors of concern, in order of decreasing importance, were the cereal crop species comprising the crop rotation for these lands (barley, wheat, and canola), livestock (cows), earthworms, small mammals and birds, waterfowl, and mammalian and avian predators. The contaminants of concern were identified as benzene, toluene, ethylbenzene, xylene (BTEX compounds), and PAHs (naphthalene, phenanthrene, pyrene, 2-methylnaphthalene, benzo[g,h,i]perylene), which are either natural constituents of crude oil and/or combustion products that formed when the residual surface oil was burned. The major routes of exposure for the ecological receptors at the site for the chemicals of concern are, in order of decreasing importance, direct uptake of contaminated soil and water, uptake or ingestion of contaminated plant or animal material, direct inhalation, and dermal adsorption through physical contact with soil and water. Measurement endpoints derived from literature data included phytotoxicity data, bioaccumulation potential, residue data, and mammalian toxicity data. Phytotoxicity could potentially result from direct toxicity to plants or, indirectly, from either N immobilization by degrading soil microorganisms or disruption of plant–soil–water relations (hydrophobicity). The risk to livestock was from the consumption of grain that might contain PAH residues resulting from bioaccumulation of PAH residues in this part of the plant, or from drinking contaminated ground water.

Studies were conducted to determine if the contaminants of concern were bioavailable to plants and earthworms and to estimate the potential for tissue residues of the contaminants in the seed and straw of the crop plants grown on the site 7 months after the crude oil spill. The results of the studies indicated that there was minimal risk to livestock associated with dietary exposure to the chemicals of concern. Although, there was evidence that contaminant levels (e.g., PAH) were slightly elevated in plant tissues such as straw, there were no significant residue levels in the grain. Because there were literature data that indicated the potential for PAHs to bioaccumulate to levels that were approximately 5 times those in the soil (Wagner and Wagner-Hering 1971), the acute and chronic effect levels were compared to 5 times the estimated exposure concentrations at the site. When this was taken into consideration, meadow voles were at risk when chronically exposed to PAHs via plant residues. Hairless meadow vole neonates, spending a significant part of their early life inhaling air with volatile hydrocarbons, having intimate contact with contaminated soil, and consuming contaminated milk from their mothers, were also likely to be at risk. These receptors and associated exposure pathways were considered to be of secondary importance for the following reasons: 1) the disruptive nature of the agriculture activities was not supportive of habitats conducive to burrowing and foraging activities; 2) the site was essentially devoid of trees, shrubs, and hedgerows, which provide refuge for animals that are likely to perform these activities; 3) the relative importance of these species is minimal in the context of agricultural activities (most are considered undesirable pest species); and 4) the berm discouraged access to the site by small mammals.

Site-specific toxicity tests with geophagus earthworms were conducted, which indicated that the level of hydrocarbon contamination in the soil was sufficiently high to cause toxicity to earthworms. However, the risk to earthworms of the current land-use practices (e.g., plowing, tilling, and application of insecticides) was comparable to the risk of mortality from site contamination. Tissue residues of petroleum constituents in earthworms might be potentially hazardous to worm-eating birds and skunks; however, the risk was considered minimal because of the lack of favorable habitat to support these species at the site.

Case Study 2: Bioavailability of diazinon to birds

This case study describes an assessment of the relative risk of diazinon to vermivorous birds. Diazinon is an organophosphate insecticide widely used in North America to control insect pests of turf grass (Fushtey and Sears 1981; OMAF 1990) and is typically applied directly to turf at rates of 7.5 kg a.i./ha. The risk of direct exposure of birds during application was considered to be minimal, as was the potential for dermal exposure from contact with diazinon residues on the grass. Most songbird deaths related to diazinon use have been associated with ingestion of sprayed vegetation or of the material itself when applied as a granular formulation to turf. However, indirect exposure from consumption of prey that contained body residues of diazinon was also of concern because it was suspected that residues in earthworms were sufficiently high to represent a potential risk to birds feeding on earthworms. The relative risk of diazinon to worm-eating birds was assessed using toxicity and bioaccumulation results from both laboratory and field studies with *Lumbricus terrestris* and the application of a pharmacokinetic model (Stephenson, Wren et al. 1997). Results of the laboratory toxicity tests indicated that the recommended application rate of 7.5 kg a.i./ha must be exceeded by 71.7-fold before acute toxicity to *L. terrestris* would be expected. The results from the field study corroborated the laboratory results.

Whole-body diazinon concentrations in *L. terrestris* increased with measured diazinon concentrations in soil in both the field study (Figure 8-2) and laboratory tests (Figure 8-3). The estimated bioaccumulation factors (BAFs) in earthworms exposed in the laboratory study are presented in Figure 8-4.

The 2 elements required for this preliminary hazard assessment were exposure and toxicity. The diazinon concentrations in earthworm tissues were used to assess exposure; literature data on feeding rates, body weights, and toxicity of diazinon to birds (house sparrow and red-winged blackbird) were used to assess toxicity. The bioaccumulation results showed that on day 7, the worm tissue concentrations of diazinon were 1.15 and 3.1g mg/kg, wet weight, at 1 and 2 times the recommended application rate (AR), respectively. The average weight of a worm used in the experiments was 5.5 g, wet weight.

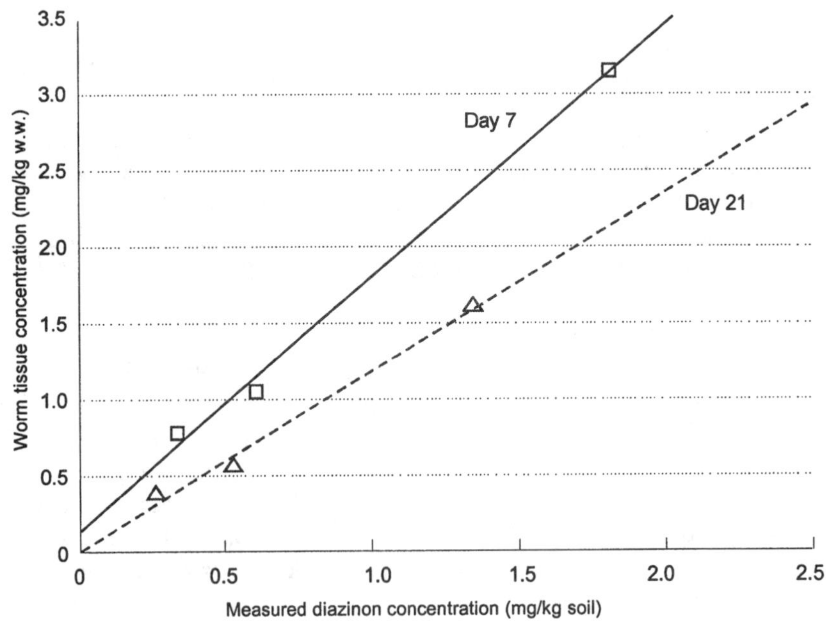

Figure 8-2 Whole-body diazinon concentrations in *L. terrestris* regressed against measured diazinon concentrations in soil in the field study

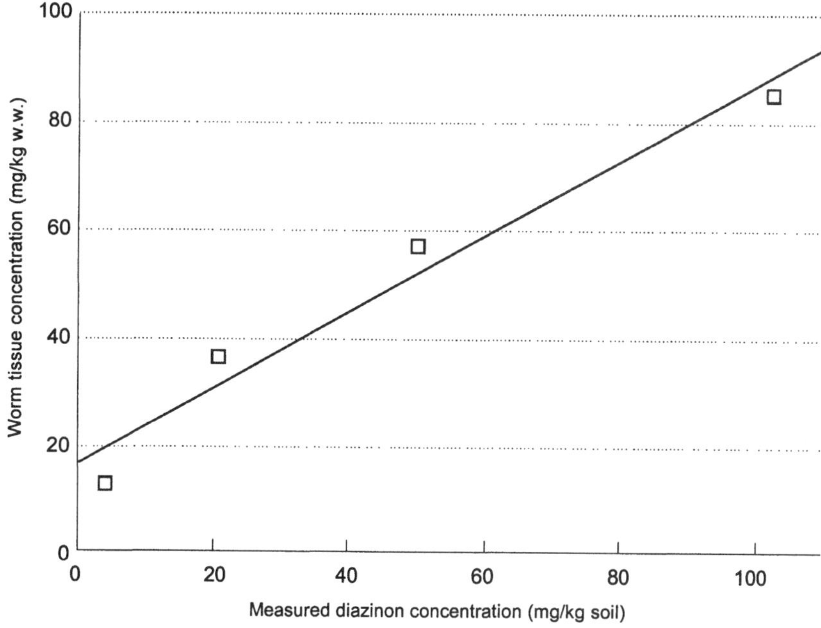

Figure 8-3 Whole-body diazinon concentrations in *L. terrestris* regressed against measured diazinon concentrations in soil in the laboratory study

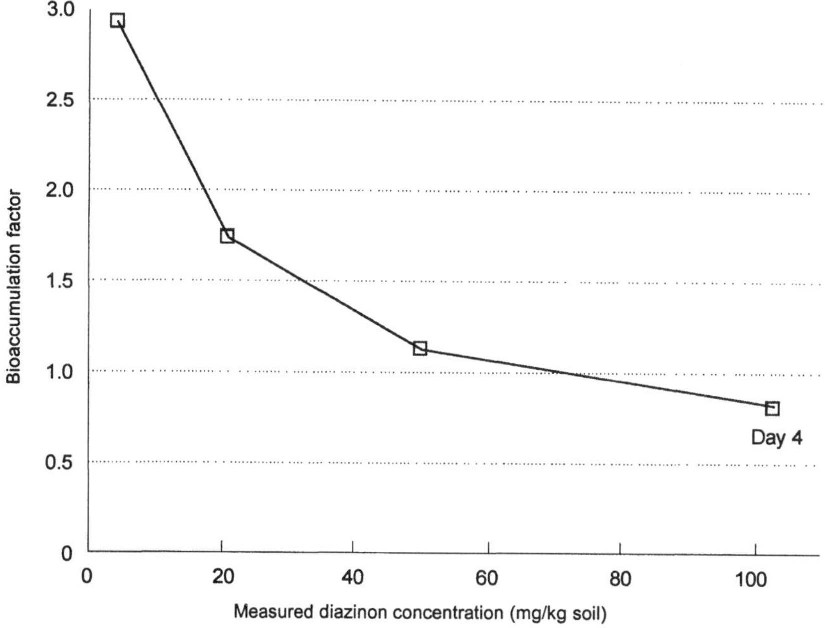

Figure 8-4 Bioaccumulation factor in *L. terrestris* plotted against measured diazinon concentrations in soil in the laboratory study

The acute LD50 for 2 bird species was estimated as follows:

A) Estimate of risk to house sparrow

LD50 = 7.5 mg/kg (USEPA 1986)

bird weight = 0.026 kg; or LD50 = 0.195 mg/bird

1) at 1X AR [worm] = 1.15 mg/bird

Therefore, the amount of food = (0.195 mg/bird)/(1.05 mg/kg)

= 0.185 kg/bird

= 185 g

= 34 worms.

2) at 2X AR, worm residue = 3.16 mg/kg

Therefore, amount of food required = 0.195/3.16

= 0.061 kg

= 61 g

= 11 worms.

B) Estimate of risk to red-winged blackbirds

LD50 = 3.2 mg/kg (USEPA 1986)

bird weight = 0.065 kg; or LD50 = 0.208 mg/bird

1) at 1X AR worm residue = 1.05 mg/kg

Therefore, amount of food = (0.208 mg/bird)/(1.05 mg/kg)

= 0.198 kg

= 198 g

= 36 worms.

2) at 2X AR, worm residue = 3.16 mg/kg

Therefore, amount of food required = 0.208/3.16

= 0.065 kg

= 65 g

= 11.8 worms.

The amount of diazinon consumed per day, or mean daily dose (MDD), was calculated by multiplying the amount of food eaten (kg) by the earthworm diazinon concentration (1.05 or 3.16 mg/kg). The margin of safety (MOS) was estimated by dividing the acute LD50 by the daily mean diazinon ingested. An MOS >1 indicates that acute adult mortality is not expected (Table 8-7).

The results indicated that at the recommended application rate of 7.5 kg a.i./ha, there was little risk of acute toxicity of diazinon to worm-eating birds under the standard conditions given above. However, there is always uncertainty associated

Table 8-7 The estimated MDD and MOS for house sparrows and red-winged blackbirds exposed to diazinon-contaminated earthworms

	AR	MDD (mg/day)	MOS (LD50 or average diazinon concentration ingested)
House sparrow	1	0.005	35.6
	2	0.016	11.9
Red-winged blackbird	1	0.010	20.3
	2	0.030	6.8

with any estimate of risk or identification of hazard. Food consumption rates were estimated to be 15% and 20% body weight/day. Higher food consumption would increase the relative risk, and higher feeding rates are reported for different bird species. For example, the food consumption rate for robins was estimated at 25% body weight/day, while a female woodcock (*Philohela minor*) may consume 100% of her own body weight daily (Reynolds et al. 1977). The diet of woodcocks is comprised largely of worms and some insects. In reality, earthworms may not constitute 100% of the diet, although other insects or materials also exposed to diazinon might be ingested. Food consumption can also differ at different times of year (i.e., breeding or prior to migration). Also, rapidly growing hatchling and fledgling birds will consume large quantities of food during the spring and early summer.

This preliminary hazard assessment was based on acute toxicity values reported in the literature. Sublethal effects may be observed at lower doses or in more sensitive life stages such as hatchlings and fledglings. For a more complete hazard assessment, a steady-state body burden model should be developed that incorporates diazinon metabolism and excretion rates for birds, as well as different application rate patterns and worm tissue concentrations. In any event, bioavailability is an essential component to the estimate of risk.

Case Study 3: Zinc in the Dutch Kempen area

Near Budel, located in the south of the Netherlands, a Zn smelter has been operating since the beginning of the 20th century. As a consequence, a large area surrounding this smelter is contaminated with heavy metals, especially Zn, Cd, Cu, and Pb. Several studies have been performed to investigate the potential human health and ecological risk of this contamination. The result of the case study presented in this section came from Posthuma et al. (1998). Soil samples were collected from 11 locations along a deposition gradient, up to a distance of 20 km northeast of the smelter. The metal concentrations and some soil characteristics are presented in Table 8-8.

Table 8-8 Main characteristics and metal concentrations (mg/kg, dry weight) in Budel gradient soils[a]

Distance (km)	pH (1M KCl)	% OM[a]	% clay	Zn	Cd	Cu	Pb
0.4	4.7	1.9	2.0	598	3.0	68	325
1.1	4.9	3.6	1.4	1787	3.9	250	424
1.6	4.0	2.8	1.3	207	0.9	29	148
2.6	4.3	2.4	1.2	219	0.8	17	91
5.0	3.1	3.7	1.3	32	0.8	10	40
6.6–20	2.9–3.4	3.5–6.4	1.2–1.6	11–28	0.14–0.4	1.9–5.4	10–61

[a]from Posthuma et al. 1998.

Soil pH was highest at the most contaminated sites (Table 8-8). This factor hampers the choice of a proper reference soil from the gradient for use in toxicity tests, because the noncontaminated soils all had lower pH values and sometimes also differed in organic matter content. Zn was the contaminant with the highest concentration, followed by Pb and Cu. Metal concentrations were compared with laboratory toxicity data available for soil invertebrates and plants. Sensitivity ratios for Zn to Cd generally were much lower than the ratios of Zn to Cd found in the field soils. It was concluded that Zn was probably the most toxic element in this area, and further research focused primarily on this metal. Concentrations of the metals in 0.01 M $CaCl_2$ extracts or in pore water of these soils generally could be described with a Freundlich-type relationship.

Several bioassays were performed on the 11 soils collected from the gradient. Results, expressed on the basis of Zn toxicity, are presented in Table 8-9.

Table 8-9 EC50 values for the effect of Zn on the growth (plants) or reproduction (invertebrates) of different test organisms exposed to soils from a contaminated gradient near a zinc smelter in Budel[a]

Organism	EC50 (mg/kg, dry weight in soil or tissue)			
	Total	Water soluble	0.01 M $CaCl_2$	Tissue residue
Trifolium pratense	347	–	321	1460 (shoots) 3700 (roots)
Folsomia candida	>1787	>10	>313	>135[b]
Eisenia andrei	1790–2533	–	300–336	>130
Enchytraeus crypticus	205	–	123	–

[a]Data taken from Posthuma et al. 1998.

[b]Value taken from Smit and Van Gestel 1998.

These data were compared with data obtained in standardized laboratory toxicity tests with both artificial and natural soils. An experimental field plot was also prepared using a soil type similar to the Budel gradient soil, and Zn chloride was mixed in with these soils. Field surveys of nematode and enchytraeid populations were conducted in the Budel gradient soils, and PICT of the microbial community was determined both in the experimental field plot and in the Budel gradient soils.

The laboratory toxicity tests generated many effect concentrations for Zn, expressed on the basis of total, water soluble, 0.01 M $CaCl_2$-extractable, and tissue residues of the different test organisms (Table 8-10).

Table 8-10 Results of laboratory toxicity test on the toxicity of Zn in freshly contaminated soils and in an aged field soil from an experimental field plot[a]

Test organism	Test soil	EC50 (mg/kg dry weight) expressed on the basis of			
		Total	Water soluble	0.01 M $CaCl_2$	Internal
Trifolium pratense	OECD[b]	356	–	–	1020 (shoots) 5300 (roots)
	Budel[c]	144	–	144	1600 (shoots) 3200 (roots)
	EFP_{fresh}[d]	73–76	–	–	1060 (shoots) 6600 (roots)
	EFP_{aged}[e]	340–890	2.5–13	99–112	200–569
Folsomia candida	OECD	473–626	8–12	62	97–150
	Budel	185	2.6	49	–
	EFP_{fresh}	184–266	13–21	63–125	146–147
	EFP_{aged}	1491–2178	19–22	255–359	328
Eisenia andrei	OECD	429–921	–	133–172	150–300
	EFP_{aged}	1277–1676	21–33	219–307	305–356
Enchytraeus crypticus	OECD	186–361	3.9	18–144	372
	EFP_{fresh}	262	–	149	–
	EFP_{aged}	844	–	102	–

[a]Data taken from Posthuma et al. 1998.

[b]OECD = artificial soil according to OECD (1984), freshly treated in the laboratory.

[c]Budel = noncontaminated Budel soil taken 20 km away from the smelter, freshly treated.

[d]EFP_{fresh} = soil used in the experimental field plot, freshly treated.

[e]EFP_{aged} = soil taken from the experimental field after aging for at least 6 months.

The following conclusions were drawn from this study:

- Effect concentrations (EC50 values) in the Budel gradient and the aged experimental field plot soil were much higher than those obtained in freshly treated soils, demonstrating the effect of aging on bioavailability.
- Effect concentrations in aged field plot soils generally were in much better agreement with those in the field gradient soils.
- For earthworms, enchytraeids, and plants, 0.01 M $CaCl_2$-extractable Zn concentrations were correlated with the toxicity regardless of soil type.
- For Collembola, water-soluble Zn concentrations correlated with effect levels for the different soils, except for the freshly treated Budel reference soil. EC50 values based on water-soluble concentrations in the laboratory tests on freshly treated and aged soils also explained the lack of effects in the Budel gradient soils.
- Internal effect concentrations were not affected by soil type for the different soils. It should be noted that Zn is an essential element, which is regulated by many organisms. Effect levels for Collembola and earthworms were only just above the regulated level encountered in unexposed animals (typically around 100 mg Zn/kg body weight) .

Apparently, any increase of the internal level immediately results in adverse effects on reproduction.

In both the Budel gradient and the experimental field plot soils, a PICT was demonstrated by a shift in the sensitivity to Zn of microbial substrate degrading activities (95 different substrates were used in Biolog plates). In both soils, PICT was significant at concentrations >60 mg Zn/kg (dry soil).

Results of field sampling of enchytraeid and nematode communities in soils along the Budel gradient were somewhat obscured by additional stress factors. Enchytraeid communities did not show a correlation with metal contamination. The variation in soil pH, vegetation type and density, soil moisture, and SOM content might have confounded concentration–effect relationships. The variation in these variables also hampered the interpretation of data on the nematode community in the Budel gradient soils. Certain taxa, however, did show a correlation with metal concentrations in these soils. Nematode communities in the experimental field plot showed a concentration-related response to Zn and also indicated the effect of aging of the Zn contamination. This was illustrated by the EC50 values for nematodes, which increased from 1067 mg/kg (dry weight) after 3 months to 1945 and 2862 mg Zn/kg (dry weight) after 10 and 22 months, respectively. The number of nematode taxa appeared to be a more sensitive endpoint than the response of individual species, with EC50 values of 574, 434, and 934 mg Zn/kg (dry weight) after 3, 10, and 22 months, respectively.

Suggested approach for using biological tools to evaluate bioavailability of chemicals from contaminated soils

It is impractical, unnecessary, and costly to quantify the biological availability of all contaminants via all exposure pathways to all the receptors of concern at a given site. The usual approach is to use a tiered assessment to focus most of the efforts on the receptor, exposure pathway, and contaminant with the highest potential for risk.

In Tier 1, measured soil contaminant levels are compared to conservative soil screening guidelines developed for specific receptor groups of concern (plants, soil invertebrates, terrestrial vertebrates, etc.). This step identifies exposure pathways, contaminants, and receptors that are of no concern and can therefore be removed from the assessment. In subsequent tiers, the remaining receptors are evaluated to see if the hypothetical risks are real. Several approaches are available to refine Tier 1 assessments.

Some of these approaches do not require site-specific estimates of biological bioavailability. These include

- simple exposure models that use conservative, literature-derived point values as input variables;
- Monte Carlo-type simulations using ranges of literature-derived input variables; and
- field surveys to assess plant, soil invertebrate, and wildlife populations and compare results to a reference site.

Other approaches comprise site-specific estimates of biological bioavailability. These include

- simple exposure models that use more realistic, field-derived values as input variables (including measures of site- and contaminant-specific BAFs);
- direct ecotoxicity testing of contaminated soils, either in the laboratory or in the field;
- estimates of species-specific and contaminant-specific critical body or tissue burdens based on measured soil concentrations and BAFs; and
- estimates of contaminant transfer to higher trophic levels based on field-collected biological samples.

Clearly, a single approach does not fit all situations. There is a need to match the appropriate tool to the problem. For smaller sites, it may not be cost-effective or relevant to determine site-specific bioavailability values. Instead, a limited ecotoxicity testing program and some targeted Monte Carlo simulations might suffice to properly characterize potential ecological risk.

At larger sites (e.g., Superfund, refinery, or mining sites), the financial liabilities are usually much higher, and more resources may be available. Determining those receptors, exposure pathways, and contaminants for which there is not an unacceptable terrestrial ecological risk under realistic and site-specific exposures can significantly reduce cleanup costs. In such cases, a relatively small initial investment to collect site-specific bioavailability estimates might yield significant cost savings.

For example, assume that a Tier 1 risk-screening approach based on generic soil guidelines suggests that a small terrestrial carnivore is potentially at risk from consuming soil invertebrates present at a site contaminated with petroleum hydrocarbons and metals. A decision is made to better quantify this potential risk. As a first step, the concentrations of hydrocarbons and metals in earthworms consumed by the carnivore are estimated using literature-derived bioavailability factors and measured soil concentrations:

$$C_{Worm} = C_{Soil}\,\mathrm{BAF} \qquad (8\text{-}1),$$

where C_{Worm} = concentration of key contaminants in earthworms,

C_{Soil} = concentration of key contaminants measured in soil, and

BAF = bioaccumulation factors for key contaminants.

The calculated concentrations of these contaminants in earthworms can then be used in exposure equations to estimate a total daily dose and, hence, risk to the carnivore. If risks are acceptable, the assessment stops. On the other hand, the estimated daily dose still might exceed a conservative critical toxicity threshold, and additional work will be required to refine the assessment.

The uncertainties associated with the calculated bioavailability values are quite large. These uncertainties include earthworms that are exposed to many contaminants in site soil even though only a few may have published BAF values, BAFs can vary depending on the duration of exposure (i.e., depending on whether steady state inside the worm has been reached), or BAFs can vary depending on the initial contaminant concentrations in the test soil.

In the next step, the assessor obtains measured bioavailability data by collecting earthworms from the site and quantifying contaminant concentrations in their tissue (other soil invertebrate prey items besides earthworms, such as crickets, millipedes, etc., could also be included in the analysis). The tissue concentrations of individual hydrocarbons and metals are measured to calculate risk to the carnivore. Toxicity values for vertebrates are available from the literature for individual hydrocarbons such as PAHs, but not for total petroleum hydrocarbons (TPHs).

At this point, representative contaminant tissue concentrations are available for the prey species consumed by the carnivore. The prey will "integrate" a number of site-specific variables, including local variations in soil contaminant levels, chemical bioavailability, and organic C content. Measured tissue values are therefore more accurate and realistic than literature values because they are specific to the site. In addition, they reflect the range of concentrations likely to be experienced by the carnivore foraging on site. These values are used once more to calculate risk to the carnivore. If those risks prove acceptable, the carnivore is removed from further consideration. Otherwise, an additional step may be required to further refine the dose estimate.

Bioavailability can be further refined by measuring contaminant concentrations in the carnivore itself to determine if contaminants are bioaccumulating or reaching a critical body or tissue residue level. In this step, a representative number of small carnivores are collected from the site. Certain tools described in Table 8-6 (e.g., assessing immunological responses or measuring metabolite concentrations in blood samples) can be used to determine if contaminant uptake via prey or incidental soil ingestion results in detectable changes in these parameters. Other tools described in Table 8-6 are more intrusive (e.g., measuring critical body or tissue burdens or contaminant bioaccumulation) but assess contaminant bioavailability and uptake more directly. Either way, these data provide quantitative information on the exposure experienced by the small carnivores at the site. These exposures may be orders of magnitude below the simplistic assessments used to develop the soil-quality guidelines in Tier 1. Such quantitative data may help show that soil contaminant levels measured at our hypothetical site are of minimal risk to the carnivore species.

The preceding example is a simplistic overview of how bioavailability measurements can provide a more realistic assessment of risk. The cost and effort involved increase with each additional refinement. How far the risk assessor proceeds to obtain site-specific bioavailability data is determined by the size of the site, the number and concentration of contaminants, the number and types of receptors, local regulatory requirements, available budgets, and other considerations (Chapter 2).

Table 8-11 provides a qualitative summary of the cost of obtaining more site-specific bioavailability values for the example outlined above. Generic values can be obtained fairly easily and quickly at lower cost. Site-specific values require more time, effort, and cost. The risk assessor must decide at what point the benefit of additional site-specific information outweighs the cost.

Table 8-11 Role of bioavailability measurements

Level of complexity	Bioavailability measured?	Estimated cost[a]
Screening-level soil guidelines	No; soil guidelines are based on published laboratory data and generic exposure assumptions	Minimal
Generic BAF values for key petroleum hydrocarbons and metals in prey species	No; BAF values are estimated using K_{ow} relationships or published data	L[b]
Site-specific BAF values for key petroleum hydrocarbons and metals in prey species	Yes; BAF values are obtained from field-collected prey tissue samples	M[c]
Site-specific bioaccumulation values for key petroleum hydrocarbons in the carnivore	Yes; bioaccumulation values are obtained from field-collected carnivore tissue samples	H[d]

[a]Costs are qualitative only and relative to the screening-level soil guidelines.
[b]L = Low.
[c]M = Medium.
[d]H = High.

Research and Development Needs

Much of the discussion in this chapter focuses on how the bioavailability of chemicals in soil systems can be assessed using living organisms, instead of just measuring different fractions of chemicals in soil and assuming these translate into differences in chemical bioavailability. Although the concept of bioavailability is used freely in regulatory and risk assessment arenas, we have relatively few techniques that can actually be used to measure bioavailability, and even fewer peer-reviewed articles that deal specifically with this issue. Some areas for research and development include the following:

- Developing quantitative linkages between chemical exposure levels in soil, chemical levels in soil that are bioavailable, and contaminant concentration in the organism. For example, if total Cu levels were determined in soil, only a portion of that total metal would be available to organisms for uptake, and only a portion of the bioavailable fraction would actually be taken up by the organism. Some of the Cu taken up by the organism could be present in the organism and measured at steady state (e.g., CBR). There is a need to correlate biological measurements and chemical measurements of bioavailability. As well, there is a need to develop biomimetic tools that are related to toxic effects and are reflective of bioavailability.
- Estimating CBRs at steady state for various toxicity endpoints, especially sublethal endpoints (e.g., reproduction) for terrestrial receptors.

- Generating terrestrial toxicity data specific to contaminated soils. Invertebrate toxicity and whole-soil plant toxicity tests with substances other than herbicides are lacking. There is a dearth of data on sublethal and chronic exposures of terrestrial organisms to contaminants in soil. Insufficient data are available for establishing ecological soil-screening levels.
- Developing methods for assessing contaminated site soils.
- Comparing or relating toxic effects to the methods of measuring bioavailability.
- Developing relationships between the different methodologies for measuring biological availability.

References

Alexander M. 1977. Introduction to soil microbiology. 2nd ed. New York NY, USA: Krieger Publishing Company. 467 p.

Alexander M. 1995. How toxic are toxic chemicals in soil? *Environ Sci Technol* 29:2713-2717.

Alexander RR, Chung N, Alexander M. 1999. Solid-phase genotoxicity assay for organic compounds in soil. *Environ Toxicol Chem* 18:420-425.

[ASTM] American Society for Testing and Materials. 1999a. Annual book of ASTM standards. Volume 11.05. West Conshohocken PA, USA: ASTM, E47, Committee on Biological Effects and Environmental Fate. 1556 p.

[ASTM] American Society for Testing and Materials. 1999b. Standard guide for conducting a laboratory soil toxicity test with the Lumbricid earthworm *Eisenia fetida*. ASTM E1676-95. In: Annual book of ASTM standards. Volume 11.05. West Conshohocken PA, USA: ASTM, E47, Committee on Biological Effects and Environmental Fate. p 1056-1074.

[ASTM] American Society for Testing and Materials. 1999c. Standard practice for conducting early seedling growth tests. ASTM E1598-94. In: Annual book of ASTM standards. Volume 11.05. West Conshohocken PA, USA: ASTM, E47, Committee on Biological Effects and Environmental Fate. p 994-1000.

[ASTM] American Society for Testing and Materials. 1999d. Standard guide for conducting terrestrial plant bioassays. In: Annual book of ASTM standards. Volume 11.05. West Conshohocken PA, USA: ASTM, E47, Committee on Biological Effects and Environmental Fate. p 1484-1500.

Bååth E, Frostegård Å, Diaz-Ravina M, Tunlid A. 1998. Microbial community-based measurements to estimate heavy metal effects in soil: The use of phospholipid fatty acid patterns and bacterial community tolerance. *Ambio* 27:58-61.

Benton MJ, Malott ML, Knight SS, Cooper CM, Benson WH. 1995. Influence of sediment composition on apparent toxicity in a solid-phase test using bioluminescent bacteria. *Environ Toxicol Chem* 14:411-414.

Bergman HL, Kimerle RA, Maki AW, editors. 1986. Environmental hazard assessment of effluents. Elmsford NY, USA: Pergamon Press. 366 p.

Bewley RJF, Stotzky G. 1983. Effects of cadmium and zinc on microbial activity in soil: Influence of clay minerals. Part II: Metals added simultaneously. *Sci Total Environ* 31:57- 69.

Beyer WN, Miller G. 1990. Trace elements in soil and biota in confined disposal facilities for dredged material. *Environ Pollut* 65:19-32.

Blair JM, Parmelee RW, Lavelle P. 1995. Influences of earthworms on biogeochemistry. In: Hendrix PF, editor. Earthworm ecology and biogeography in North America. Boca Raton FL, USA: CRC Press. p 27-158.

Blanck H, Wängberg S-Å, Molander S. 1988. Pollution-induced community tolerance: A new ecotoxicological tool. In: Cairns Jr J, Pratt JR, editors. Functional testing of aquatic biota for estimating hazards of chemicals. Philadelphia PA, USA: ASTM. STP 988. p 219-230.

Bulich AA. 1979. Use of luminescent bacteria for determining toxicity and bioavailability in aquatic environments. In: Markings LL, Kimerle RA, editors. Aquatic toxicology. Philadelphia PA, USA: ASTM. STP 667. p 97-106.

Bulich AA. 1982. A practical and reliable method for monitoring the toxicity and bioavailability of aquatic samples. *Process Biochem* 17:45-47.

Bulich AA. 1984. Microtox: A bacterial toxicity and bioavailability test with several environmental applications. In: Liu D, Dutka BJ, editors. Toxicity and bioavailability screening procedures using bacterial systems. New York NY, USA: Marcel Dekker. p 55-64.

Bulich AA. 1986. Bioluminescent assays. In: Bitton G, Dutka BJ, editors. Toxicity and bioavailability testing using microorganisms. Volume 1. Boca Raton FL, USA: CRC Press. p 57-74.

Callahan CA, Menzie CA, Burmaster DE, Wilborn DC, Ernst T. 1991. On-site methods for assessing chemical impact on the soil environment using earthworms: A case study at the Baird and McGuire Superfund site, Holbrook, Massachusetts. *Environ Toxicol Chem* 10:817-826.

Chaineau CH, Morel JL, Oudet J. 1997. Phytotoxicity and plant uptake of fuel oil hydrocarbons. *J Environ Qual* 26:1478-1483.

Chang FH, Broadbent FE. 1982. Influence of trace metals on some soil nitrogen transformations. *J Environ Qual* 11:1-4.

Chapman PM, Allen HE, Godtfredsen K, Zgraggen MN. 1996. Evaluation of bioaccumulation factors in regulating metals. *Environ Sci Technol* 30:A448-A452.

Chen SC, Fitzpatrick LC, Goven AJ, Venables BJ, Cooper EL. 1991. Nitroblue tetrazolium dye reduction by earthworm (*Lumbricus terrestris*) coelomocytes: An enzyme assay for nonspecific immunotoxicity of xenobiotics. *Environ Toxicol Chem* 10:1037-1043.

Coles EH. 1986. Veterinary immunology. 4th ed. New York NY, USA: W.B. Sanders Company. 486 p.

Connell DW, Markwell RD. 1990. Bioaccumulation in the soil to earthworm system. *Chemosphere* 20:91-100.

Corke CT, Thompson FR. 1970. Effects of some phenylamide herbicides and their degradation products in soil nitrification. *Can J Microbiol* 16:567-571.

Corp N, Morgan AJ. 1991. Accumulation of heavy metals from polluted soils by the earthworm *Lumbricus rubellus*: Can laboratory exposure of control worms reduce biomonitoring problems? *Environ Pollut* 74:39-52.

Crommentuijn T, Doodeman CJAM, Doornekamp A, Van der Pol JJC, Bedaux JJM, Van Gestel CAM. 1994. Lethal body concentrations and accumulation patterns determine time-dependent toxicity of cadmium in soil arthropods. *Environ Toxicol Chem* 13:1781-1789.

Curtis C, Lima A, Lorano SJ, Veith GD. 1982. Evaluation of a bacterial bioluminescence bioassay as a method for predicting acute toxicity and bioavailability of organic chemicals to fish. In: Pearson JG, Foster RB, Bishop WE, editors. Aquatic toxicity and bioavailability and hazard assessment. Philadelphia PA, USA: ASTM. STP 766. p 170-178.

Davis N, Ho M. 1976. Quantitation of immunoglobulins. In: Rose N, Friedman H, editors. Manual of clinical immunology. Washington DC, USA: American Society of Microbiology. p 4-16.

Exon JH, Koller LD, Talcott PA, O'Reilly CA, Henningsen GM. 1986. Immunotoxicity testing: An economical multiple-assay approach. *Fund Appl Toxicol* 7:387-397.

Eyambe GS, Goven AJ, Fitzpatrick LC, Venables BJ, Cooper EL. 1990. A noninvasive technique for sequential collection of earthworm (*Lumbricus terrestris*) leukocytes during subchronic immunotoxicity studies. *Lab Animals* 25:61-67.

Fitzgerald DG, Warner KA, Lanno RP, Dixon DG. 1996. Assessing the effects of modifying factors on pentachlorophenol toxicity to earthworms: Applications of body residues. *Environ Toxicol Chem* 15:2299-2304.

Folsom Jr BL, Price RA. 1991. A plant bioassay for assessing plant uptake of contaminants from freshwater soils or dredged material. In: Gorsuch JW, Lower WR, Lewis MA, Wang W, editors. Plants for toxicity assessment. Volume 2. Philadelphia PA, USA: ASTM. STP 1115. p 172-177.

Fushtey SG, Sears MK. 1981. Turfgrass diseases and insect pests (descriptions, illustrations, and controls). OMAF Publication 162. Toronto ON, Canada: Ontario Ministry of Agriculture and Food.

Gerhardson B, Clarholm M. 1986. Microbial communities and plant roots. In: Jensen V, Kjoller A, Sorensen LH, editors. Microbial communities in soil. New York NY, USA: Elsevier. p 19-34.

Giashuddin M, Cornfield AH. 1979. Effects of adding nickel (as oxide) to soil on nitrogen and carbon mineralization at different pH levels. *Environ Pollut* 19:67-70.

Glaister J. 1986. Principles of toxicological pathology. London, UK: Taylor and Francis. 223 p.

Graumann W, Drukker J, editors. 1991. Histochemistry and cytochemistry as a tool in environmental toxicology. Progress in histochemistry and cytochemistry, Volume 23. Copenhagen, DK: Urban & Fisher Verlad. 415 p.

Greene JC, Bartels CL, Warren-Hicks WJ, Parkhurst BR, Linder GL, Peterson SA, Miller WE. 1989. Protocols for short term toxicity screening of hazardous waste sites. Corvallis OR, USA: USEPA, Office of Research and Development, Environmental Research Laboratory. EPA-600-3-88-029. 102 p.

Haimi J, Salminen J, Huhta V, Knuutinen J, Palm H. 1992. Bioaccumulation of organochlorine compounds in earthworms. *Soil Biol Biochem* 24:1699-1703.

Hamelink JL, Landrum PF, Bergman HL, Benson WH, editors. 1994. Bioavailability: Physical, chemical, and biological interactions. Chelsea MI, USA: Lewis Publishers. 239 p.

Holland AF. 1990. Near coastal program plan for 1990: Estuaries. Narragansett RI, USA: USEPA, Environmental Research Laboratory. EPA-600-4-90-033. 182 p.

Hopkin SP, Jones DT, Dietrich D. 1993. The isopod *Porcellio scaber* as a monitor of the bioavailability of metals in terrestrial ecosystems: Towards a global 'woodlouse watch' scheme. *Sci Total Environ Suppl* (Part 1):357-365.

Horvath B, Gruiz K. 1995. Ecotoxicological testing of contaminated soil. In: Van den Brink WJ, Bosman R, Arendt F, editors. Contaminated soil '95. Dordrecht, NL: Kluwer Academic Publishers. p 619-620.

Huggett RJ, Kimerle RA, Mehrle Jr PM, Bergman HL, editors. 1992. Biomarkers: Biochemical, physiological, histological markers of anthropogenic stress. Chelsea MI, USA: Lewis Publishers. 347 p.

[ISO] International Standards Organization. 1998a. Soil quality: Effects of pollutants on earthworms (*Eisenia fetida*) Part 2: Determination of effects on reproduction. Geneva, CH: ISO. ISO-11268-2. 16 p.

[ISO] International Standards Organization. 1998b. Soil quality: Inhibition of reproduction of Collembola (*Folsomia candida*). Draft International Standard. Geneva, CH: ISO. ISO-11267. 16 p.

Jones TC, Hunt RD, King NW, editors. 1997. Veterinary pathology. 6th ed. Baltimore MD, USA: Lippincott, Williams and Wilkins. 1392 p.

Kapustka LA, La Point TW, Fairchild JF, McBee K, Bromenshenk JJ. 1989. Field assessments. In: Warren-Hicks W, Parkhurst B, Baker Jr S, editors. Ecological assessments at hazardous waste sites. Seattle WA, Corvallis OR, USA: USEPA, Corvallis Environmental Research Laboratory. EPA-600-3-89-013. p 8-1-8-47.

Keddy CJ, Greene JC, Bonnell MA. 1995. Review of whole-organism bioassays: Soil, freshwater sediment, and freshwater assessment in Canada. *Ecotoxicol Environ Saf* 30:221-251.

Knecht JA, Van Dillen M, Koevoets P, Schat LM, Verkleij JAC, Ernst WHO. 1994. Phytochelatins in cadmium-sensitive and cadmium-tolerant *Silene vulgaris*. *Plant Physiol* 104:255-261.

Kochwa S. 1976. Immunoelectrophoresis. In: Rose N, Friedman H, editors. Manual of clinical immunology. Washington DC, USA: American Society of Microbiology. p 17-35.

Kratz W. 1998. The bait-lamina test: General aspects, applications and perspectives. *Environ Sci Pollut Res* 5:94-96.

Lanno R, Leblanc S, Knight B, Tymoswski R, Fitzgerald D. 1998. Application of body residues as a tool in the assessment of soil toxicity. In: Sheppard S, Bembridge J, Holmstrup M, Posthuma L, editors. Advances in earthworm ecotoxicology. Pensacola FL, USA: SETAC. p 41-53.

Lanno RP, McCarty LS. 1997. Earthworm bioassays: Adopting techniques from aquatic toxicity testing. *Soil Biol Biochem* 29:693-697.

La Point T, Fairchild J. 1989. Aquatic field methods. In: Warren-Hicks W, Parkhurst B, Baker Jr S, editors. Ecological assessments at hazardous waste sites. Seattle WA, Corvallis OR, USA: USEPA, Corvallis Environmental Research Laboratory. EPA-600-3-89-013. p 8-14–8-28.

Lawlor K, Duncan K, Levetin E, Buck P, Wells H, Jennings E, Hettenbach S, Bailey S, Fisher JB, Todd T. 1997. Effects of crude oil contamination and bioremediation in a soil ecosystem. In: (Proceedings) 4th International In Situ and On-site Bioremediation Symposium, Volume 4; 1997 Apr 28-May 1; New Orleans, LA, USA. Columbus OH, USA: Battelle Press. p 383-398.

Linder G, Bollman M, Callahan C, Gillette C, Nebeker A, Wilborn D. 1998. Bioaccumulation and food-chain analysis for evaluating ecological risks in terrestrial and wetland habitats. I. availability-transfer factors (ATFs) in "soil macroinvertebrate amphibian" food chains. In: Hoddinott K, editor. Superfund risk assessment in soil contamination studies; 3rd Volume. West Conshohocken PA, USA: ASTM. STP 1338. p 51-65.

Linder G, Ingham E, Brandt J, Henderson G. 1992. Evaluation of terrestrial indicators for use in ecological assessments at hazardous waste sites. Corvallis OR, USA: USEPA, Environmental Research Laboratory. EPA-600-R-92-183. 206 p.

Linz DG, Nakles DV, editors. 1997. Environmentally acceptable endpoints in soil: Risk-based approach to contaminated site management based on availability of chemicals in soil. Annapolis MD, USA: American Academy of Environmental Engineers. 630 p.

Løkke H, Van Gestel CAM, editors. 1998. Handbook of soil invertebrate toxicity tests. West Sussex, England: Wiley and Sons, Ltd. 281 p.

Ma W-C, Van Kleunen A, Immerzeel J, De Maagd PGJ. 1998. Bioaccumulation of polycyclic aromatic hydrocarbons by earthworms: Assessment of equilibrium partitioning theory in in situ studies and water experiments. *Environ Toxicol Chem* 17:1730-1737.

Marigómez I, Kortabitarte M, Dussart GBJ. 1998. Tissue-level biomarkers in sentinel slugs as cost-effective tools to assess metal pollution in soils. *Arch Environ Contam Toxicol* 34:167-176.

Marquenie JM, Simmers JW, Kay SH. 1987. Preliminary assessment of bioaccumulation of metals and organic contaminants at the Times Beach confined disposal site, Buffalo, NY. Vicksburg MS, USA: Department of the Army, Waterways Experimental Station, Corps of Engineers. USAWES Miscellaneous Paper EL-87-6.

McBee K. 1989. Terrestrial field methods. In: Warren-Hicks W, Parkhurst B, Baker Jr S, editors. Ecological assessments at hazardous waste sites. Seattle WA, Corvallis OR, USA: USEPA, Corvallis Environmental Research Laboratory. EPA-600-3-89-013. p 8-5-8-14.

McCarthy JF, Shugart LR, editors. 1990. Biomarkers of environmental contamination. Chelsea MI, USA: Lewis Publishers. 457 p.

McCarty LS, Mackay D, Smith AD, Ozburn GW, Dixon DG. 1992. Residue-based interpretation of toxicity and bioconcentration QSARs from aquatic bioassays: Neutral narcotics. *Environ Toxicol Chem* 11:917-930.

Menzie CA, Burmaster DE, Freshman JS, Callahan CA. 1992. Assessment of methods for estimating ecological risk in the terrestrial component: A case study at the Baird and McGuire superfund site in Holbrook, Massachusetts. *Environ Toxicol Chem* 11:245-260.

Mohrig W, Kanschke E, Ehleers M. 1984. Rosette formation by coelomocytes of earthworm *Lumbricus terrestris* L. with sheep erythrocytes. *Dev Comp Immunol* 8:471-476.

Morgan JE, Morgan AJ. 1988. Earthworms as biological monitors of cadmium, copper, lead and zinc in metalliferous soils. *Environ Pollut* 54:123-138.

[OECD] Organization for Economic Cooperation and Development. 1984. OECD guidelines for testing of chemicals: Earthworm acute toxicity test. Paris, France: OECD. Guideline Nr 207.

[OMAF] Ontario Ministry of Agriculture and Food. 1990. Recommendations for turfgrass management. Toronto ON, Canada: Ontario Ministry of Agriculture and Food. OMAF Publication 384.

Paine JM, McKee MJ, Ryan MF. 1993. Toxicity and bioaccumulation of soil PCBs in crickets: Comparison of laboratory and field studies. *Environ Toxicol Chem* 12:2097-2103.

Parkhurst B, Linder G, McBee K, Bitton G, Dutka B, Hendricks C. 1989. Toxicity testing. In: Ecological assessments at hazardous waste sites. Seattle WA, Corvallis OR, USA: USEPA, Corvallis Environmental Research Laboratory. EPA-600-3-89-013. p 6-1–6-5.

Parmelee RW, Bohlen PJ, Blair JM. 1998. Earthworms and nutrient cycling processes: Integrating across the ecological hierarchy. In: Edwards CA, editor. Earthworm ecology. Boca Raton FL, USA: St. Lucie Press. p 123-143.

Paton GI, Campbell CD, Glover LA, Killham K. 1995. Assessment of the bioavailability of heavy metals using lux modified constructs of *Pseudomonas fluorescens*. *Lett Appl Microbiol* 20:52-56.

Paton GI, Palmer G, Burton M, Rattray EAS, McGrath SP, Glover LA, Killham K. 1997. Development of an acute and chronic ecotoxicity assay using lux-marked *Rhizobium leguminosarum* biovar *trifolii*. *Lett Appl Microbiol* 24:296-300.

Paton GI, Rattray EAS, Campbell CD, Meussen H, Cresser MS, Glover LA, Killham K. 1997. Use of genetically modified microbial biosensors for soil ecotoxicity testing. In: Pankhurst CS, Doube B, Gupta V, editors. Bioindicators of soil health. Wallingford, UK: CAB International. p 397-418.

Pawert M, Triebskorn R, Graff S, Berkus M, Schulz J, Köhler HR. 1996. Cellular alterations in collembolan midgut cells as a marker of heavy metal exposure: Ultrastructure and intracellular metal distribution. *Sci Total Environ* 181:187-200.

Peakall DB, Walker CH. 1994. The role of biomarkers in environmental assessment (3). Vertebrates. *Ecotoxicology* 3:173-179.

Pinkart HC, Ringelberg DB, Stair JO, Sutton SD, Pfiffner SM, White DC. 1995. Phospholipid analysis of extant microbiota for monitoring in situ bioremediation effectiveness. In: Hinchee RE, Douglas GS, Ong SK, editors. Monitoring and verification of bioremediation. Columbus OH, USA: Battelle Press. p 49-57.

Posthuma L, Van Gestel CAM, Smit CE, Bakker DJ, Vonk JW, editors. 1998. Validation of toxicity data and risk limits for soils: Final report. Bilthoven, NL: National Institute of Public Health and the Environment. Report Nr 607505004. 230 p.

Presnell JK, GL Humason, Schreibman M. 1997. Humason's animal tissue techniques. 5th ed. Baltimore MD, USA: Johns Hopkins University Press. 536 p.

Reynolds JW, Krohn WB, Jordon G. 1977. Proceedings of Woodcock Symposium 6:135-147.

Ringelberg DB, Davis JD, Smith GA, Pfiffner SM, Nichols PD, Hensen JM, Wilson JT, Yates M, Kampbell DH, Reed HW, Stocksdale TT, White DC. 1988. Validation of signature polar lipid fatty acid biomarkers for alkane-utilizing bacteria in soils and subsurface aquifer materials. *FEMS Microbiol Ecol* 62:39-50.

Rodriguez-Grau J, Venables BJ, Fitzpatrick LC, Cooper EL. 1989. Suppression of secretory rosette formation by PCBs in *Lumbricus terrestris*: An earthworm assay for humoral immunotoxicity of xenobiotics. *Environ Toxicol Chem* 8:1201-1207.

Rose NR, Friedman H, editors. 1976. Manual of clinical immunology. Washington DC, USA: American Society for Microbiology. 932 p.

Rutgers M, Van't Verlaat I, Wind B, Posthuma L, Breure AM. 1998. Rapid method to assess pollution-induced community tolerance in contaminated field soil. *Environ Toxicol Chem* 17:2210-2213.

Sanchez PS, Sato MIZ, Paschoal CMRB, Alves MN, Furlan EV, Martins MT. 1988. Toxicity and bioavailability assessment of industrial effluents from Sao Paulo State, Brazil, using short-term microbial assays. *Toxic Bioavail Assess* 3:55-80.

Sheppard SC, Evenden WG, Abboud SA, Stephenson M. 1993. A plant life-cycle bioassay for contaminated soil, with comparison to other bioassays: mercury and zinc. *Arch Environ Contam Toxicol* 25:27-35.

Simenstad CA, Tanner CD, Thom RM, Conquest LL. 1991. Estuarine habitat assessment protocol. Seattle WA, USA: USEPA, Region 10, Office of Puget Sound. EPA-910-9-91-037. 206 p.

Smit CE, Van Beelen P, Van Gestel CAM. 1997. Development of zinc bioavailability and toxicity for the springtail *Folsomia candida* in an experimentally contaminated field plot. *Environ Pollut* 98:73-80.

Smit CE, Van Gestel CAM. 1997. Influence of temperature on the regulation and toxicity of zinc in *Folsomia candida* (Collembola). *Ecotoxicol Environ Saf* 37:213-222.

Smit CE, Van Gestel CAM. 1998. Effects of soil type, pre-percolation and ageing on bioaccumulation and toxicity of zinc for the springtail *Folsomia candida*. *Environ Toxicol Chem* 17:1132-1141.

Stephenson GL. 1995. Proposal to develop terrestrial toxicity test methods for assessment of contaminated soils. Prepared for Environment Canada. Orton ON, Canada: Aquaterra Environmental. 40 p.

Stephenson GL. 1998a. Development of plant toxicity tests for assessment of contaminated soils. Ottawa ON, Canada: Environment Canada, Method Development and Application Section. 75 p.

Stephenson GL. 1998b. Development of a reproduction test with *Onychiurus folsomi* for assessment of contaminated soils. Ottawa ON, Canada: Environment Canada, Method Development and Application Section. 253 p.

Stephenson GL. 1998c. Development of earthworm toxicity tests for assessment of contaminated soils. Ottawa ON, Canada: Environment Canada, Method Development and Application Section. 52 p.

Stephenson GL, Solomon KR, Hale B, Greenberg BM, Scroggins RP. 1997. Development of suitable test methods for evaluating the toxicity of contaminated soils to a battery of plant species relevant to soil environments in Canada. In: James Dwyer F, Doane TR, Hinman ML, editors. Environmental toxicology and risk assessment: Modeling and risk assessment. 6th Volume. West Conshohocken PA, USA: ASTM. STP 1317. p 474-489.

Stephenson GL, Wren CD, Middelraad IJC, Warner JE. 1997. Exposure of the earthworm, *Lumbricus terrestris*, to diazinon, and the relative risk to passerine birds. *Soil Biol Biochem* 29:717-720.

Stephenson GL, Kuperman R, Linder GL, Visser S. 2002. Toxicity tests for assessing contaminated soils and ground water. In: Sunahara GI, Renoux AY, Thellen C, Gaudet CL, Pilon A, editors. Environmental analysis of contaminated sites. Sussex, UK: John Wiley and Sons, Inc. p 25-43.

Stokke K, Stenersen J. 1993. Non-inducibility of the glutathione transferases of the earthworm *Eisenia andrei*. *Comp Biochem Physiol* 106C:753-756.

Stroomberg GJ, Reuther C, Kozin I, Van Brummelen TC, Van Gestel CAM, Gooijer C, Cofino WP. 1996. Formation of pyrene metabolites by the terrestrial isopod *Porcellio scaber*. *Chemosphere* 33:1905-1914.

Svendsen C, Weeks JM. 1997a. Relevance and applicability of a simple earthworm biomarker of copper exposure. I. Links to ecological effects in a laboratory study with *Eisenia andrei*. *Ecotoxicol Environ Saf* 36:72-79.

Svendsen C, Weeks JM. 1997b. Relevance and applicability of a simple earthworm biomarker of copper exposure. II. Validation and applicability under field conditions in a mesocosm experiment with *Lumbricus rubellus*. *Ecotoxicol Environ Saf* 36:80-88.

Triebskorn R, Köhler H-R. 1992. Plasticity of the endoplasmic reticulum in three cell types of slugs poisoned by molluscicides. *Protoplasma* 169:120-129.

Tizard IR, Kersey R, editors. 1996. Veterinary immunology. 5th ed. New York NY, USA: W.B. Sanders Company. 531 p.

[USEPA] U.S. Environmental Protection Agency. 1985. Toxic Substances Control Act test guidelines; Final rules. Washington DC, USA: USEPA. 40 CFR parts 796, 797, and 798.

[USEPA] U.S. Environmental Protection Agency. 1986. Diazinon: Position document 4. Intent to cancel registrations and denial of applications for registration of pesticide products containing diazinon; Conclusion of special review. Washington DC, USA: USEPA.

Van Beelen P, Doelman P. 1997. Significance and application of microbial toxicity tests in assessing ecotoxicological risks of contaminated soils and sediments. *Chemosphere* 34:455-499.

Van Brummelen TC, Verweij RA, Wedzinga SA, Van Gestel CAM. 1996. Polycyclic aromatic hydrocarbons in earthworms and isopods from contaminated forest soils. *Chemosphere* 32:315-341.

Van Gestel CAM. 1997. Scientific basis for extrapolating results from soil ecotoxicity tests to field conditions and the use of bioassays. In: Van Straalen NM, Lokke H, editors. Ecological risk assessment of contaminants in soils. London, England: Chapman & Hall, p 25-50.

Van Gestel CAM, Adema DMM, De Boer JLM, De Jong P. 1988. The influence of soil clean-up on the bioavailability of metals. In: Wolf K, Van den Brink WJ, Colon FJ, editors. Contaminated soil '88. Hamburg, DE: Kluwer Academic Publishers. p 63-65.

Van Gestel CAM, Dirven-Van Breemen EM, Kamerman JW. 1993. The influence of soil clean up on the bioavailability of heavy metals for earthworms and plants. In: Eijsackers HJP, Hamers T, editors. (Proceedings) Integrated soil and sediment research: A basis for proper protection. Hamburg, DE: Kluwer Academic Publishers. p 345-348.

Van Gestel CAM, Hensbergen PJ. 1997. Interaction of Cd and Zn toxicity for *Folsomia candida* Willem (Collembola: Isotomidae) in relation to bioavailability in soil. *Environ Toxicol Chem* 16:1177-1186.

Van Straalen NM. 1994. Biodiversity of ecotoxicological responses in animals. *Neth J Zool* 44:112-129.

Van Straalen NM. 1996. Critical body-residues: Their use in bioindication. In: Van Straalen NM, Krivolutsky DA, editors. Bioindicator systems for soil pollution. Dordrecht, NL: Kluwer Academic Publishers. p 5-16.

Van Wensem J, Vegter JJ, Van Straalen NM. 1994. Soil quality criteria derived from critical body concentrations of metals in soil invertebrates. *Appl Soil Ecol* 1:185-191.

Verhoef HA, Brussaard L. 1990. Decomposition and nitrogen mineralization in natural and agro-ecosystems: The contribution of soil animals. *Biogeochemistry* 11:175-211.

Wagner KH, Wagner-Hering E. 1971. The cycle of cancer-causing substances—polycyclic aromatic hydrocarbons—in plants and humans. *Protectio Vitae* 6:260.

Weeks JM, Svendsen C. 1996. Neutral red retention by lysosomes from earthworm (*Lumbricus rubellus*) coelomocytes: A simple biomarker for exposure to soil copper. *Environ Toxicol Chem* 15:1801-1805.

Wells JB, Lanno RP. 2001. Passive sampling services (PSDs) as biological surrogates for estimating the bioavailability of organic chemicals in soil. In: Greenberg BM, Hull RN, Roberts Jr MH, Gensemer RW, editors. Environmental toxicology and risk assessment: Science, policy, and standardization - implications for environmental decisions. 10th Volume. West Conshohocken PA, USA: ASTM. STP 1403. p 253-270.

Zanger M, Gräff S, Braunbeck T, Alberti G, Köhler H-R. 1997. Detection and induction of cytochrome P450IA (CYP 1A)-like proteins in *Julus scandinavius* (Diplopoda) and *Oniscus asellus* (Isopoda): A first analysis. *Bull Environ Contam Toxicol* 58:511-517.

Chapter 9

Chemical Measures of Bioavailability

Martin Alexander, Scott D. Cunningham, Rufus R. Chaney, Joseph B. Hughes, Joop Harmsen

Introduction

The ability to accurately determine the effects of pollutants on populations of individual species or on ecosystems is hampered by an enormous uncertainty in the quantification of receptor exposure. This uncertainty has resulted in a conservative approach in exposure assessment that assumes that the entire concentration or mass of a contaminant present in a given soil is potentially available for uptake by possible receptors. This assumption is in marked contrast to laboratory studies and field observations. Numerous studies have demonstrated that the responses of at-risk populations are not affected by the total concentration of a contaminant in the soil where a receptor resides, but instead by only that fraction that is biologically available at that time and soil condition. This is particularly true in soils that contain contaminants that undergo "poorly reversible sorption" (sometimes referred to as "sequestration" or "irreversible sorption"). An essential step in conducting realistic risk assessments in these situations is the development of analytical methods to measure the concentrations of bioavailable contaminants and relate those concentrations to biological responses or accumulation. This should greatly help in the determination of site-specific remediation goals in relation to present and future land use, as well as possible endpoints for a biologically based remediation technique.

This chapter addresses the development and use of analytical procedures to assess the bioavailability of soil contaminants. The soil–contaminant–organism–ecosystem continuum is complex; however, the bioavailability of a few compounds from soil to organisms has been successfully modeled, and that information is used sometimes on a predictive basis. It is very likely that appropriate procedures can and will be devised for using chemical assays for assessing the bioavailability of toxicants in soils. The existing base of information on the reactions of toxicants in and with soil and the range of analytical approaches is sufficiently large that suitable procedures can be devised and validated to be useful in estimating the bioavailability of contaminants for risk assessment.

Contaminated Soils: From Soil-Chemical Interactions to Ecosystem Management. Roman P. Lanno, editor.

Background

The premise behind assays covered in this chapter is that chemical assays can be developed to reflect a biologically relevant fraction of soil contaminants. These protocols would be more readily adopted if they were based on techniques and equipment familiar to existing environmental laboratories. Current analytical protocols developed for environmental matrices have 4 components: extraction, separation, identification, and quantification. The focus of many of the new analytical protocols is primarily the first step, the development of procedures that extract biologically available fractions. Much of the separation, identification, and quantification relies on traditional approaches and instrumentation. Since the 1960s, most accepted extraction protocols used for the analysis of toxicants in soils have relied on exhaustive techniques intended to recover the total mass of toxicant present. More recently, extraction techniques have been developed to focus on nonexhaustive extractions that can be correlated with observed biological responses. Two approaches have been taken. The first is an empirical approach in which soil is subjected to a wide variety of extraction conditions. The results are compared to biological assays, and the extraction protocol with the highest correlation coefficient is chosen. In contrast, in a biomimetic approach, an extractant is used that has characteristics that correspond to exposure pathways or organisms (i.e., inhalation, ingestion, dermal contact, and plant and microbial uptake). The extraction conditions then are adjusted to give the highest possible correlation with bioassay results. Both approaches have shown some success in predicting bioavailable fractions.

Figure 9-1 presents a conceptual model of the distribution of soil toxicants and their transport into a medium in contact with an organism. As depicted, only a fraction of the total mass of soil-bound toxicants will be readily available to an organism. Also, it is recognized that the presence of organisms may result in a modification of the composition of the aqueous solution contacting the soil particle. Specific examples of how an organism affects the properties of soil in contact with an organism include pH changes (either alkaline or acidic) and redox changes in the gastrointestinal tract and the production of chelating agents or surfactants near plant roots and by microorganisms. Regardless of the nonexhaustive extraction techniques developed (i.e., empirical or biomimetic), it is important that the range of conditions appropriate for the exposure pathway be considered in the correlation of chemical procedures with biological assays. The heterogeneous distribution of many contaminants in soils and uncertainties in analytical assays can be limiting factors in establishing good correlations between the chemical assays and a biological response. Standard deviations of 20% or higher can be expected, and sample-pooling methods may be appropriate (Lamé et al. 1996).

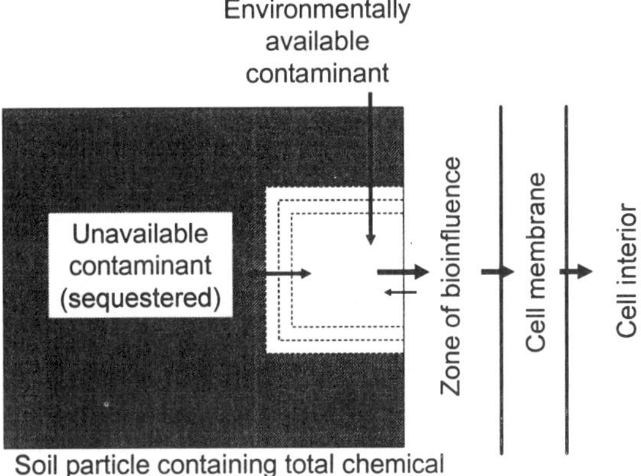

Figure 9-1 Conceptual diagram of the distribution of soil contaminants (i.e., available and sequestered form) and the surrounding microenvironment that links soil contaminants with biological membranes

What is Required for Chemical Assays to be Applicable?

It is essential that the results of chemical assays for the bioavailable fraction of contaminants in soils of various types be closely correlated with the results of biological assays. Some of the chemical procedures that have been proposed, or are being developed, appear to be promising, but until the data they provide are correlated with biological responses, even the initial assessment of the utility of those chemical approaches cannot be considered adequate.

Comparisons of precision, sensitivity, and time and cost per test need to be made between chemical and biological assays. It is expected that the chemical test will be more precise, more sensitive, and less costly, but this assumption must be verified. The length of time for a chemical assay needs to be compared to the length needed for bioassays to assess whether there also is a savings in time. It is of critical importance that the chemical, or any surrogate test of bioavailability, be accepted by regulators.

Many laboratories currently use chemical procedures to determine the total concentration of toxic chemicals and pollutants in soil. They have large investments in equipment, and their staffs are trained to use vigorous solvent-extraction procedures or conduct analyses of total metal concentrations. Thus, it is desirable to have chemical assays that allow existing laboratories to conduct the testing, and for those laboratories to use their existing equipment and personnel. From this viewpoint, less vigorous extraction procedures appear to be advantageous. The purchase of new or expensive equipment and the requirement for personnel with different expertise should be minimized or avoided. Nevertheless, the key decision in selecting one or more chemical assays must be based on the correlation with biological response, and this may require approaches that involve different types of testing laboratories, instrumentation, or personnel.

Established Methods

Inorganic ions

More than 50 years ago, agronomists and soil scientists began to search for chemical methods to determine the concentration of individual plant nutrients in agricultural soils. The impetus for this search was the need to establish a basis for making recommendations or decisions about the quantity of these nutrients that had to be added to soil to achieve maximum yields of crops. As a result of this research, chemical methods were devised that give data that are reasonably predictive of the bioavailability of inorganic ions necessary for plant development. Indeed, such procedures are now widely used and fully accepted. For example, extraction techniques are used to predict the levels of P available to plants. Soil water can be directly analyzed to determine the immediately available phosphate. Certain extractants can then be employed to assess the availability of P with time. Total P concentration, combining available and unavailable forms, can be determined after complete digestion of a soil sample with strong acids (Pierzynski 1998). These widely used methods have been correlated with crop response to phosphate fertilizers in different soils.

Chemical extractions are used to evaluate the available levels of micronutrients in soils because total concentrations are often inadequate to predict deficiencies for specific crops. Extraction methods were established by correlation between the results of extraction and the response of crop species to the addition of fertilizers. An example of a routine soil test for plant-available (deficient versus adequate) Zn, Cu, Fe, and Mn is the diethylenetriamine pentaacetic acid (DTPA) test of Lindsay and Norvell (1978). When any single extractant is used, it can be well correlated with adequacy for one, but not all, crops. Agricultural experience has identified the crops and soils in particular regions that are likely to be deficient in a particular element. Growers are advised to test when they expect to grow a crop on a soil that has not been fertilized with the nutrient for years.

Chemical assays of bioavailability also are considered in agriculture to predict the toxicity of elements such as Zn, Cd, Ni, Cu, Co, Mn, and Al to higher plants. Soils differ in the concentration of metals required to cause phytotoxicity, and the soil differences are usually related to the ability of the soil to sorb or precipitate the element. Sorption capacity of soils is commonly related to pH, organic matter (which affects the strength of sorption by these sorbing components), and levels of hydrous Fe and Mn oxides. Because these soil properties strongly affect solubility and phytotoxicity, analyzing total concentration provides little useful information.

Extractions used to characterize levels of plant-available elements in soils are of 3 general types: 1) those that extract the total amount or a fraction of the total, 2) those that extract from the "labile" pool (i.e., the potentially root-absorbable pool), and 3) those that reflect the instantaneous plant-available concentration. The last extraction technique is useful in determining whether the soil in a field will cause toxicity to a particular crop. However, if soil pH or organic matter concentrations change through management practices, both the phytotoxic and extractable concentrations can be altered. Thus, an "instantaneous plant-available" test is useful for only a short time after it is conducted. To predict the solubility or availability to plants of the element at the existing soil pH, neutral salt extractions are commonly used, rather than chelating agents or acids. The dilute salt solution extracts amounts proportional to uptake; 0.01 M $CaCl_2$ and 0.1 M NH_4NO_3 have been shown to extract amounts of Zn, Cd, or Ni proportional to plant uptake (Prüe et al. 1991; Sauerbeck and Hein 1991). Similar observations have been made with respect to soil metal availability by soil invertebrates. Smit et al. (1997) were able to correlate the EC50 of the springtail *Folsomia candida* to both the $CaCl_2$-available and water-available extraction in both a freshly and a historically polluted soil. Conder and Lanno (2000) found $Ca(NO_3)_2$-extractable Cd, Pb, and Zn were better predictors of toxicity to earthworms than total metal measurements.

Calibration of phytotoxicity tests is complicated because roots can grow through surface soil, which may be rich in metal contaminants, and obtain nutrients and water from deeper soil horizons. For this reason, the response observed in pot studies is commonly much greater than the response measured in the field (deVries and Tiller 1978). This leads to a situation in which analytical protocols give a better predictive measure than laboratory bioassays.

In general, it is recognized that metals in soil are either in the available or the occluded fractions. The "labile pool" of an element is a good measure of the fraction that is potentially bioavailable to plants and is defined by the specific activity of an element in plant shoots when the soil is amended with a carrier-free isotope (Tiller 1979). It is clear that some elements can be occluded within Fe or Mn oxides and be unavailable to plants—even to the extent that plants or animals grazing on the plants are deficient in the element. In particular, Co is occluded within MnO_2, and Ni is occluded within Fe oxides (Tiller et al. 1969; McKenzie 1970; McLaren et al. 1985; Bruemmer et al. 1988). Cd is poorly occluded in either oxide, while Zn and Pb may be occluded (Bruemmer et al. 1988; Ford et al. 1997).

Organic chemicals

Pesticide bioavailability has been of major interest for many years. In particular, work has been done on the development of assays to determine efficient application rates as well as to predict or measure the potential for pesticide carryover to the next growing season. Among the pesticides, especially with herbicides with long residual efficacy in soil, the biological distinction between total levels of compound detectable in the soil and plant-available compounds has long been recognized and mirrors findings with plant nutrients. With many of the modern pesticides that are used at very low rates (g/ha), the development of chemical assays to measure bioavailable residues has required the parallel development of analytical capabilities with vastly improved sensitivity. Total concentrations in soil often can be measured with harsh extraction and conventional analysis; however, detection of bioavailable fractions, often in water-phase extractions, requires analytical modifications such as column-switching techniques. This further highlights the dichotomy between total chemical analysis and biologically available fractions.

Numerous types of assays have been evaluated to determine whether the results of extractions correlate well with phytotoxicity. Initial results demonstrated that plant damage was poorly correlated with total extractable compound, particularly in soils containing aged pesticides. As a result, biological assays were employed, and these assays were shown to be predictive but cumbersome and expensive. Eventually, many of these bioassays were replaced by porewater assays (analogous to techniques developed for sediments [Adams et al. 1985; Di Toro et al. 1991]) that were capable of correlating herbicide concentration in soil with plant injury. The method involves the centrifugation of a sample of soil to obtain the water that is trapped within the soil matrix, and the soil water is then subjected to chemical analysis. The results of such experiments indicate that for some compounds that are acutely toxic (e.g., herbicide to a sensitive plant or an insecticide to an appropriate insect), porewater analysis provides a good approximation of bioavailability (Ronday et al. 1997). These chemical tests have correlated well with observed plant injury (Duffy et al. 1993) and allowed the agricultural industry to decrease its reliance on bioassays. If similar extraction-analysis approaches can be developed for the evaluation of the bioavailable fraction of other groups of compounds, it may be possible to systematically and quantitatively measure the bioavailability of, and ultimately the risk associated with, contaminants in a specific soil matrix.

Promising New Methods

A number of chemical methods have been described that may be useful for assessing or predicting bioavailability. These are usually empirically based; however, some are based upon attempts to mimic conditions of biological uptake. These "biomimetic" extractions are intellectually attractive, but until good correlations with actual field data are available, they should not be counted on for any inherent predictive advantage over empirically derived methods.

Aqueous extractions

The transfer of many contaminants to organisms involves a release of the contaminant from a soil particle into the aqueous phase and then a transfer to a surface membrane or cuticle of an organism. The aqueous phase may be modified by the organism present, but obtaining information on the quantity of contaminant available in a water phase is often especially important as a measurement of availability to leaching, potential biological uptake, or subsequent resorption. A number of techniques are being developed to sample this water phase within the soil.

Pore water

The concentrations of contaminants present in the water phase in pores between soil particles have been determined. The pore water can be obtained by centrifugation of soil samples using tubes consisting of 2 compartments separated by a filter or by using porous cups. It is important that the contaminant to be analyzed does not react with the equipment being used for the analysis. Preliminary studies suggest that this method is useful for correlation with the response of invertebrates to toxicants (Houx and Aben 1993; Ronday et al. 1997).

Solid-phase extractions from soil water

Solid-phase extraction (SPE) is a technique in which a bead, fiber, or membrane is introduced into a wet soil, soil suspension, or water phase above the soil and withdrawn after a period of time. This method of contaminant sampling relies on the capacity of the solid matrix to sample the concentrations in solution and the potential capacity of the soil to maintain that equilibrium concentration. Contaminants that have partitioned into these materials are then extracted from the solid materials for separation analysis, or in some cases, the solid-phase material can be placed directly in the injection port of an instrument (e.g., gas chromatograph). This technique appears to offer considerable promise and currently is being used to assess the bioavailability of several organic compounds in soil to earthworms, the solid-phase extractants being Tenax TA beads and C18 membranes. Preliminary data show good correlations between DDT, DDE, DDD, and polycyclic aromatic hydrocarbon (PAH) uptake by *Eisenia fetida* from a range of soils and the quantity sorbed by the SPEs (Tang and Alexander 1999; Morrison et al. 2000). Tenax beads also have been used for measuring the desorption kinetics of some chlorobenzenes, polychlorinated biphenyls (PCBs), and PAHs (Cornelissen et al. 1997, 1998). Solid-phase microextraction (SPME) fibers have been used to measure phenanthrene exposure in earthworm lethality tests in artificial soil with varying amounts of organic matter (Wells and Lanno 2001). Results of SPE tests can also be correlated with the results of bioassays, the biodegradable fraction of PAHs in the field, and bioaccumulation tests (Harmsen and Ferdinandy 1999). The SPE tests described above are used to measure solution concentrations of contaminants in a specific mass of soil. Semipermeable membrane devices (SPMDs) offer another alternative and can be used to measure gas phase or porewater nonpolar organic contaminants in a relatively large volume of soil, integrated over an extended period of time. However, measurements were more variable than with other SPE techniques and were not found to predict earthworm lethality of phenanthrene as well as SPME measurements (Wells and Lanno 2001).

Nonaqueous extractions

Extraction with single solvents

Some approaches rely on the use of a single extractant. The approach is based on the assumption that the quantity of chemical removed from soil by a mild or nonexhaustive extraction procedure will be similar to the amount that is bioavailable to one or more groups of organisms. The mild extraction is quite different from the Soxhlet extraction now commonly used for analyses of soils to determine the total concentration of organic compounds. The vigorous extraction by the Soxhlet procedure does not provide values that reflect bioavailability and usually overestimates, sometimes grossly, the bioavailable concentration. Mild extractability by several organic solvents has been correlated with the availability of phenanthrene and atrazine to earthworms and microorganisms (Kelsey et al. 1997), and extractability with tetrahydrofuran water has been correlated with the bioavailability of several organic compounds to earthworms (Tang and Alexander 1999). It has also been used to determine the portion of PAHs in sediments that are biodegradable (Doddema et al. 1998).

Automated, sequential, and multiple extractions

For ease of optimization and sample handling, as well as providing additional adjustable parameters, automated extraction equipment is increasingly being employed as a more flexible alternative to single extractions. Two possibilities, which are currently under development for the extraction of biologically relevant organic materials, may be of value: supercritical-fluid extraction (SFE) and accelerated solvent extraction (ASE). Both techniques are familiar to the regulatory community as total extraction methods; however, with minor instrumental adjustments, they may be made to differentiate relatively available and sequestered materials. Both ASE and SFE can be conducted as a harsh extraction designed to obtain the maximum amount of contaminant possible, a mild extraction condition designed to extract only an amount equivalent to that available in the pore water, or in a sequential mode using increasingly stringent conditions on a single soil sample. The latter is depicted in Figure 9-2 and parallels sequential extraction protocols that have been available for many years for metals in soils.

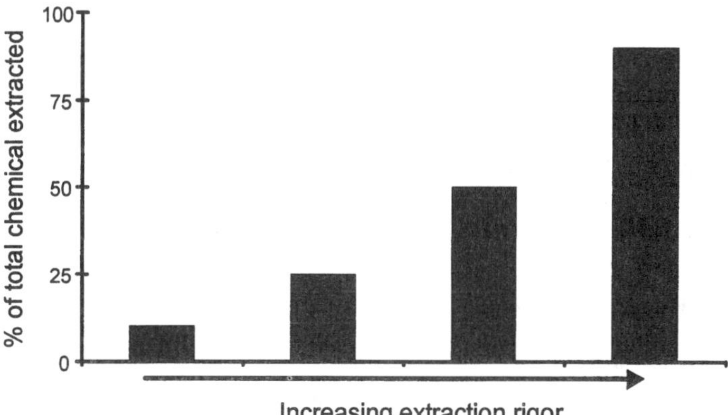

Figure 9-2 Hypothetical example of the influence of extraction conditions on the percent recovery of the total amount of a soil contaminant. In the case of metals, this might reflect a series of water, mild acid, strong acid, and finally strong acid with heat. In the case of hydrophobic organics, this might reflect a series of solvents ranging in polarities (hydrophilic to hydrophobic).

The use of SFE as a harsh extractant for removal of total organics in soil is increasingly accepted (Fahmy et al. 1993; Hawthorne et al. 1995), but its use as either a tool to explore desorption kinetics (Rochette and Koskinen 1996; Weber and Young 1997) or as an extractant for available fractions (Rochette and Koskinen 1998) is still being optimized. The influence of aging on extractability also has been demonstrated through the use of milder SFE conditions with PAHs, petroleum hydrocarbons, and pesticides (Camel et al. 1995; JR Shann, M Singh, and SD Cunningham, unpublished data). These promising analytical results have not yet been sufficiently paired with appropriate bioassays.

Like SFE, ASE is being developed by numerous investigators for the total extraction of organic contaminants from soil. ASE, under harsh extraction conditions, has even replaced Soxhlet extraction as an acceptable technique for some applications (USEPA 1995). For logistical reasons, sample handling, and ease of use, ASE is currently favored over SFE by many investigators. Like SFE, ASE under mild conditions has also been used to demonstrate that increasingly stringent extractions are required to remove compounds as they age in soils (JR Shann, M Singh, and SD Cunningham, unpublished data). ASE methodologies have not yet been applied to mimic a biological response nor have correlations with biological availability been studied to an appreciable extent.

Accelerated solvent extraction and SFE have many variables that can be used to optimize extractions to reflect the results of bioassay. These include the strength of an extract (polarity, pH, etc.), temperature (higher temperatures generally increase amounts extracted), time (increased time increases extraction harshness), and energy input (pressure, shaking, microwave, and sample grinding all increase the stringency of extraction). Correlation coefficients can sometimes be markedly improved in these multiparameter systems by adjusting one or more of these variables.

Direct analysis techniques

Thermal desorption

Recently, a number of investigators have begun to explore the potential for thermal desorption to differentiate the relative availability of chemical contaminants in soil. The premise behind this technique is that sequestered compounds should require more time and/or energy to be removed from soil samples. Thermal desorption experiments (Uzigiris et al. 1995; Werth and Reinhard 1997) have shown that thermal desorption spectra give biphasic desorption patterns like solution-based desorption profiles. The initial experimental designs were not appropriate to differentiate between relatively available and sequestered compounds in a single sample of soil; hence, more recent experiments have been conducted on single soil samples.

A coupled thermal desorption-mass spectrometer also has been used to study desorption of aged hydrocarbons from soil. This coupled arrangement is possible only when the desorbed contaminants are ionized for introduction into the mass spectrometer in a gentle manner (e.g., by photoionization). Initial results suggest that the effects of aging, clay content, and compound type are reflected in the shape and placement of the desorption curve (Zoller personal communication).

Metal speciation

With many metal contaminants, the use of analytical techniques that help speciate a metal in situ provides a mechanistic basis for laboratory-based metal desorption. Alternative analytical methods offer insight into measurements of metal bioavailability made in solutions (e.g., the reduction of Pb bioavailability observed in the extraction solution due to the Pb cation being complexed with P or iron oxides). This kind of information suggests possible conditions where the observed reduction in Pb bioavailability may be compromised. An example of the latter case would be Pb bound to iron oxides that might become available under iron-reducing conditions.

Extractions to predict contaminant bioavailability of soil ingested by infants

Because Pb in soil may pose a significant risk to children and animals following ingestion, research has been conducted to find methods to reduce the bioavailability of Pb in soils. Feeding tests showed that soils differ appreciably in the relative bioavailability of the Pb they contain (Freeman et al. 1992; Chaney and Ryan 1994; Schoof et al. 1996). The formation of very insoluble phosphate compounds (Ma et al. 1993; Zhang et al. 1998) and increasing soil adsorption strength for Pb by additions of phosphate and hydrous Fe oxide have effectively reduced the bioavailable fraction of Pb in soils (Chaney and Ryan 1994; Berti and Cunningham 1997).

Progress with in situ remediation of soils rich in Pb requires a method that is well correlated with Pb absorbed by mammals. Pigs and rats have been used in these studies, and both have shown that some treatments can reduce bioavailability of Pb in soil (Chaney and Ryan 1994; Brown et al. 1997). Chemical methods to evaluate the potential bioavailability of Pb in soil have been under development for several years. One of the early attempts used a bioavailability test developed for dietary Fe, and it was found that the measurement of dissolved Pb after 1 hour of stomach-like treatment of the soil, or after 1 hour of small intestine-like treatment of the soil from the stomach-like treatment, gave a good correlation with the fraction bioavailable to rats (Ruby et al. 1993). This method was further improved, simplified, and standardized as the physiologically based extraction technique (PBET) (Ruby et al. 1996). A similar degree of correlation between the stomach-phase extracted Pb and absorbed Pb (blood or bone) was observed in pig-feeding studies (Medlin 1997). Rodriguez et al. (1999) have also developed an in vitro gastrointestinal (IVG) method for examining the bioavailability of As from contaminated soils.

Although Pb in soils is not as soluble or bioavailable as Pb in foods or drinking water, soil can be an important source of absorbed Pb. Unfortunately, children absorb much more Pb from paint dust in their environment, and replacing Pb-contaminated soil has little effect on blood Pb in urban children exposed to housepaint Pb (Weitzman et al. 1993).

It has been shown that the in vitro method was well correlated with Pb absorbed from mine waste and other contaminated soils when fed to pigs. The good correlation between blood or bone Pb and extracted Pb (r approximately 0.8 to 0.9) suggests that the extraction method may replace feeding studies. Feeding 1 soil sample to pigs costs about \$60,000 to \$100,000 with appropriate quality assurance–quality control (QA–QC) practices, whereas the extraction method may cost only \$100 per sample. The method has been examined further to simplify it as much as possible, and the original inclusion of enzymes, organic acids, and some other substances was found to have no apparent effect on extraction of Pb. As long as the pH of the extraction fluid is approximately 2.0 to 2.5, the simulated stomach-phase extraction remains correlated with absorption. It is not yet clear whether this extraction reflects the effect of treatments to reduce the bioavailability of soil Pb. Questions also remain about the reliability of both the pig and rat bioassays for estimating Pb risk to children consuming soil, but children cannot be tested. Adults have been tested, and weanling primates could be tested with different soils to determine whether remediation reduces soil Pb bioavailability effectively and whether different Pb-rich soils have different Pb risk due to different bioavailabilities. Similar correlations with Pb absorbed by birds or other organisms may be needed for other risk assessment models.

Techniques derived from molecular biology

One possible approach to analyzing the bioavailable fraction of contaminants in soil is to use an organism, or a system derived from an organism, that responds to a biologically available fraction of a toxicant. Such organisms may be engineered to contain a gene sequence that, in the presence of a toxicant or substrate, triggers the transcription of a reporter gene. The gene product then may be measured either directly (e.g., light from a *lux* cassette or fluorescent-green protein) or indirectly through an enzyme assay (e.g., β-galactosidase from a lac gene).

To date, relatively few "sensor" organisms have been developed and used to differentiate bioavailable compounds from those that are sequestered, although this appears to be a fruitful area for research. Nevertheless, bacteria that are responsive to the presence of aromatic hydrocarbons such as toluene and naphthalene (Applegate et al. 1998), heavy metals (Corbisier 1996), and certain genotoxic agents (Van der Lelie et al. 1997; Alexander et al. 1999) are now being used. Combining the organism, sensor function, reporter function, and assay system remains to be done efficiently and the information correlated with other receptors and environmental endpoints, and results from the approach must be shown to correlate with biological assays. Certain novel biosensor approaches are beginning to find their way into use (Simpson et al. 1998). Work remains to be done before such systems can be designed, implemented, and correlated with human and ecological exposures; however, the biosensor approach appears promising. These biosensors may one day replace chemical analysis as an integrative tool to determine biologically relevant fractions of contaminants; however, they are currently subject to too many confounding factors in real soil to be reliably used in a risk-based regulatory framework.

Confounding Factors

Several factors may confound chemical assays or may lead to the lack of correlation of an otherwise suitable chemical assay with biological response. For example, in a soil with mixtures of toxicants, there may be synergistic or antagonistic effects so that the influence of several toxicants may be greater or less than predicted from their effects when tested individually; chemical procedures will not reveal these synergisms or antagonisms. Toxicants in soil sometimes exist in nonaqueous-phase liquids (NAPLs) or as crystalline materials, and the correlation between biological availability of compounds in NAPLs or present as crystals and the quantities detected by chemical procedures is presently unknown. The properties of some soils (e.g., pH, texture) may result in an inhibition of the growth or activity of organisms independently of the presence or absence of toxicants, and the chemical assays thus will not be related to the relevant biological response. Furthermore, although the concentration–response curve for the chemical assay may parallel the dose–response curve for a test organism (and thus represent a useful surrogate test) at low concentrations of a toxicant, the same may not be true at high concentrations (Figure 9-3).

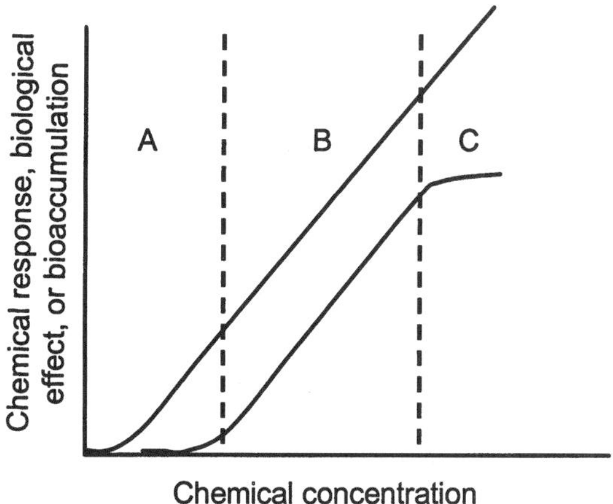

Figure 9-3 Comparison of a hypothetical response of chemical assays and bioassays to an increasing concentration of a toxicant in soil. In region A, the toxicant is detected by chemical analysis but there is no detectable biological effect. In region B, a parallel response is observed and correlation is possible. In region C, assays become nonlinear and correlation is lost.

Contaminants are not homogeneously distributed in soil, and this heterogeneous distribution may be represented by highly localized, small sites of contamination or large zones containing the toxic chemical. A bioassay organism will be affected differently by the small sites or by the large zones, or it may move away from the locus of contamination. In contrast, the chemical assay will be based on analysis of a small sample that, hopefully, is well mixed but does not show locally high or low concentrations, and it will neither reflect the capacity of animals to move away from zones of contamination nor the ability of organisms to integrate exposure over time.

Implementation

The use of chemical analyses to gain information on the bioavailability of toxicants in soil may often require analytical testing in addition to what is currently employed for the determination of total concentrations of contaminants. With the increase in information gained through these tests, one can anticipate an increase in the total cost of soil analyses. Thus, it will be advantageous to implement appropriate protocols that minimize this potentially significant economic burden. This can be done through a tiered decision-making approach whereby redundant testing could be avoided and by the selection of an appropriate test organism for which the chemical test is to be correlated early in the decision-making process.

An example of how a tiered approach may be used is in the case of a sample that contains significant levels of a contaminant that readily desorbs from the soil into an aqueous medium. Under this scenario, contaminants are available for biological uptake (and for transport), and it would not be necessary to perform an extensive evaluation of potential bioavailability. Conversely, if a soil contains contaminants (as determined in a test for total concentrations of soil contaminant) that are resistant to desorption, the information gained through chemical analysis correlated to bioavailability is warranted. Current testing of total concentrations of soil contaminants could be augmented by either porewater analysis or a stepwise desorption process as described previously. The additional information may act as a basis for initial decision-making, or it may modify the current regulatory paradigm.

It is possible, and indeed it may be highly likely, that the percentages of the total concentration of a compound that are available will vary with different biological species. This could lead to the assumption that the results of chemical assays must be correlated with a wide range of dissimilar species. Such an approach would be expensive, time consuming, and confusing, and it would be difficult to integrate into a regulatory system. Two alternatives can be considered. First, a conservative approach may be used in which the test organism is one that is able to acquire the highest percentage of the total concentration of the toxicant in soil, that is, it would have more of the compound bioavailable to it than any other species. This approach would offer the highest degree of protection to ecological receptors, but it also may be overprotective to the natural functioning of ecosystems. A second approach would rely on the use of a test organism (for correlating biological and chemical assays) that is most relevant or has special importance for the particular land use associated with the contaminated soil. Neither approach needs to be considered if humans are the species of concern for exposure at the site; in this case, results of the chemical assay need to be correlated with some measure of bioavailability to humans by the dermal, ingestion, or inhalation routes of exposure.

Conclusion

Chemical assays offer considerable promise for providing meaningful information that will be useful in developing realistic information for risk analysis of contaminated soils and for decision-making in regard to polluted sites. Nevertheless, the development of such methods as applied to contaminants in soil is still in its infancy. Additional research and development are clearly necessary. With the results from this further research and development, however, society will greatly benefit by having a firmer base for making decisions and attaining a better and cleaner environment.

References

Adams WJ, Kimerle RA, Mosher RG. 1985. Aquatic safety assessment of chemicals sorbed to sediments. In: Cardwell RD, Purdy R, Bayner RC, editors. Aquatic toxicology and hazard assessment. Philadelphia PA, USA: ASTM. STP 854. p 429-453.

Alexander R, Chung IY, Alexander M. 1999. Solid-phase genotoxicity assay for organic compounds in soil. *Environ Toxicol Chem* 18:420-425.

Applegate BM, Kehrmeyer SR, Sayler GS. 1998. A chromosomally based tod-lux CDABE whole-cell reporter for benzene, toluene, ethylbenzene and xylene (BTEX) sensing. *Appl Environ Microbiol* 64:2730-2735.

Berti WR, Cunningham SD. 1997. In-place inactivation of Pb in Pb-contaminated soils. *Environ Sci Technol* 31:1359-1364.

Brown SL, Xue Q, Chaney RL, Hallfrisch JG. 1997. Effect of biosolids processing on the bioavailability of Pb in urban soils. In: Biosolids management innovative treatment technologies and processes. (Proceedings) Water Environment Research Foundation Workshop #104; 1997 Oct 4; Chicago, IL, USA. Water Environment Research Foundation. p 43-54.

Bruemmer GW, Gerth J, Tiller KG. 1988. Reaction kinetics of the adsorption and desorption of nickel, zinc, and cadmium by goethite. I. Adsorption and diffusion of metals. *J Soil Sci* 39:37-52.

Camel V, Tambute A, Claude M. 1995. Influence of aging on the supercritical fluid extraction of pollutants in soils. *J Chromatogr* 693:101-111.

Chaney RL, Ryan JA. 1994. Risk based standards for arsenic, lead and cadmium in urban soils. In: Deutsche Gesellschaft für Chemisches Apparatewesen. Frankfurt, DE: Chemische Technik und Biotechnologie. 130 p.

Conder JM, Lanno RP. 2000. Evaluation of surrogate measures of cadmium, lead, and zinc bioavailability to *Eisenia fetida*. *Chemosphere* 41:1659-1668.

Corbisier P, Thiry E, Diels L. 1996. Bacterial biosensors for the toxicity assessment of solid wastes. *Environ Toxicol Water Qual* 11:171-177.

Cornelissen G, Richterink H, Ferdinandy MMA, Van Noort PCM. 1998. Rapidly desorbing fractions of PAHs in contaminated sediments as a predictor of the extent of bioremediation. *Environ Sci Technol* 32:966-970.

Cornelissen G, Van Noort PCM, Parsons JR, Govers HAJ. 1997. Temperature dependence of slow adsorption and desorption kinetics of organic compounds in sediments. *Environ Sci Technol* 31:454-460.

deVries MPC, Tiller KG. 1978. Sewage sludge as a soil amendment, with special reference to Cd, Cu, Mn, Ni, Pb, and Zn—Comparison of results from experiments conducted inside and outside a greenhouse. *Environ Pollut* 16:213-240.

Di Toro DM, Zarba CS, Hansen DJ, Berry WJ, Swartz RC, Cowan CE, Pavlou HE, Allen NA, Paquin PR. 1991. Technical basis for establishing sediment quality criteria for nonionic organic chemicals using equilibrium partitioning. *Environ Toxicol* 10:1541-1583.

Doddema HJ, Cuypers MP, Derksen GB, Grotenhuis JTC, Harkes MP, Harmsen J, Rulkens WH, Zweers AJ. 1998. Characterization of dredged sediments for bioremediation, Report STOWA. Utrecht, NL.

Duffy MJ, Carski TH, Hanafey MK. 1993. Conceptually and experimentally coupling sulfonylurea herbicide sorption and degradation in soil. In: Proceedings, 9th Symposium on Pesticide Chemistry; Lucca, IT.

Fahmy TM, Paulaitis D, Johnson DM, McNally MEP. 1993. Modifier effects in the supercritical fluid extraction of solutes from clay, soil, and plant materials. *Anal Chem* 65:1462-1469.

Ford R, Bertsch PM, Farley KJ. 1997. Changes in transition and heavy metal partitioning during hydrous iron oxide aging. *Environ Sci Technol* 31:2028-2033.

Freeman GB, Johnson JD, Killinger JM, Liao SC, Feder PI, Davis AO, Ruby MV, Chaney RL, Lovre SC, Bergstrom PD. 1992. Relative bioavailability of lead from mining waste soil in rats. *Fund Appl Toxicol* 19:388-398.

Harmsen J, Ferdinandy M. 1999. Measured bioavailability as a tool for managing clean-up and risks on landfarms. In: In situ and on-site bioremediation, 4th Symposium. Columbus OH, USA: Battelle Press.

Hawthorne S, Galy A, Schmitt V, Miller D. 1995. Effect of SFE flow rate on extraction rates: Classifying sample extraction behavior. *Anal Chem* 67:2723-2732.

Houx NWH, Aben WJM. 1993. Bioavailability of pollutants to soil organisms via the soil solution. *Sci Total Environ* Suppl Part 1:387-395.

Kelsey JW, Kottler BD, Alexander M. 1997. Selective chemical extractants to predict bioavailability of soil aged chemicals. *Environ Sci Technol* 31:214-217.

Lamé PJ, Kroes PJ, Defize PR, Frapporti G. 1996. Soil protocol for support of building material. Appledorn, NL: Report TNO-MEP R961009.

Lindsay WL, Norvell WA. 1978. Development of a DTPA soil test for zinc, iron, manganese, and copper. *J Soil Sci Soc Am* 42:421-428.

Ma QY, Traina SJ, Logan TJ, Ryan JA. 1993. In situ lead immobilization by apatite. *Environ Sci Technol* 27:1803-1810.

McKenzie RM. 1970. The reaction of cobalt with manganese dioxide minerals. *Aust J Soil Res* 8:97-106.

McLaren RG, Lawson DM, Swift RS, Purves D. 1985. The effects of cobalt additions on soil and herbage cobalt concentrations in some S.E. Scotland pastures. *J Agric Sci* 105:347-363.

Medlin EA. 1997. An in vitro method for estimating the relative bioavailability of lead in humans [MS thesis]. Boulder CO, USA: University of Colorado.

Morrison DD, Robertson BK, Alexander M. 2000. Bioavailability to earthworms of aged DDT, DDE, DDD, and dieldrin in soil. *Environ Sci Technol* 34:709-713.

Pierzynski G. 1998. Methods of phosphorous analysis for soils, sediments, residuals and waters. Southern Cooperative Series Bulletin nr 30. SERA-IEG17. USDA-CSREES Regional Committee.

Prüe SS, Turian G, Schweikle V. 1991. Ableitung kritischer Gehalte an NH_4NO_3-extrahierbaren ökotoxikologisch relevanten Spurenelementun in Böden SW-Deutschlands. Mitt Dtsch Bodenk Gesel 66.

Rochette E, Koskinen W. 1996. Supercritical carbon dioxide for determining atrazine sorption by field moist soils. *Soil Sci Soc Am J* 60:453-460.

Rochette E, Koskinen W. 1998. Atrazine sorption in field-moist soils: Supercritical carbon dioxide density effects. *Chemosphere* 36:1825-1839.

Ronday R, Kammen-Polman AMM, Dekker A, Houx NWH, Leistra M. 1997. Persistence and toxicological effects of pesticides in top soil: Use of equilibrium partitioning theory. *Environ Toxicol Chem* 16:601-607.

Rodriguez RR, Basta NT, Casteel SW, Pace LW. 1999. An in vitro gastrointestinal method to estimate bioavailable arsenic in contaminated soils and solid media. *Environ Sci Technol* 33:642-649.

Ruby MV, Davis A, Link TE, Schoof R, Chaney RL, Freeman GB, Bergstrom PD. 1993. Development of an in vitro screening test to evaluate the in vivo solubility of ingested mine-waste lead. *Environ Sci Technol* 27:2870-2877.

Ruby MV, Davis A, Schoof R, Eberle S, Sellstone CM. 1996. Estimation of lead and arsenic bioavailability using a physiologically based extraction test. *Environ Sci Technol* 30:422-430.

Sauerbeck DR, Hein A. 1991. The nickel uptake from different soils and its prediction by chemical extractions. *Water Air Soil Pollut* 57-58:861-871.

Schoof RA, Butcher MK, Sellstone C, Bell RW, Fricke JR, Keller V, Keehn B. 1996. An assessment of lead absorption from soil affected by smelter emissions. *Environ Geochem Health* 17:189-199.

Simpson ML, Sayler GS, Applegate BM, Ripp S, Nivens DE, Paulus MJ, Jellison Jr GE. 1998. Bioluminescent-bioreporter integrated circuits from whole cell biosensors. *Trends Biotechnol* 16:332-338.

Smit CE, Van Beelen P, Van Gestel CAM. 1997. Development of zinc bioavailability and toxicity for the springtail *Folsomia candida* in an experimentally contaminated field plot. *Environ Pollut* 98:73-80.

Tang J, Alexander M. 1999. Mild extractability and bioavailability of polycyclic aromatic hydrocarbons in soil. *Environ Toxicol Chem* 18:2711-2714.

Tiller KG. 1979. Applications of isotopes to micronutrient studies. In: Isotopes and radiation in research on soil-plant relationships. Vienna, AT: International Atomic Energy Agency. p 359-372.

Tiller KG, Honeysett JL, Hallsworth EG. 1969. The isotopically exchangeable form of native and applied cobalt in soils. *Aust J Soil Res* 7:43-56.

[USEPA] U.S. Environmental Protection Agency. 1995. Test methods for evaluating solid waste, Method 3540,3541,3545. 3rd ed. Update III. Washington DC, USA: USEPA. EPA-SW-846.

Uzigiris EE, Edelstein WA, Philipp HR, Iben IET. 1995. Complex thermal desorption of PCBs from soil. *Chemosphere* 30:377-387.

Van der Lelie D, Regniers L, Borremans B, Provoost A, Verschaeve L. 1997. The Vitotox test, an SOS bioluminescence *Salmonella typhimurium* test to measure genotoxicity kinetics. *Mutation Res* 389:279-290.

Weber WJ, Young TM. 1997. A distribution reactivity model for sorption by soils and sediments. 6. Mechanistic implications of desorption under supercritical fluid conditions. *Environ Sci Technol* 31:1686-1691.

Weitzman M, Aschengrau A, Bellinger D, Jones R, Hamlin JS, Beiser A. 1993. Lead-contaminated soil abatement and urban children's blood lead levels. *J Am Med Assoc* 269:1647-1654.

Wells JB, Lanno RP. 2001. Passive sampling sevices (PSDs) as biological surrogates for estimating the bioavailability of organic chemicals in soil. In: Greenberg BM, Hull RN, Roberts Jr MH, Gensemer RW, editors. Environmental toxicology and risk assessment: Science, policy, and standardization-implications for environmental decisions: 10th Volume. West Conshohocken PA, USA: ASTM. STP 1403. p 253-270.

Werth CJ, Reinhard M. 1997. Effects of temperature on trichloroethylene desorption from silica gel and natural sediments. 2. Kinetics. *Environ Sci Technol* 31:697-703.

Zhang P, Ryan JA, Bryndzia LT. 1998. Pyromorphite formation from goethite adsorbed lead. *Environ Sci Technol* 31:2673-2678.

Chapter 10

Assessing Contaminated Soils: Lessons Learned and Future Directions

Roman P. Lanno

> The land ethic simply enlarges the boundaries of the community to include soils, waters, plants, and animals, or collectively: the land.
>
> —Leopold 1949

The notion of managing terrestrial systems from a holistic perspective, as suggested by Aldo Leopold (1949), has not been a focus of current environmental management practices. Leopold warned against conservation and wildlife management practices based solely upon economic motives because "they ignore, and eventually eliminate, many elements of the land community that lack commercial value, but that are essential to the healthy functioning of the ecosystem" (Leopold 1949). The management of aquatic systems has experienced a movement toward management at the watershed level, trying to examine the cumulative effects of deforestation, altered flow regimes, introduction of exotic species, and inputs of toxic contaminants and nutrients. This approach may be more difficult in terrestrial systems, especially since land in disturbed terrestrial ecosystems (e.g., agricultural, industrial) is treated as a commodity to be sold, not a resource to be nurtured and maintained. Although reductionistic approaches for the management of contaminated sites are currently practiced, each site or parcel of land is a functional, or dysfunctional, part of a bigger landscape. Aquatic toxicology began with simple toxicity tests that were used to monitor end-of-pipe effluents, and the tremendous reduction in acute toxicity of effluents was an important achievement in improving the health of aquatic ecosystems. It took many years to adopt sublethal toxicity tests and in situ effects assessment of aquatic organism populations. Do we need to make similar errors in managing terrestrial systems by looking at only a small portion of the larger management picture? Do we have the appropriate tools available to assess contaminated soils over many orders of magnitude in spatial, temporal, and biological scales? We hope this book brings together some of the tools necessary to accomplish holistic assessment and management of contaminated sites, or at least point us in the right direction. This final chapter attempts to coalesce some of the more important concepts and ideas as presented by the authors of the chapters into a cohesive whole, linking major

Contaminated Soils: From Soil–Chemical Interactions to Ecosytem Management. Roman P. Lanno, editor.
 ISBN 1-880611-31-7

issues associated with assessing contaminated soil sites. The summary also poses some of the many questions that still need to be addressed if we are to push forward our understanding of assessing and managing contaminated soils.

Exposure Assessment

Exposure is the interaction or contact of a contaminant with a biological receptor and cannot be measured using chemical analytical techniques alone. Yet, the naive assumption is made that measuring total chemical levels in soil provides an accurate measure of exposure. Vigorous extraction techniques (e.g., Soxhlet extraction for organic chemicals, strong acid digestion for metals) used to determine the total chemical concentrations in soil overestimate, sometimes by orders of magnitude, the bioavailable chemical concentration (Alexander 2000). Many abiotic modifying factors in the soil external to the organism (e.g., pH, organic matter), and biotic modifying factors such as behavior, size, and ability to metabolize contaminants, all act to alter the bioavailability of chemicals and reduce the usefulness of total soil chemical levels as a measure of chemical exposure. Initially, we presented a simple model relating total chemical to bioavailability for soil organisms (see Figure 1-2). After the discussions presented in various chapters in this book, the model can be expanded to include more detail (Figure 10-1). At this

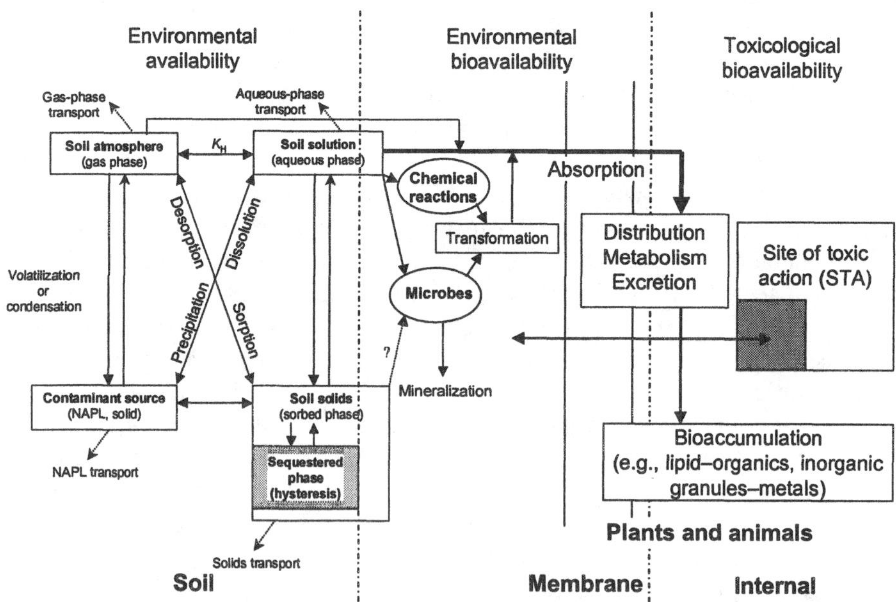

Figure 10-1 Detailed schematic model relating chemical fate and transport to the bioavailability of a chemical to a soil organism via dermal exposure

point, the model becomes more complex, and measuring all the parameters to fit a complete model would become a very expensive and time-consuming task. So which parameters would be best to measure as an estimate of exposure for ecological risk assessment (ERA)? From a toxicological perspective, the ideal measure of exposure would be the number of moles of chemical that are actually present in the organism interacting with receptor molecules at the site of toxic action. Any chemical measurement other than that is only a surrogate for chemicals present at the site of toxic action. From a risk assessment perspective, a chemical measure of exposure must be not only well correlated with a biological effect, but it must be precise, rapid, simple, and inexpensive. Although total chemical levels are often not well correlated with bioaccumulation or toxicity, they are relatively precise, rapid, simple, and inexpensive. Also, few alternatives to total chemical measures have been available, with almost none correlated with any type of biological response.

Correlation with a biological response is a very important point, since chemical analytical methods for assessing various fractions of chemicals in soils have typically been developed independently of any biological exposure, especially with organisms other than microbes. Assumptions are made that since the measured chemical fraction is not the total chemical mass present, this somehow equates with bioavailability. If bioavailability is determined, it has usually been associated with the bioavailability of organic compounds for microbes during bioremediation, often determined indirectly as evolution of carbon dioxide or dissipation or degradation of the parent compound. There is a great need for the development of chemical measures of different fractions of chemicals in soil that are correlated with biological responses, either toxic effects or bioaccumulation. Recent research has increased the number of chemical analytical tools available for determining environmentally available fractions of chemical. In many cases, these chemical measurements have been correlated with biological responses, providing data that may be useful in ERA.

Toxicity and bioaccumulation are an integration of the environmentally bioavailable concentration of a chemical (Figure 10-1) and exposure duration, not simply the chemical concentration. Yet, very few measures of exposure include a temporal component. This may not be as important in aquatic systems where an organism is completely immersed in a medium with relatively rapid advective transport mechanisms, where chemical concentration often can be assumed to be fairly constant. However, in soil systems, where larger organisms are not immersed in water, a temporal component may become very important. Exposure may become the net flux into the organism as determined by what an organism takes up from soil pore water and atmosphere and desorption kinetics of chemicals from soil particles. This would suggest that the most relevant chemical measures of bioavailability would include a temporal component and not simply be a one-point-in-time chemical concentration.

Chemical methods for measuring bioavailability can be grouped into methods that measure the bioavailable fraction of chemicals in soil solution (e.g., aqueous and weak solvent extractions) and methods that mimic chemical uptake by a soil organism (i.e., biomimetic techniques). Biomimetic techniques can also be grouped into those that measure chemical concentrations at a point in time and those that integrate chemical exposure over time.

Current hypotheses suggest that the bioavailable fraction of organic chemicals in soil resides in the soil pore water (Figure 10-1), with the partitioning of contaminant from soil particles into pore water prior to actual uptake by an organism (Belfroid et al. 1996). The chemical concentration in the aqueous phase in soil solution or pore water has been suggested as a suitable surrogate measure, better correlated with biological endpoints in soil invertebrates than total chemical measurements (Houx and Aben 1993; Belfroid et al. 1996; Conder and Lanno 2000). However, the collection of pore water for determining chemical levels is expensive and time-consuming. One alternative would be to estimate porewater concentrations of organic chemicals based upon the partitioning of chemicals between different phases in the soil based upon equilibrium partitioning (Di Toro et al. 1991). While this model works for some organic chemicals in some cases, there are also many exceptions. Equilibrium partitioning does not account for the effects of nonaqueous-phase liquids (NAPLs), effects of hysteresis, routes of uptake other than dermal contact with aqueous phases, exclusionary mechanisms (e.g., plant cell walls), and for differences in the behavior of soil organisms, which may result in great differences in chemical exposure. Equilibrium partitioning has been proposed for estimating porewater concentrations of organic chemicals in sediments (Di Toro et al. 1991) but has not been applied extensively in soil systems. Additionally, equilibrium partitioning assumes that chemicals are in equilibrium between solution and solid phases, an assumption that may not be valid in many exposure scenarios in soil systems. In sediments, sediment-dwelling organisms are completely immersed in pore water and some may modify their exposure environments slightly by building tubes in sediments (e.g., polychaetes). In soil systems, there may be allometric reasons that account for differences in organism exposure to porewater levels of chemicals. Many of the smaller organisms inhabiting soil, such as bacteria, yeasts, protozoa, rotifers, nematodes, and enchytraeids, essentially live in an aquatic environment, immersed in free pore water or the porewater films surrounding soil particles. Larger organisms, such as earthworms, may be exposed to chemicals only at points where their dermal surfaces come in contact with porewater films and are not immersed in water at all. They may also ingest soil particles and associated organisms, as well as being exposed to chemicals in the soil atmosphere, providing additional routes of chemical exposure. Soil arthropods vary in their soil water requirements, but all have a cuticle that may limit uptake of chemicals from pore water across the dermal surface, with dietary and inhalation pathways the primary routes of exposure.

These differences in exposure pathways would suggest that in order to determine bioavailability, various chemical measurements would need to be correlated with the uptake and/or responses of many different organisms. An alternative approach would be to estimate bioavailability using an organism presumed to have the greatest fraction of soil chemicals bioavailable to it. This organism often has been assumed to be the earthworm since both dermal and dietary routes of exposure may be important. However, one has to be careful not to confuse bioavailability and toxicity. When comparing toxicity among various species, increased bioavailability does not always equate to increased toxicity due to differences among various organisms in their capacity to metabolize and detoxify chemicals.

Regardless of the chemical measure used to estimate chemical availability, the chemical measure must be correlated with either the accumulation or metabolism of the chemical by an organism or an observed effect in the organism. Alternatives to porewater concentrations of chemicals include nonaqueous extraction methods and solid-phase extraction techniques. Nonaqueous extraction methods are used assuming that the quantity of chemical removed from soil by a mild or non-exhaustive extraction reflects the amount of chemical bioavailable to soil organisms. Mild extractions with organic solvents have been correlated with the bioavailability of nonpolar organic chemicals to earthworms and microorganisms (Kelsey et al. 1997; Tang and Alexander 1999). Supercritical-fluid extraction (SFE) and accelerated solvent extraction have been used to measure total organic chemical levels in soil. Physical conditions associated with extraction (e.g., temperature, pressure) allow the extraction of different chemical fractions, including sequestered chemicals. This has allowed SFE to be adapted for examining desorption kinetics (Rochette and Koskinen 1996; Weber and Young 1997; Hawthorne and Grabanski 2000) and the effect of aging on chemical availability (Camel et al. 1995). Thermal desorption techniques also have been used to assess the availability of organic chemicals in soil, with sequestered compounds requiring more time and/or energy to be removed from soil samples than easily desorbable compounds (Uzigiris et al. 1995; Werth and Reinhard 1997). All of these techniques appear promising in detecting different available fractions of organic chemicals in soil, but until they are correlated with bioaccumulation or toxicity, they will remain only promising techniques.

Another group of methods for measuring the environmentally available fraction of a chemical in soil is solid-phase extraction (SPE) techniques in which a solid phase is placed in wet soil or a soil suspension and withdrawn after a period of time. These methods allow partitioning of the chemicals to occur between soil solution and the solid phase, with the amount of chemical that partitions to the solid phase determined by the capacity and polarity of the solid phase and the ability of the soil to maintain porewater chemical concentrations. In this respect, SPE devices mimic partitioning of chemicals from soil to organisms, such as earthworms, and may be termed "biomimetic." After deployment, contaminants that have partitioned into the solid phase are eluted and quantified, or in some cases, the solid-phase material can be placed directly in the injection port of an instrument (e.g., gas chromatograph). Various SPE devices that have been used to assess the availability of organic compounds in soil include Tenax beads, C18 membranes, and solid-phase microextraction (SPME) fibers. Good correlations have been observed between the uptake of DDT, DDE, DDD, and polycyclic aromatic hydrocarbon (PAH) by *Eisenia fetida* from a range of soils and the quantity of chemical sorbed by C18 membranes (Tang and Alexander 1999; Morrison et al. 2000). Solid-phase microextraction fibers have also been used to assess the bioavailability of phenanthrene in artificial soils differing in organic matter content (Wells and Lanno 2001). This research also defined a level of phenanthrene measured using a SPME fiber that was associated with the toxicity endpoint of mortality in earthworms, thus adding a predictive capacity to this measurement. Variations in this technique appear very promising as a measure of chemical availability in soils.

All of the preceding techniques only measure chemical availability from a soil subsample, although they may measure partitioning from that sample over time. This exposure scenario may be adequate for smaller organisms such as microbes and nematodes. However, larger, more mobile organisms such as earthworms and arthropods are exposed to a continuously changing chemical gradient as they move through the soil, integrating exposure over the heterogeneity of chemical distribution in soil and via diffusion of chemicals in soil air from sites distal to the organism. A recent method developed to mimic both spatial and temporal distribution of chemicals in the soil is the application of semipermeable membrane devices (SPMDs) in soil (Wells and Lanno 2001). Semipermeable membrane devices have been used effectively to assess chemical availability in water and sediment (Huckins et al. 1990, 1993). Semipermeable membrane device application in soils is limited by slow uptake kinetics and high variability between SPMDs, precluding good correlations with earthworm uptake of phenanthrene (Wells and Lanno 2001). However promising SPE methods may seem, continued development must be coupled with biological responses. To date, responses have been correlated only with microbes and earthworms, with good results. More experiments are needed to assess the relationship between SPE techniques and chemical bioavailability of toxicologically significant chemical tests for plants, arthropods, and soil-dwelling mammals.

Rather than measuring chemical availability using various chemical analysis methods as an indirect measure of bioavailability, bioavailability can be measured directly by determining the residues of chemicals in tissues or whole organisms. Critical body residues (CBRs) are tissue, organ, or whole body residues of specific chemicals or groups of chemicals that are correlated with adverse effects (e.g., reduced survival, decreased reproductive success). Results of bioaccumulation tests or tissue residues measured in organisms exposed to contaminated soil in toxicity tests or in the field can be compared with the CBR to obtain an indication of bioavailability and potential risk. Although the CBR concept has been well established for organic chemicals in aquatic ecotoxicology (McCarty et al. 1992; Van Straalen 1996; Lanno et al. 1998), only a few CBR values for organic chemicals have been derived for terrestrial soil-dwelling organisms (Fitzgerald et al. 1996; Wells and Lanno 2001). Crommentuijn et al. (1994) determined lethal body concentrations for cadmium in six different soil arthropod species. CBRs for cadmium and/or zinc on the growth and/or reproduction of plants, earthworms, and Collembola have also been reported (Van Gestel and Hensbergen 1997; Smit and Van Gestel 1997, 1998; Posthuma et al. 1998).

All the preceding methods for determining exposure include some chemical measurement of the environmentally available fraction of chemical coupled with a biological response. These measurements are useful in determining the amount of chemical that is available during the present set of environmental conditions, but do not serve as useful predictors of the environmentally available chemical under future conditions. In order to predict how much environmentally available chemical will be present in the future, it becomes necessary to understand the mechanisms involved in the interactions between soil and chemicals (Figure 10-1). As seen in Chapters 5 through 7, there is a wealth of research in this area. In this manner, determination of the environmentally available fraction of a chemical is the first step in determining the environmentally bioavailable fraction of a chemical. Central to understanding the effects of abiotic modifying factors on the environmentally available fraction of a chemical are the mechanisms of sorption and desorption. These mechanisms determine the readily desorbable and irreversibly sorbed (sequestered) fractions of chemicals in soil. Currently, the rates of sorption and desorption of chemicals from soil particles are poorly predicted based upon parameters that are typically provided in the soil toxicology literature, such as total contaminant concentration and typical soil properties (e.g., pH, organic matter content, texture).

Part of the problem is that the contaminant is not rapidly equilibrated between the stationary and fluid phases on time scales appropriate to the exposure period of test organisms (Pignatello and Xing 1996). Contaminants exist in a continuum of sorbed states, from instantaneously in equilibrium to very slowly in equilibrium with the fluid phase to a fraction that may be completely immobilized. This suggests that for each type of soil organism there exist "available" and "sequestered" states of sorbed molecules (Alexander 1995). Contaminant bioavailability is therefore highly dependent on the specific contaminant, soil, receptor, mode of uptake, and duration of uptake, resulting in great differences in bioavailability between organisms. As an example, metals sequestered in soil may not be available for dermal absorption by earthworms but may be readily taken up in a mammalian digestive system where low pH conditions solubilize the metal. It is this specificity that results in difficulty in the development of predictive models.

In cases where contaminant uptake by the organism does not significantly alter the fluid-phase concentration, such as when uptake is very slow or when the organism moves rapidly through the contaminated medium, bioavailability may be controlled by the fluid-phase concentration. In all other cases, kinetics describing the flux of contaminant through the particle, across the particle-bulk fluid interface, and across the fluid–membrane interface will be important. When the organism alters the fluid-phase chemical concentration during uptake, the concentration gradient between the particle and fluid and between the receptor and fluid are both altered. Since sorption and desorption rates from particles are governed primarily by molecular diffusion, and since diffusion depends on the concentration gradient, such biologically induced changes in concentration gradient may influence contaminant flux out of the particle. The presence of an organism may accelerate the flux of contaminant from soil particles in a variety of ways: 1) through increasing the concentration gradient across the particle–fluid interface as a result of depletion of contaminant in the fluid, thereby accelerating particle–fluid flux; 2) through organism-induced changes in the composition of the fluid phase that affect the chemical potential of the solute, and as a result, its sorption coefficient; and 3) through organism-caused alteration of soil properties via changes in fluid-phase composition. Therefore, an accurate bioavailability model will require linking biological uptake and depuration kinetics with sorption and desorption kinetics, but models coupling these principles are essentially absent from the literature.

In order to develop such models, critical information is needed in a number of areas. Although a lot of information is available on K_{oc} values, most of these data are based on isotherms constructed over short equilibration times (<48 hours), and their relevance to highly aged contaminated soils is questionable. The apparent K_{oc} for historically contaminated soils may be as much as 2 orders of magnitude greater than laboratory-determined values with clean soils that are freshly spiked with the contaminant (Pignatello and Xing 1996).

A major reason for reexamining the relevance of existing K_{oc} values is the observation that desorption is not simply the opposite of sorption. A fraction of the total sorbed compound is not removable, as would be expected based upon standard sorption isotherms, and this fraction increases with aging times, from days to weeks and longer. When equilibrium desorption is vastly different from sorption, the process is termed "hysteresis." Hysteresis implies the existence of a time-independent process, that is, desorption points do not approach the sorption isotherm curve even after long equilibration times, yet the original compound can be recovered quantitatively. The mechanisms of hysteresis are not completely understood, but there are a number of theories. There may be molecular rearrangement after sorption so desorption may take place from a different molecular environment than sorption. Diffusion through the fixed intraparticle pore system of soil particles may be an important retardation process, especially in micropores (<2 nm) and nanopores (<1 nm). Exclusion of even the smallest cells (microbes approximately 200 nm in diameter) from the intraparticle micropore system may provide some explanation for reduced microbial bioavailability of contaminants with aging time.

Two site model desorption rate constants have been estimated and were found to depend inversely on the K_d, K_{ow}, or molecular connectivity index, a measure of topological size and degree of chain branching in the molecule (Piatt and Brusseau 1998). Thus, for a given class of compounds, large molecules desorbed more slowly than small, and branched more slowly than linear. Polar compounds appear to desorb more slowly than apolar compounds having the same K_{ow}, as expected due to the additional drag of functional group interactions (e.g., hydrogen bonding) occurring at each molecular jump through the matrix. It should be noted that these relationships have been developed from sorption experiments occurring over only minutes to hours; although the relative relationships are likely to hold over longer times, it is still not possible to predict desorption rates in aged systems with any certainty.

In addition to hysteresis, sorption and desorption models should also consider compounds associated with nonsettling particles (NSPs) in soil solution. Nonsettling particles are colloidal size particles made up of humic substances often associated with inorganic materials and are often referred to as "dissolved organic matter" (DOM). While sorption mechanisms are assumed to be the same as in bulk soil, NSP sorption becomes important as its K_{ow} exceeds approximately 10^4. Sorption to NSPs may affect the apparent sorption distribution coefficient by giving an artificially high aqueous-phase concentration. The bioavailability of molecules associated with NSPs is unknown, but recent studies suggest that some molecules associated with DOM may also become sequestered (Schlebaum et al. 1998).

An important question for setting soil-quality criteria and establishing environmentally safe remediation endpoints is the potential for release of sequestered fractions through sudden changes in environmental conditions at some time in the future. Desorption of sequestered fractions may be accelerated by abrasion, application of heat, introduction of solvents and oils, or for metals, changes in pH. It is possible that competing cosolutes (either natural or anthropogenic) introduced to the soil could displace sequestered fractions. Theoretically, natural microbial decomposition of soil organic matter (SOM) could lead to the release of sequestered chemical fractions, but such a process is not likely to occur suddenly and without renewal of SOM. However, the younger SOM is likely to have a more rubbery structure (Weber et al. 1999), which has different sorptive properties than older SOM.

Contaminated sites are spatially heterogeneous with respect to many properties, including contaminant concentration, mineralogy, fraction of organic carbon, moisture, etc. Chemical concentrations may vary up to 1000-fold or more over a site (Keith 1996). Thermodynamic coefficients and rate parameters may vary correspondingly with these disperse properties and conditions. For example, sorption coefficients and diffusion coefficients are concentration dependent. Diffusion coefficients are highly dependent on soil geometry and other properties. It is quite possible that the "bioavailability factor" of a given compound varies spatially over a site from near zero to near unity. Research is needed to establish the site-scale heterogeneity of thermodynamic and kinetic parameters of desorption if we are to have any real confidence in predicting bioavailability.

Sorption and desorption thermodynamic and kinetic models are presently inadequate for predicting bioavailable fractions using only the total contaminant concentration and soil properties typically available. Literature K_{oc} values generally overestimate risk because short-term sorption studies underestimate K_{oc} values of aged contaminants. Therefore, measurements are required on a site-specific basis in order to obtain the minimal thermodynamic and kinetic information necessary to make bioavailability predictions. Measurements that should be routinely made to help assess bioavailability include prevalent porewater concentration and desorption rate profile for the contaminant that will give the maximum desorbable contaminant mass over a given period.

Interaction of heavy metals with soils reduces the toxicity of the metals, the same phenomenon that is seen when organic compounds interact with soils. Addition of metal salts to soils results in a greater bioavailability of the metal than for metal that has been allowed to age in the soil or for metal that is added together with components with which it can react, such as sewage sludge. Metals added to natural or artificial soils have been used to ascertain dose–response relationships. When the concentrations of chemicals in field samples are compared to these, the level of biological effect is overestimated. If metal concentrations are determined based on aqua regia digests rather than on milder procedures, this effect is exacerbated. The importance of considering bioavailability in setting soil-quality standards has been discussed by Peijnenburg et al. (1997).

The most critical parameter for supporting bioavailability assessments would be reliable information on the porewater concentration of the contaminant of interest. This can come from direct measurements of porewater concentrations and/or from measurements of chemical-specific and soil-specific parameters that can be used in some model to predict porewater concentrations under a variety of soil/site conditions, as well as over a reasonable time frame. The desired chemical-specific parameters include partition coefficients, rate constants, and diffusion coefficients. Soil-specific parameters would include organic carbon content, pH, clay content, and cation-exchange capacity (see list in *Soil Properties*, Chapter 7). This focus on porewater concentration is not meant to detract from the likelihood that other, artificial means might provide a more direct basis for estimating bioavailability in specific circumstances. These other means might involve, for example, specialized leaching with liquids other than natural waters and/or uptake by solid-phase extractants that mimic biological systems. Regardless of whether direct porewater measurements of chemical concentrations or the measurement of model-related parameters are made, standardized protocols are desperately needed.

For the direct measurement of porewater chemical concentrations, protocols would include the number of samples needed for characterization, porewater contact time, porewater recovery technique, and similar issues. For model-related parameters, the initial question is which model to choose? Both mechanistic and empirical models are needed. Mechanistic models that accurately describe sorption and desorption mechanisms are needed to better understand the underlying processes, and they will allow more realistic assessments at high priority sites where such detailed assessments are warranted. However, the problems with this "mechanistic" modeling approach are that

- there is currently no clear understanding of the mechanisms;
- there is insufficient consensus among experts on which mechanisms and models are the most representative;
- they require too many inputs and are, consequently, not easy to use; and
- there is little likelihood of regulatory near-term acceptance.

Empirical models are needed in order to deal with near-term problem sites, that is, those where management decisions will be made in the next few years. While these models may be based on strictly empirical correlations, or on simplistic or incorrect assumptions regarding the sorption and desorption process, they are all there is today that are relatively easy to use and generally acceptable to the regulatory agencies (at the level that most sites are given regulatory oversight). Properly applied and qualified, the use of such models should provide reasonable estimates of porewater concentrations under many conditions of concern. Perhaps what a protocol could spell out is the conditions of chemicals under which the simplistic models do and do not provide a reasonably reliable basis for assessment. By conditions, it is meant soil and contaminant types, spatial and temporal scales, exposure scenarios, and receptors. In many cases, it may well be that the uncertainty in the outputs from the simplistic or empirical models is acceptable.

Effects Assessment

Regardless of which chemical method of measuring environmental availability is selected, these measurements will likely have greater precision than biological responses, suggesting that the selection of the appropriate biological effects are of paramount importance in reducing uncertainty in risk characterization.

Characterization of bioavailability requires, by definition, an interaction between some fraction of chemical in the environment and an organism (Figure 10-1). However, bioavailability will be expressed differently by different organisms, so which measurements of bioavailability are most appropriate? Biological measures of bioavailability may be indirect or direct. Indirect measures of bioavailability (e.g., toxicity tests, biomarkers) include some response of organisms to a presumed bioavailable level of a contaminant. For example, a soil sample from a contaminated site may be lethal to earthworms. However, the sample contains a mixture of many chemicals and the resources to conduct a complete chemical analysis of the many chemicals present in the soil may not be available, yet it is a reasonable assumption that some bioavailable fraction of those chemicals resulted in lethality to earthworms. Nonspecific biomarkers (e.g., stress proteins, mixed-function oxidase [MFO] enzyme induction) also will respond to bioavailable chemicals, yet we cannot say with certainty exactly which chemicals were responsible for the observed response. Direct measures of bioavailability include bioaccumulation and CBRs, where the amount of chemical taken up by an organism is estimated directly from residues in the organism. This represents a complete integration of abiotic and biotic modifying factors affecting the bioavailability of the chemical for that specific organism. Both direct and indirect measures of bioavailability are tools that serve specific purposes in the assessment of contaminated soils.

Biological measures of bioavailability can range from community-level measures to suborganismal responses. Unlike previous discussions of the hierarchial nature of responses dealing with specific organisms in specific environments (e.g., fish as indicators of environmental stressors in aquatic systems), the assessment of contaminated soils involves both macroscopic and microscopic organisms. As such, biological responses to contaminated soils involve macroscales and microscales in both space and time. Soil organisms such as earthworms and arthropods often are used to assess chemical bioavailability and effects at the suborganismal, organismal, and population levels, while microbial assessments measure bioavailability and effects at the population and community levels. Many of the most important processes in soil are carried out by microbes and can be used as measurement endpoints to examine the effects of chemicals on soil functions and processes. These include various microbial processes such as nitrification, respiration, and decomposition that are involved in the essential process of nutrient cycling. Community-level adaptations to contaminated soils can also be assessed using pollution-induced community tolerance (PICT) procedures. Macroscopic organism responses can be examined both in the lab and the field and include both indirect bioavailability (e.g., toxicity tests) and direct bioavailability measurements (e.g., bioaccumulation, CBRs). Suborganismal parameters such as the measurement of chemical metabolites, chemical-specific biomarkers, and histochemistry provide direct measures of bioavailability, while indirect evidence of chemical bioavailability can be assessed using histological, physiological, and immunological techniques.

Although many techniques, both direct and indirect, exist for assessing chemical bioavailability in soil, the question remains of when to apply these techniques and which ones are most useful at what times? While chemical measures of bioavailability may be useful for modifying Tier 1 soil-screening levels, on a site-specific basis, a test battery comprising ecologically relevant soil-dwelling organisms, and measurement endpoints and standard test protocols should be used for a Tier 2 assessment of site-specific contaminant bioavailability. The biological tests used to assess bioavailability should be selected after considering a number of selection criteria including these:

- Are the organisms representative of the ecosystem that is being evaluated?
- Are the organisms in the test battery representative of the range of species that could potentially inhabit a site?
- Do the organisms in the test battery reflect the diversity that exists for organism-specific sensitivity to contaminants?
- Are the organisms selected in consideration of the current and future land use, the soil type, and the ecoregion in which the site is located?
- Are the selected organisms able to survive and thrive in the control or reference soils used as experimental controls or as diluents in a test where the organisms are being exposed to sequential or proportional dilutions of the contaminated soils?

The test battery recommended for biological measures of bioavailability at a particular site will depend to a large extent on the purpose of the assessment, the site characteristics, and the magnitude and extent of the contamination problem that has been identified. However, at a minimum, it should include tests with plants, soft-bodied soil invertebrates, and soil arthropods, and tests designed to assess decomposition and/or a microbial process such as nitrification.

The most common tests conducted to examine contaminant bioavailability in soils are toxicity tests, bioaccumulation analyses, and metabolite analyses (plus biomarkers in earthworms). Biomarkers (with plants and arthropods) and CBR analyses have received less attention. The remaining tests (microbial, histological, physiological, and immunological assays) have not been applied to any significant extent.

This breakdown in practice is not surprising. The organisms used in toxicity testing integrate the entire bioavailable dose from the contaminants in soil into a toxic response. This may reflect toxicity from cocontaminants, parent compound of concern, intermediate breakdown products, or internally generated metabolites. As such, toxicity testing can provide an unambiguous answer to the question "are soil contaminants bioavailable?" There are, however, some significant limitations with this approach:

- For soils containing complex mixtures, the measured toxicity does not determine which of the contaminants are bioavailable or responsible for the observed response.
- Significant contaminant uptake from the soil matrix can take place even if no toxic responses are observed in the exposed organisms; this will occur if

the total uptake does not exceed a critical body burden for the duration of the test.

- "Bioavailability" will be a function of the toxicity endpoints measured. Short-term, acute exposures may not result in a toxic response (such as mortality), whereas a longer-term, chronic exposure may result in significant toxicity (such as adverse effects on reproduction, survival, growth, etc.). Decisions about bioavailability of contaminants based only on the results of acute tests might be inaccurate.
- Clean control soils collected in the field, or prepared in the laboratory, might not adequately support the test organism. This, however, does not reflect an adverse biological effect associated with contaminant bioavailability, but rather, the result of unsuitable physicochemical characteristics of the reference soil.

In any terrestrial habitat, the physical, chemical, and biological properties of the soil are partially responsible for determining its productivity. Methods for evaluating the bioavailability and toxicity of chemicals in soil must distinctly separate the adverse effects associated with soil physicochemical characteristics from the adverse effects of contaminants in the soil. In general, chemicals spilled in soil primarily cause changes in soil chemical properties, but their release, or activities associated with their release (e.g., increased traffic or physical disruption of habitat), may also influence physical and biological soil properties. Therefore, the physicochemical reference conditions of a soil must be adequately described if deviations in biological parameters of organisms exposed to contaminated soils are to be attributed to the contaminant itself. In addition, a quantitative knowledge of certain soil properties is often necessary for modeling the fate and transport of the contaminant in question. For example, the conversion of contaminant concentration to loading rates on an areal basis requires knowledge of the bulk density and depth of the contaminated soil horizon. An adequate description of soil properties involves both field observations and laboratory measurements. Field observations include variables that are commonly recognized as traditional agronomic site characteristics (e.g., texture, fertility), while laboratory measurements are those commonly associated with toxicity testing (e.g., pH, organic matter content, cation exchange capacity). Both types of measurements are important in defining soil productivity.

As with chemical measures of contaminant availability and the determination of sorption and desorption coefficients, there is a dire need for additional standardized protocols for conducting toxicity tests with organisms in contaminated soils, especially in North America. Ecological risk assessments at most contaminated sites in the U.S. rely extensively on chemical measurements and ratios of measured chemical concentrations and literature-based toxicity thresholds (i.e., hazard quotients). If biological testing is done at all, it is often limited to a scant few methods such as earthworm toxicity or phytotoxicity tests (e.g., lettuce seed germination and root elongation tests) (Greene et al. 1988; Edwards and Bohlen 1992). Several other terrestrial tests have been standardized in the past 10 years and are available for use. Compendia of test methods are available, allowing an increased flexibility with respect to test species selection, such that greater ecological relevance can be achieved during biological site assessment (Linder et al. 1992; Kapustka 1997; ASTM 1999; Løkke and Van Gestel 1998). Standardization of in situ invertebrate sampling techniques and toxicity tests, as well as microbial process measurements, would improve the utility of many more tools germane to assessing soil ecology parameters. However, as tests become more standardized, they become further and further removed from the variability and complexity observed in real ecosystems. It is important to understand that standardized laboratory tests are tools that can be used to address questions regarding the relative toxicity of chemicals in various soils. Such tests are not designed, nor were they intended, to replace actual field tests or sampling, even though they are often applied in this context. Caution must be applied when extrapolating laboratory test results to field conditions.

During the assessment and cleanup of contaminated sites, management decisions are ultimately made as to whether or not active remediation is warranted, and if so, which remediation technology best achieves project goals. Most often, such decisions are based on the measurement of contaminants in soil and whether or not the contaminant level exceeds some sort of predetermined value (e.g., hazard quotient). However, the information generated from ecological effects measurements can also be used effectively to both select the appropriate technology and evaluate performance of the technology, rather than just measuring contaminant levels in soil. In a simplified sense, the choice of effects measurements should reflect as closely as possible the potential or actual harm contaminants have on valued resources. As remediation technologies are considered, each should be examined, at least on a theoretical basis, in terms of how the technology would influence those parameters used to measure ecological effects. For example, if soil microbial processes were measured to understand effects on soil fertility and ultimately forest growth potential, the following questions could be framed: What are the consequences to microbial processes of excavating the top 50 cm of soil? What are the consequences of incinerating the soil? What are the consequences of soil amendments that promote the biodegradation of specific contaminants by microbes? What are the predictions for the future if a "no remediation" or "natural attenuation" option is selected? This approach recognizes that there are potential ecological risks associated with all remedial options. Although digging and hauling contaminated soil may be a simple approach to reducing the concentration of a contaminant in a confined area, the actual removal of the upper horizons of soil and the compaction of remaining soil by heavy equipment used in soil removal may have a greater impact on soil processes over the long term than leaving contaminants in place. Posing questions from this perspective would shift the focus of remediation to preserving the ecological resource to be protected, rather than simply reducing the concentration of contaminants. Maintaining the ecological integrity of soil would certainly be beneficial to the overall goal of preserving soils and soil services for future generations. With the advent of new methods for assessing soil as a valued biological resource, rather than as a matrix for contaminant deposition, it is possible to provide quantitative biological data alongside the chemical analysis of soil for the management of contaminated sites.

Soil-Quality Guidelines and Risk Assessment

Humans are almost entirely dependent on soils for existence. We are physically supported by it, it is the source of most of our food, and it ultimately provides many of the materials used for our clothing and shelter. Protection of the soil resource is thus not an option or frill, it is a societal imperative. Contamination by toxic substances is one of several stresses on the soil resource that must be responded to by society.

A significant and seriously underestimated challenge in the development of a numerical soil-quality guideline (NSQG) and other risk-based tools lies in collecting and interpreting intelligence on societal values for and expectations of the soil resource. Protection of the soil resource and mitigation where protection has failed is complicated by the fact that soil is bought, sold, used, and abused as "land," simply viewed as a commodity rather than a precious, living, biological resource to be nurtured. Systems of land tenure are used to organize use patterns and define rights and responsibilities of parties tied to the land. These rights and responsibilities are expressed and seen in various ways, depending on the form of tenure (e.g., owned, leased, held in public trust, etc.). In addition, differing patterns of use, and potential or actual conflicts among these uses, have compelled societies to establish zoning systems, within which the range of uses and activities is restricted.

Land stewardship is a concept that, when followed, serves to minimize impacts on soils. However, also revered is the concept of property rights that individuals should be free to conduct whatever activities they see fit, so long as adjacent property owners are not adversely affected. This concept extends to care and control of contaminated lands and restricts the extent that government-based management and regulation apply, thus making land stewardship a difficult task.

In Western Europe and North America, the management of soils is primarily the responsibility of individuals and private entities. Despite an assumed overall appreciation for the importance of soils in sustaining human societies, the specific benefits and services provided by or associated with soils are often not clearly articulated to the public. Soils are created by processes that operate on geological time scales, yet can be depleted and irreversibly changed in a short amount of time (Daily et al. 1997). Remediation of soils can be expensive and ecologically disruptive (Milloy 1995). Because of the large implications in terms of money and sustainability, and the dispersed responsibility for soil management, it is particularly important to discuss and articulate the benefits and expectations associated with soils.

A clear understanding of soil services can direct the development, application, and underlying science supporting NSQGs. Examples of services provided by soils and soil ecosystems include support of plant and wildlife communities and biodiversity, buffering and modulation of the hydrological cycle, protection of groundwater and surface water quality, decomposition, regulation of element and nutrient cycling, and simply physical support. By articulating soil services and the expectations associated with soils in the context of different land uses, risk assessors and managers ensure that the right questions are being addressed and that NSQGs are applied appropriately. The best science cannot contribute to guidance for protecting soil if we are not sure what it is about soil we are trying to protect and why.

Articulating soil services provides the basis for deriving management goals that can provide a strong foundation for risk assessment activities. Management goals are desired characteristics of ecological values that the public wants to protect (USEPA 1998). In the development of NSQGs, management goals are defined in generic ways that are intended to encompass most scenarios, yet must be specific enough to identify clear objectives for protection. Objectives more specific than "protect the environment" require considerable thought and discussion.

Natural areas such as wilderness areas and wildlife refuges are expected to support all of the soil services. We also have high expectations that managed forests will support most categories. Expectations regarding biodiversity and wildlife communities are somewhat less because these areas are often managed for a particular plant species or community structure. However, this should not preclude the notion that biodiversity of other species (e.g., microbes, soil invertebrates) is important in the ecological support of targeted management species.

In many European countries, land parcels are often expected to support a variety of land uses (i.e., multifunctional land use policy in the Netherlands and Denmark) so NSQGs do not vary by land use. However, larger European countries such as Germany and France do provide for different land uses within their NSQG frameworks. In the U.S. and Canada, most NSQGs are used to identify sites or chemicals that can be eliminated from further analysis and are used with the limited data available at the early tiers of risk assessment. The uncertainty associated with limited data availability is partially addressed through attempts to bias sampling toward contaminated areas but is also addressed by using a high level of protection. The high level of protection minimizes the chances that sites or chemicals that truly pose a significant risk are not eliminated.

Issues of scale in the ecological risk assessment of soils

Many issues related to various scales exist when ERAs of contaminated soils are conducted. These include spatial scales (distribution of contaminants and receptors), temporal scales (chemical release, transformation, and sequestration rates), biological processes (e.g., bioaccumulation, location and life stage of receptor), and effects scales (type and magnitude of effect).

Risk assessments of contaminants in ecosystems begin with a conceptual model of how an ecosystem works in order to identify potential pathways for exposure to contaminants. Such a conceptual model is also required during the development of NSQGs. This model is usually formulated in terms of the macrospatial scale. However, what is often not recognized is that these models involve an implicit assumption concerning ecosystem stability at the macrotemporal scale. When most people look out upon a landscape, they assume they observe a representation of what the ecosystem should look like, what it has always looked like, and what it should look like in the future. Such a view of a "stable" ecosystem assumes that when the system is stressed, it will either return to its original state when the stress is removed or will change unalterably. However, other models of ecosystem stability include adaptation to the stress in a new stable state (Holling 1986; Watt and Craig 1986; Wiens et al. 1986), as well as predictable or unpredictable cycling between stable states. The impact of contaminant stress on ecosystem stability may be even more complex in successional ecosystems or ecosystems that are adapted to periodic disturbance (Rice and Jain 1985; Walter and Chapin 1987; George et al. 1992). In these cases, it is unclear how ecosystem response to the additional stress induced by contaminants should be measured. Depending upon one's view of ecosystem stability, the goal may be to protect (preserve) the ecosystem in its state as observed at time = 0 or to preserve its ability to move between states. It is by far more tractable to assume ecosystem stability when developing NSQGs. One can assume for the purposes of screening that if ecosystem structure and function are preserved at time = 0, the potential for natural dynamics is also preserved. However, with the use of population-level and community-level methods of determining ecosystem response to contaminants in latter tiers of an ERA, assumptions concerning ecosystem stability must be revisited.

Related to the issue of ecosystem stability over temporal and spatial scales is the issue of selecting which organisms to use in measurement endpoints. Organisms in "lower" taxonomic groups are considered on a microscale (i.e., microorganisms, soil invertebrates) and are often simply lumped together, while organisms in higher taxonomic groups are often further subdivided (e.g., resident and migratory birds). There are two reasons for this. First is the recognition that society tends to value individuals and populations of higher organisms to a greater degree than lower organisms. The trophic structure and presence of higher organisms is an important assessment endpoint with respect to public perception. Second, lower organisms are often perceived to be more important with respect to ecosystem function rather than ecosystem structure. The concept of structure without functioning components is often difficult to convey. For "lower" organisms, species substitution or functional redundancy may play a major role in maintaining the functioning of an ecosystem (Moore and DeRuiter 1993). Certain soil invertebrates or microorganisms may be susceptible to soil contamination and may be reduced in total number or biomass, but the ecosystem process of concern may not be impacted due to the ability of other soil species to compensate and perform the function of the reduced or eliminated species. Soil microorganisms and invertebrates represent a substantial portion of the world's biodiversity, but due to the focus on function instead of community or ecosystem structure for these organisms, biodiversity of these species is not often considered an important endpoint, and there is always the potential that the loss of species, even if function is momentarily retained, may have future implications for ecosystem stability (Tilman 1997).

Organisms in higher taxonomic groups are often assumed to reside in the contaminated area for 100% of the time. This may be unrealistic with respect to sites with a small areal extent of contamination compared to the size of home ranges of many vertebrate species. In addition, with the exception of endangered species, it is largely recognized that the unit of protection for most plant and vertebrate species is the population, not the individual (Moriarty 1988; USEPA 1998; Suter 1993). For this reason, the elimination of sites based solely on size is not necessarily protective. Should the ability to eliminate such small sites be desired, an infrastructure would need to be developed and put into place to identify and track such sites to ensure that cumulative population impacts are not significant.

Risk assessors must define both spatial and temporal scales of exposure appropriate to the particular contaminated site or chemical use. Spatial scales are relevant to particular endpoints and fate and transport processes. Sampling of chemical contaminants in soil and mathematical models of exposure should reflect these scales. Because of the diversity of potential receptors and our low-to-moderate understanding of their activities and locations, there is often a compromise between what is known, what is assumed, and what can actually be accomplished. For example, it is not feasible to measure chemical concentrations in soil from the surface to 1 cm, 2 cm, 3 cm, etc., just because certain soil invertebrates or plants have their primary exposures at all of these depth intervals. Nor is it feasible to measure these concentrations at 1-m horizontal intervals. It also is not desirable to expend resources on this level of sampling if most assessment endpoints are aggregations of ecological receptors (e.g., communities).

A few or many of these scales may be considered, depending on the needs of the assessment for supporting management decisions. Simple screening-level analyses consider only a few scales (e.g., areas where contamination is suspected), while more comprehensive assessments may evaluate the spatial distribution of populations or receptors at the site in relation to the site-wide exposure field for soil contaminants and other stressors.

The depth of sampling should be related to the assessment endpoint. Sampling depths chosen to reflect exposure to the soil invertebrate community should be based on an average estimate of depth intervals where these organisms reside. Similarly, the sampling depth for the plant community should be an average unless a particular threatened and endangered population is of interest, in which case the rooting depth of the particular species is applicable, if known. A default sampling depth of 30 cm for plant communities may be identified, based primarily on rooting depths in Jackson et al. (1996). However, this default depth should be used with caution because rooting depth will vary greatly with plant species and climatological conditions. A reasonable default sampling for soil invertebrates may also be 30 cm, though the depth of the organic soil horizon would also be appropriate. Default sampling depths should be adjusted if the vertical gradient of contamination is known. If ideal sampling depths are different for different assessment endpoints, a compromise depth may be required. Where contaminant concentrations are measured (e.g., at Superfund sites), exposure depth is not treated as variable. Therefore, any uncertainty in the use of a particular default depth should be stated. The choice of a 10-cm sampling depth versus a 1-m sampling depth could mean a difference of an order of magnitude (e.g., plutonium in Litaor et al. 1994).

In selecting a statistical representation of concentrations for organisms exposed to chemicals in soils, a fundamental distinction must be made between receptors that average their exposure over space and time and those that have essentially constant exposure. For plants and soil invertebrates that are immobile or nearly immobile, the exposure concentration is best represented by the maximum detected concentration (Sample et al. 1997). Mobility is irrelevant if the organism does not move the distance between two sampling points. However, for mobile ecological receptors (terrestrial wildlife consuming soil, vegetation, or animal foods) that do not experience their environment on a "point" basis, it is necessary to convert measured data from single sample points into an estimate that represents some relevant spatial area, such as their habitat (Sample et al. 1997).

A number of methods are used to consider spatial area when computing exposure concentrations, including assuming that contaminants are evenly distributed within a habitat and that ecological receptors forage randomly with respect to contamination within that habitat. Consideration can be given to the size and quality of habitat on and near the site utilized by mobile species in relation to species foraging areas and contaminant levels in these habitats (Sample et al. 1997). Area-weighted averaging methods, which take into account the relative differences in contamination levels per unit area, may be used to estimate exposure concentrations for both individual species and populations (Freshman and Menzie 1996). More sophisticated techniques are available, including bootstrapping and probabilistic methods, to enhance area-weighted approaches by allowing for quantitation of variance in both the area and contaminant concentration estimates (Burmaster and Thompson 1997). Finally, geographic information system approaches may be used to simultaneously integrate spatial information on contaminant concentrations, habitat distribution, species foraging area, and topographic features that may affect exposure routes (Clifford et al. 1995).

Temporal scales have received little consideration in the assessment of risks of contaminated soils to ecological receptors. These factors include seasonal patterns of behavior and contaminant uptake and changes in exposure in future scenarios. For example, in northern temperate climates, few organisms are exposed to contamination in the winter because many species are dormant, others are migratory, and uptake of contaminants by food items (vegetation, invertebrates) is reduced or halted. Changes in chemical concentrations in soil occur due to biodegradation or other chemical transformations, and exposure estimates are more accurate if reasonable representations of fate and transformations are included in the relevant models.

Measures of effects need to be coordinated with measures of exposure to help establish a causal relationship between responses and amount of contact with soil contaminants and other stressors. Temporal analysis provides information on the persistence of the exposure, as well as on the rate at which ecological receptors may respond to changes in the exposure regime. In general, temporal dimensions of ecological risk have been poorly described for contaminants in soil. However, such information is useful for determining the duration of exposure to chemicals at toxic levels as well as potential rates for natural or enhanced recovery.

Lab-to-field extrapolation

Numeric soil-quality guidelines are derived by extrapolating results from single-species toxicity tests in the laboratory to soil concentrations in the field which, with a certain level of protection, aim to prevent adverse effects in soil ecosystems. However, in many cases there are not sufficient data available to evaluate the toxicity of a chemical and develop an NSQG. Even when sufficient data exist, they may not be of a form that allows a refined assessment (e.g., missing data on soil type and soil chemistry such as organic matter, clay content, and pH). In other cases, insufficient information is available regarding relevant bioaccumulation factors (BAFs), depreciating the assessment of risk to wildlife as a result of bioaccumulation or biomagnification. BAFs may be very site specific, changing according to chemical concentrations and soil physicochemical characteristics. Typically, the highest BAFs are found at concentrations close to background levels. Therefore, site-specific measurement of BAFs may lead to an adjustment of the NSQGs.

Applications of bioavailability

Developing methods for the measurement and understanding of the mechanisms of bioavailability is important for assessing the toxicity of contaminated soils, but an understanding of how this information will be applied in ERA is also needed. First, a functional relationship between toxicity (the assessment endpoint) and bioavailability must be identified. Second, estimates of bioavailability are needed for both the bioassays supporting the NSQGs and the site soils under consideration. With these in hand, a site manager could estimate expected effects without detailed site-specific bioassay information.

Exposure

Movement of contaminants from soils through the food chain to higher trophic-level receptors is a common concern to risk managers. These higher trophic-level receptors are often perceived as having greater societal ("charismatic soil megafauna," e.g., earthworms) or economic (e.g., earthworms as fish bait) value. They also may be located some distance from the soil contaminant source area. To evaluate such exposures, consideration must be given to both direct contact with or incidental ingestion of soils, and indirect exposures through ingestion of food items contaminated by contact with soils. The relative importance of these pathways depends on the bioavailability of a contaminant from a soil medium and its bioaccumulation potential through the food chain. If contaminants with bioaccumulation potential are found to be present, the risk assessment can be designed to include appropriate measures (e.g., bioaccumulation models, tissue analysis) of potential food web exposure and risk. The risk manager should also recognize that land use patterns (and changes in these patterns) influence both exposure pathways and routes for a given receptor.

Measures of exposure available to terrestrial ecological risk assessors include

- concentrations of chemicals in soil, which may be adjusted for bioavailability,
- doses to wildlife that are derived from these concentrations or are adjusted for bioavailability using knowledge of digestion, and
- concentrations of chemicals in tissues of organisms (food items or consumers).

The measured concentration of a chemical in an organism or particular tissue is often the most direct estimate of exposure. However, trapping vertebrates and sampling plants may not be feasible, given the funding level and/or timing of a risk assessment. In addition, few existing exposure–response models for wildlife relate tissue concentrations to toxicity.

Total chemical concentrations in soil are the most common measure of exposure. These measures are also used to derive BAFs or regression relationships with chemical residues in organisms (BJC 1998; Sample et al. 1999). In addition, most of the available exposure–response thresholds, screening benchmarks, and other dose–response models are derived from studies in which only total concentrations of chemicals in soil were measured. Concentrations and doses based upon total chemical measurements are used in both screening and higher-tier risk assessments.

Although methods for adjusting total chemical concentrations for the fraction that is bioavailable to organisms (e.g., plant available nutrients in soil) have been available for some time, the application of these methods to measuring toxicant bioavailability have only recently been developed, and their application is not common. Experience is being gained on how to use bioavailability measures in ERA. It is advisable that methods for measuring bioavailability continue to be developed and tested for ranges of organisms, soils, and chemicals. In addition, it would be useful for risk assessors if soil chemists and toxicologists state the ranges of uncertainty and limits of applicability for exposure estimates derived from these methods. For example, if a particular organic solvent can be used to estimate the bioavailable fraction of phenanthrene to earthworms, can the same organic solvent be used to estimate the bioavailable fraction of other PAHs in different soils to other soil invertebrates? Could this extraction technique be adapted for use with all nonionic, semivolatile organic compounds?

Many estimates of bioavailability are measures of chemical concentrations in soil pore water, and this may be an appropriate exposure pathway for receptors such as plants and earthworms, where direct contact with dissolved contaminants may be an important uptake pathway. However, for exposure pathways involving the incidental ingestion of soil or transfer between solid phases, soil water measures may not be correlated with the relevant biologically available fraction.

In predictive risk assessments, estimates of exposures are conducted without the benefit of direct measures of bioavailability. Instead, estimates of bioavailability are made based on chemical properties, soil characteristics, and biological receptors of concern. It would be useful to have national or international databases that relate chemical concentrations in contaminated soil to levels of bioaccumulation and/or toxicity.

Bioavailability is a factor when considering comparisons between site chemical levels and background chemical concentrations. Many risk assessments use such comparisons when evaluating exposure conditions, and in some cases, background chemical levels may actually exceed screening levels. The bioavailability of compounds is rarely considered in such situations because comparisons are made strictly on the basis of total chemical levels. The comparison of total chemical levels between different soils and sites is meaningless because bioavailability is governed by the form of the chemical present, the physicochemical structure of the soil, and the duration of the soil–chemical interaction period (aging and weathering). For these reasons, background concentrations of chemical may pose very different exposures than chemicals associated with relatively recent additions of chemicals. In such cases, determination of the bioavailable fraction of chemical present would facilitate a true comparison of exposure levels.

The implications of incorporating bioavailability measures into ERA include

- the decreased uncertainty in site-specific exposure estimates for the relevant assessment endpoints;
- the requirement for site-specific soil tests from which to develop exposure–response relationships;
- the increased precision of exposure–response relationships developed from site-specific data (e.g., laboratory toxicity tests, field studies) where the measure of exposure is the bioavailability-adjusted concentration or dose;
- the relegation of generic empirical exposure–response models (where exposure is represented by total contaminant concentrations) to a position of low importance in the weight-of-evidence analysis;
- the possibility of developing empirical bioaccumulation models that relate bioavailability-adjusted contaminant concentrations in soil to concentrations in plants and other biological tissues and wildlife foods;
- the possibility of using estimates of soil water concentrations in available exposure–response models, where exposure is represented by the concentration in nutrient solution or soil water (i.e., the method of Sauvé et al. [1998] estimates exposure from total chemical concentration in soil and soil pH); and
- the increased cost of assessment, to the extent that different extraction methods are required for different chemicals or chemical families and different endpoints.

The use of bioavailability-adjusted measures of exposure would reduce uncertainty but increase costs of risk assessments. Decisions to proceed with more expensive assessments are usually balanced by the possibility that better estimates of exposure lead to more cost-effective response actions and remedial decisions. Menzie et al. (2000) list 3 questions that a risk assessor or manager should ask when considering the use of information on bioavailability:

1) Do the governing site-management regulations or policies allow for risk-based approaches that can incorporate bioavailability information?
2) Within these risk-based approaches, when is it appropriate to consider information on bioavailability?
3) Will the information provide value for decision-making?

When judging the value of bioavailability information, Menzie et al. (2000) suggest the following be asked:

- Is the chemical in a form that is considered, for all practical purposes, to be physically and chemically unavailable as supported by a regulatory determination, previous studies, or simple tests accepted by the regulatory agency?
- Are risks associated with site soils being driven by chemicals for which the acquisition of bioavailability information is being considered?

- Are other factors associated with the presence of the chemicals (e.g., the presence of a free product or visibly stained soils) likely to determine site management options?
- Will incorporating bioavailability information change the risk estimate by an amount that affects the way in which the overall site risk is managed?

Incidental ingestion of soil by wildlife can result in exposure to contaminants in the soil. The relative importance of this exposure route depends on the relative amount of soil in the diet as well as the degree to which food items (plants and animals) accumulate the contaminant in soils (Chapter 2, Table 2-4). Assuming that the contaminant in the soil is 100% bioavailable, incidental soil ingestion becomes relatively important (50% or more of the oral dose) at soil ingestion rates of 1% and greater in the diet and where the soil BAF into food items is approximately 0.1 and less. For chemicals that fit this category, information on the bioavailability of the chemical incidentally ingested with soil could result in significant modifications in dose estimates.

General Comments on Ecological Risk Assessment of Soils

In general for ERA of soils, a logical analysis of the data should proceed from more realistic (i.e., site-specific) to more precise and controlled (e.g., single chemical and species toxicity tests). Field surveys indicate the actual state of the receiving environment, so other lines of evidence that contradict the field surveys, after allowing for limitations of the field data, are clearly incorrect. For example, the presence of plants that are growing and not visibly injured indicates that lethal and gross pathological effects are not occurring but does not preclude reductions in reproduction or growth rates.

Managing sites with contaminated soils requires considerations of a number of technical and nontechnical factors. The site manager relies on various sources of information, including input from risk assessors and stakeholders, to reach decisions concerning appropriate courses of action. Technically sound and well focused assessments of ecological risk are important for insuring that ecological resources are adequately considered and appropriately protected. The impact that this information has on site decisions is also related to how often and how well communications occur between the assessors and site managers. It is helpful for risk assessors to understand this process and to be aware of the diverse factors that influence decisions. This helps the assessor develop a broader vision that enables the identification of appropriate assessment approaches, information needs, and communication procedures.

Selection of appropriate levels of protection is part of the risk management process and often occurs after the risks associated with a situation have been assessed. In the case of NSQGs, however, a level of protection must be selected a priori so that a specific number can be generated.

Summary

The discussions presented in each of the chapters of this book are only the beginning of the development of tools and methods for assessing contaminated soils. Some of the methods appear to be well developed and codified, such as the earthworm toxicity testing protocol (ASTM 1999). Upon closer inspection, it becomes obvious that even standard guidance such as this is lacking in many aspects, and its application should be conducted with caution and thought, not as a simple recipe to be followed. Other methods are currently at an empirical level, and the mechanisms behind the observed results are not completely understood. As an example, consider hysteresis. This phenomenon has been observed for a number of organic compounds and metals using aging and rate-of-release studies, yet the exact mechanisms that govern the slow release or irreversible binding of chemicals in soil particles are not understood. Understanding these mechanisms is as vital for the development of models predicting long-term toxicity of chemicals in soil as measuring the bioavailable fraction of chemical in soil that causes a toxic response, with the two measurements certainly linked. Although understanding dose–response relationships is an integral part of assessing contaminated soils, it is also extremely important to have a framework into which these data can be incorporated to allow the proper application of scientific data in the management of contaminated soils.

Although specific areas of research have been discussed, there are still many general questions that need to be answered in the context of contaminated soil management. Some of these questions are presented and discussed below.

- Will we be forever chained to the chemist's bench? Most of the current environmental-quality guidelines are based upon meeting some chemical criterion that is based upon total chemical concentrations in laboratory studies. If we are to truly develop a "land ethic" in the management of contaminated terrestrial environments, it is important to bring more biology and ecology into the decision-making process. This requires shifting the management focus from simply chemical concentration monitoring to actively include actual biological effects monitoring, both in the laboratory and in the field.
- How can we integrate biocomplexity into risk assessment? The biocomplexity of microbiological and soil invertebrate communities form the foundation of preserving "healthy ecosystems," yet we have only begun to scratch the surface of understanding the relationships between these communities

and the "charismatic megafauna" we are often trying to protect and manage. It becomes necessary for scientists and risk managers to become more proficient at conveying the importance of the biocomplexity of these often invisible communities to the normal functioning of soil ecosystems.

- How do we link the impact of organism uptake of a chemical to desorption phenomena and its relation to the flux of bioavailable contaminant? This question forms the basis of one of the fundamental needs in soil ecotoxicology: the marriage of expertise from numerous disciplines. This question cannot be answered by soil chemists, environmental engineers, or ecotoxicologists alone but requires dialogue and the development of combined approaches from these disciplines.
- What is the difference between statistical and risk assessment hypotheses, and what implications does this have for hypothesis testing?
- How can we use a cost–benefit approach to risk management of remedial options? All actions have effects on ecosystems, though this is rarely recognized, and typically, investigators focus on chemical concentrations rather than biological sustainability.
- How do we apply the concept of bioavailability to the risk assessment of contaminated soils? Although some progress has been made in addressing this question, much more understanding is needed. Tools for measuring the bioavailable fraction of chemicals in soils are being developed, but we need to understand the limitations of these tools as well as their applications. Which tool to use may very well depend upon the question being asked: Is this soil toxic? Is a specific chemical toxic? What chemical is causing the toxicity? Which measure of bioavailability do we use to assess chemical bioavailability to which organism? Can we use the same tool to measure bioavailability from different routes of exposure (e.g., dermal, dietary)?

The assessment of contaminated soils is an extremely complex process, encompassing issues ranging from biological to chemical; microscale to macroscale in time, space, and biology; and ethical to regulatory. It is clear from the contributions in this book that many individuals are addressing these issues in research and in policy. The number of tools available for addressing the many facets of assessing contaminated soils is continually growing, presenting us with an ever-increasing number of choices. It is important that we understand when and where to apply various tools to answer specific questions, and understand that no single technique or test will adequately assess a system as complex as contaminated soils. It is our hope that this book will begin to assist in understanding the assessment of contaminated soils.

References

Alexander M. 1995. How toxic are toxic chemicals in soil? *Environ Sci Technol* 29:2713-2717.

Alexander M. 2000. Aging, bioavailabiity, and overestimation of risk from environmental pollutants. *Environ Sci Technol* 34:4259-4265.

[ASTM] American Society for Testing and Materials. 1999. Standard guide for conducting a laboratory soil toxicity test with the Lumbricid earthworm *Eisenia foetida*. In: Annual Book of ASTM Standards. Volume 11.05. E47. Philadelphia PA, USA: ASTM, Committee on Biological Effects and Environmental Fate. ASTM E1676-95. p 1056-1074.

Belfroid AC, Sijm DTHM, Van Gestel CAM. 1996. Bioavailability and toxicokinetics of hydrophobic aromatic compounds in benthic and terrestrial invertebrates. *Environ Rev* 4:276-299.

[BJC] Bechtel Jacobs Company. 1998. Empirical models for the uptake of inorganic chemicals from soil by plants. Oak Ridge TN, USA: U.S. Department of Energy. BJC/OR-133.

Burmaster DE, Thompson KM. 1997. Estimating exposure point concentrations for surface soils for use in deterministic and probabilistic risk assessments. *Human Ecol Risk Assess* 3:363-384.

Camel V, Tambute A, Claude M. 1995. Influence of aging on the supercritical fluid extraction of pollutants in soils. *J Chromatogr* 693:101-111.

Clifford PA, Barchers DE, Ludwig DF, Sielken RL, Klingensmith JS, Graham RV, Banton MI. 1995. An approach to quantifying spatial components of exposure for ecological risk assessment. *Environ Toxicol Chem* 14:895-906.

Conder JM, Lanno RP. 2000. Weak-electrolyte extractions and ion-exchange membranes as surrogate measures of cadmium, lead, and zinc bioavailability to *Eisenia fetida* in artificial soils. *Chemosphere* 41:1659-1668.

Crommentuijn T, Doodeman CJAM, Doornekamp A, Van der Pol JJC, Bedaux JJM, Van Gestel CAM. 1994. Lethal body concentrations and accumulation patterns determine time-dependent toxicity of cadmium in soil arthropods. *Environ Toxicol Chem* 13:1781-1789.

Daily GC, Matson PA, Vitousek PM. 1997. Ecosystem services supplied by soil. In: Daily GC, editor. Nature's services: Societal dependence on natural ecosystems. Washington DC, USA: Island Press. p 113-132.

Di Toro DM, Zarba CS, Hansen DJ, Berry WJ, Sqartz RC, Cowan CE, Pavlou SP, Allen HE, Thomas NA, Paquin PR. 1991. Technical basis for establishing sediment quality criteria for nonionic organic chemicals using equilibrium partitioning. *Environ Toxicol Chem* 10:1541-1583.

Edwards CA, Bohlen PJ. 1992. The effects of toxic chemicals on earthworms. *Rev Environ Contam Toxicol* 125:23-99.

Fitzgerald DG, Warner KA, Lanno RP, Dixon DG. 1996. Assessing the effects of modifying factors on pentachlorophenol toxicity to earthworms: Applications of body residues. *Environ Toxicol Chem* 15:2299-2304.

Freshman JS, Menzie CA. 1996. Two wildlife exposure models to assess impacts at the individual and population levels and the efficacy of remedial actions. *Human Ecol Risk Assess* 2:481-498.

George MR, Brown JR, Clawson WJ. 1992. Application of nonequilibrium ecology to management of Mediterranean grasslands. *J Range Manag* 45:436-440.

Greene JC, Bartels CL, Warren-Hicks WJ, Parkhurst BR, Linder GL, Peterson SE, Miller WE. 1988. Protocols for short-term toxicity screening of hazardous waste sites. Corvallis OR, USA: USEPA. EPA-600-3-88-029.

Hawthorne SB, Grabanski CB. 2000. Correlating selective supercritical fluid extraction with bioremediation behavior of PAHs in a field treatment plot. *Environ Sci Technol* 324:4103-4110.

Holling CS. 1986. Resilience of ecosystems; local surprise and global change. In: Clark WC, Munn RE, editors. Sustainable development of the biosphere. Cambridge, UK: Cambridge University Press. p 292-317.

Houx NWH, Aben WJM. 1993. Bioavailability of pollutants to soil organisms via the soil solution. *Sci Total Environ Suppl* Part 1:387-395.

Huckins JN, Tubergen MW, Manuweera GK. 1990. Semipermeable membrane devices containing model lipid: A new approach to monitoring the bioavailability of lipophilic contaminants and estimating their bioconcentration potential. *Chemosphere* 20:533-552.

Huckins JN, Manuweera GK, Petty JD, Mackay D, Lebo JA. 1993. Lipid containing semipermeable membrane devices for monitoring organic contaminants in water. *Environ Sci Technol* 27:2489-2496.

Jackson RB, Canadell J, Ehrlender JR, Mooney HA, Sala OE, Schulze ED. 1996. A global analysis of root distributions for terrestrial biomes. *Oecologia* 108:489-411.

Kapustka LA. 1997. Selection of phytotoxicity tests for use in ecological risk assessments. In: Wang W, Gorsuch J, Hughes JS, editors. Plants for environmental studies. Boca Raton FL, USA: Lewis Publishers. p 515-548.

Keith LH. 1996. Principles of environmental sampling. 2nd ed. Washington DC, USA: ACS. 848 p.

Kelsey JW, Kottler BD, Alexander M. 1997. Selective chemical extractants to predict bioavailability of soil-aged chemicals. *Environ Sci Technol* 31:214-217.

Lanno R, Leblanc S, Knight B, Tymoswski R, Fitzgerald D. 1998. Application of body residues as a tool in the assessment of soil toxicity. In: Sheppard S, Bembridge J, Holmstrup M, Posthuma L, editors. Advances in earthworm ecotoxicology. Pensacola FL, USA: SETAC. p 41-53.

Litaor MI, Thompson ML, Barth GR, Molzer PC. 1994, Plutonium-239+240 and Americium-241 in soils east of Rocky Flats, Colorado. *J Environ Qual* 23:1231-1239.

Leopold A. 1949. A Sand County Almanac. Oxford, UK: Oxford University Press. 295 p.

Linder G, Ingham E, Brandt CJ, Henderson, G. 1992. Evaluation of terrestrial indicators for use in ecological assessments at hazardous waste sites. Corvallis OR, USA: USEPA. EPA-600-R-92-183.

Løkke H, Van Gestel CAM. 1998. Handbook of soil invertebrate toxicity tests. New York NY, USA: John Wiley & Sons. 281 p.

McCarty LS, Mackay D, Smith AD, Ozburn GW, Dixon DG. 1992. Residue-based interpretation of toxicity and bioconcentration QSARs from aquatic bioassays: Neutral narcotics. *Environ Toxicol Chem* 11:917-930.

Menzie C, Burke AM, Grasso D, Harnois M, Magee B, McDonald D, Montgomery C, Nichols A, Pignatello J, Price B, Rose J, Shatkin JA, Smets B, Smith J, Svirsky S. 2000. An approach for incorporating information on chemical availability in soils into risk assessment and risk-based decision making. *Human Ecol Risk Assess* 6:479-510.

Milloy SJ. 1995. Science-based risk assessment: A piece of the Superfund puzzle. Washington DC, USA: National Environmental Policy Institute.

Moore JC, DeRuiter PC. 1993. Assessment of disturbance in soil ecosystems. *Vet Parasitol* 48:75-85.

Moriarty F. 1988. Ecotoxicology: The study of pollutants in ecosystems. 2nd ed. London, UK: Academic Press. 289 p.

Morrison DD, Robertson BK, Alexander M. 2000. Bioavailability to earthworms of aged DDT, DDE, DDD, and dieldrin in soil. *Environ Sci Technol* 34:709-713.

Piatt JJ, Brusseau ML. 1998. Rate-limited sorption of hydrohobic organic compounds by soils with well characterized organic matter. *Environ Sci Technol* 32:1604-1608.

Pignatello JJ, Xing B. 1996. Mechanisms of slow sorption of organic chemicals to natural particles. *Environ Sci Technol* 30:1-10.

Peijnenburg WJGM, Posthuma L, Eijsackers HJP, Allen HE. 1997. A conceptual framework for implementation of bioavailability of metals for environmental management purposes. *Ecotoxicol Environ Saf* 37:163-172.

Posthuma L, Van Gestel CAM, Smit CE, Bakker DJ, Vonk JW, editors. 1998. Validation of toxicity data and risk limits for soils: Final report. Bilthoven, NL: National Institute of Public Health and the Environment. Report Nr 607505004. 230 p.

Rice K, Jain S. 1985. Plant population genetics and evolution in disturbed environments. In: Pickett STA, White PS, editors. The ecology of natural disturbance and patch dynamics. Orlando FL, USA: Academic Press. 472 p.

Rochette E, Koskinen W. 1996. Supercritical carbon dioxide for determining atrazine sorption by field moist soils. *J Soil Sci Soc Am* 60:453-460.

Sample BE, Aplin MS, Efroymson RA, Suter II GW, Welsh CJE. 1997. Methods and tools for estimation of the exposure of terrestrial wildlife to contaminants. Oak Ridge TN, USA: Oak Ridge National Laboratory, Environmental Sciences Division. ESD Nr 4650, ORNL/TM-13391.

Sample BE, Suter II GW, Beauchamp JJ, Efroymson RA. 1999. Literature-based bioaccumulation models for earthworms: Development and validation. *Environ Toxicol Chem* 18:2110-2120.

Sauvé S, Dumestre A, McBride M, Hendershot W. 1998. Derivation of soil quality criteria using predicted chemical speciation of Pb^{2+} and Ca^{2+}. *Environ Sci Technol* 17:1481-1489.

Schlebaum W, Badora A, Schraa G, Riemsdijk WHV. 1998. Interactions between a hydrophobic organic chemical and natural organic matter: Equilibrium and kinetic studies. *Environ Sci Technol* 32:2273-2277.

Smit CE, Van Gestel CAM. 1997. Influence of temperature on the regulation and toxicity of zinc in *Folsomia candida* (Collembola). *Ecotoxicol Environ Saf* 37:213-222.

Smit CE, Van Gestel CAM. 1998. Effects of soil type, pre-percolation and ageing on bioaccumulation and toxicity of zinc for the springtail *Folsomia candida*. *Environ Toxicol Chem* 17:1132-1141.

Suter GW. 1993. Ecological risk assessment. Ann Arbor MI, USA: Lewis Publishers. 538 p.

Tang J, Alexander M. 1999. Mild extractability and bioavailability of polycyclic aromatic hydrocarbons in soil. *Environ Toxicol Chem* 18:2711-2714.

Tilman D. 1997. Biodiversity and ecosystem function. In: Daily GC, editor. Nature's services, societal dependence on natural ecosystems. Washington DC, USA: Island Press. p 93-112.

[USEPA] U.S. Environmental Protection Agency. 1998. Guidelines for ecological risk assessment; Notice. *Federal Register* 63:26846-26924.

Uzigiris EE, Edelstein WA, Philipp HR, Iben IET. 1995. Complex thermal desorption of PCBs from soil. *Chemosphere* 30:377-387.

Van Gestel CAM, Hensbergen PJ. 1997. Interaction of Cd and Zn toxicity for *Folsomia candida* Willem (Collembola: Isotomidae) in relation to bioavailability in soil. *Environ Toxicol Chem* 16:1177-1186.

Van Straalen NM. 1996. Critical body-residues: Their use in bioindication. In: Van Straalen NM, Krivolutsky DA, editors. Bioindicator systems for soil pollution. Dordrecht, NL: Kluwer Academic Publishers. p 5-16.

Walter LR, Chapin FS. 1987. Interactions among processes controlling successional change. *Oikos* 50:131-135.

Watt KEF, Craig PP. 1986. System stability principles. *Systems Res* 3:191-201.

Weber WJ, Young TM. 1997. A distribution reactivity model for sorption by soils and sediments. 6. Mechanistic implications of desorption under supercritical fluid conditions. *Environ Sci Technol* 31:1686-1691.

Weber Jr WJ, Huang W, LeBoeuf EJ. 1999. Soil/sediment organic matter geochemistry and its relationship to the binding and sequestration of organic contaminants. *Colloids Surfaces A: Physiochem Eng Aspects* 151:167.

Wells JB, Lanno RP. 2001. Passive sampling services (PSDs) as biological surrogates for estimating the bioavailability of organic chemicals in soil. In: Greenberg BM, Hull RN, Roberts Jr MH, Gensemer RW, editors. Environmental toxicology and risk assessment: Science, policy, and standardization- Implications for environmental decisions. Tenth Volume. West Conshohocken PA, USA: ASTM. STP 1403. p 253-270.

Werth CJ, Reinhard M. 1997. Effects of temperature on trichloroethylene desorption from silica gel and natural sediments. 2. Kinetics. *Environ Sci Technol* 31:697-703.

Wiens JA, Addicott JF, Case TJ, Diamond J. 1986. Overview: The importance of spatial and temporal scale in ecological investigations. In: Diamond J, Case T, editors. Community ecology. New York NY, USA: Harper and Row. 665 p.

Glossary

A

Adsorption
Surface retention of molecules, atoms, or ions by a solid or liquid; as opposed to absorption, penetration of substances into the bulk of the solid or liquid.

AF **Availability factor**

AFLP **Amplified fragment length polymorphism**

Aged compound
Compound that has persisted in soil or sediment for a period of time, becoming less available to organisms or to removal by a mild solvent extraction.

Agricultural soils
Topsoils with or without soil amendments; likely to contain residues of pesticides, fertilizers.

ALAD **Aminolevulinic acid dehydratase**

ANOVA **Analysis of variance**

AR **Application rate**

ARAR **Applicable or relevant and appropriate requirement**

ARDRA **Amplified ribosomal DNA restriction analysis**

Artificial soil
Soil formulated by mixing together naturally occurring particulates, according to a recipe/formulation, producing a soil with specific characteristics that the organisms can burrow or reside; used as control soil in toxicity tests with reference toxicants or as a diluent of contaminated site soils to expose test organisms to a range of exposure dilutions.

ASE **Accelerated solvent extraction**

Assessment endpoint
Explicit expression of a specific ecological receptor and an associated function or quality to be maintained or protected.

ASTM **American Society for Testing and Materials**

ATP **Adenosine triphosphate**

B

BAF **Bioaccumulation factor**
Ratio of contaminant residues in whole body or specific tissue to soil contaminant concentrations at steady state.

Contaminated Soils: From Soil–Chemical Interactions to Ecosytem Management. Roman P. Lanno, editor.

Batch

Quantity of the same kind of material made, handled in an identical manner at the same time and considered as one group (batch of seeds refers to seed with the same lot number and batch of soil refers to soil with the same origin [i.e., originating from the same bucket]).

Bioaccumulation

Net accumulation of a substance by an organism as a result of uptake from all environmental sources and routes of exposure (e.g., dermal, dietary).

BCE **British Columbia Environment**

BCF **Bioconcentration factor**

Ratio of contaminant residues in whole body or specific tissue to aqueous contaminant concentrations at steady state.

Bioavailability

Measure of the potential of a chemical for entry into ecological or human receptors; specific to the receptor, route of entry, time of exposure, and matrix containing the contaminant.

Bioconcentration

Net accumulation of a substance by an organism as a result of uptake from an aqueous phase (e.g., soil pore water).

Biodegradation

Biological transformation of a contaminant compound.

Biomimetic

Object that mimics an organism in structure or function.

Bound residue

Chemical species in soil, plant, or animal tissue originating from a contaminant, unextracted by standard, exhaustive methods that do not significantly change the chemical nature of the residues.

BSAF **Biota-to-soil accumulation factor**

Ratio of lipid-normalized tissue residue to organic carbon-normalized soil contaminant concentrations at steady state.

BTEX **Benzene toluene ethylbenzene xylene**

C

Capacity

Quantity of metal contained in an exchangeable pool that is capable of interchanging with metal in the solution phase.

CBR **Critical body residue**

Tissue, organ, or whole body residue concentrations for specific chemicals or groups of chemicals that are correlated with adverse effects such as reduced survival or decreased reproductive success.

CCME **Canadian Council of Ministers of the Environment**

CERCLA **Comprehensive Environmental Response Compensation and Liability Act**

Chromophore

Functional group or conjugated system of double bonds that absorb UV/Vis light (e.g., aromatic ring).

COC **Chemical of concern**

CMOS **Center for Subsurface Modeling Support**

Cometabolism

Contaminant degradation in the presence of another compound (primary substrate) necessary for enzyme induction and carbon energy source.

Complexation

Formation of a group of compounds in which a part of the molecular bonding between compounds is of the coordinate type (i.e., bond formed by a shared pair of electrons).

Congener

One of a group of similar compounds produced during the same reactions, such as individual PCB molecules.

Control

Treatment in an investigation, experiment, or study that duplicates the condition and factors that might affect the results of the investigation, except the specific condition that is being studied; is used to determine the absence of measurable toxicity due to basic test conditions (e.g., particle size, temperature, viability of test organisms, or effects due to their storage or handling).

Control soil

Natural (e.g., field collected), artificial, or formulated, clean soil or substrate of known physicochemical composition and of consistent quality; must not contain concentrations of contaminants that cause discernible stress to the test organisms or reduce their survival (e.g., trace levels may be present).

CSM **Conceptual site model**

Model that describes and integrates information on ecological receptors, contaminants of interest, exposure routes, potential effects, assessment endpoints, risk hypotheses, and measurement endpoints (measures of exposure, effect, and characteristics) used by risk assessors to guide the design of the ERA, as well as by risk managers to convey information about potential site-related risks to stakeholders.

D

DAF **Dilution-attenuation factor**

Ratio of porewater concentration of a chemical at its source to the groundwater concentration at the receptor.

DDD **1,1-Dichloro-2,2-bis-(*p*-chlorophenyl)ethane**

DDE **1,1-Dichloro-2,2-bis-(*p*-chlorophenyl)ethylene**

DDT **1,1,1-Trichloro-2,2-bis-(*p*-chlorophenyl)ethane**

Direct measure of bioavailability

Quantification of chemical residues in an organism to determine chemical bioavailability.

Dredged aquatic sediments

Aquatic sediments that may have spent years in an upland disposal area or near-shore confined disposal area, then selected for beneficial reuse as fill or cover material.

DMSO **Dimethylsulfoxide**

DOM **Dissolved organic matter**

DQO **Data quality objective**

DTPA **Diethylenetriamine pentaacetic acid**

E

EC50 **Median effective concentration**

The concentration (mg/kg soil) or proportion (% by dry weight) of material in the soil that is estimated to cause a discernible sublethal toxic effect to 50% of the test organisms. The EC50 (together with its 95% fiducial limits) is statistically derived by analysis of an observed sublethal response for various exposure concentrations or dilution, after a fixed period of exposure.

EcoSSL **Ecological soil screening level**

EDTA **Ethylene diamine tetraacetate**

Endpoint

Variables used to measure or detect an effect, adverse or otherwise; either a measurement endpoint (e.g., shoot or root length) or a calculated or derived value that characterizes the results of the test (e.g., EC50).

Equilibrium

Common chemical concentration between separate states reached after prolonged contact.

Exposure

Co-occurrence of or contact between a contaminant and a receptor.

ERA **Ecological risk assessment**

ERT **Enchytraeid reproduction test**

F

Fate

Processes that distribute contaminants among various compartments in the soil and transform the original contaminant to other chemical forms.

Fill
Wide variety of materials from both virgin ("borrow" areas, i.e., sand and gravel pits) and secondary sites (residential sites, dredged aquatic sediments, and material from sites undergoing demolition, abandonment, or redevelopment).

Fugacity
Tendency for a substance to transfer from one environmental phase to another; analogous to chemical potential as it pertains to the tendency of a chemical to escape from a phase (e.g., from water to air).

FETAX **Frog embryo teratogenesis assay—*Xenopus***

G

GIS **Geographic information system**

H

Half-life
Time required for the concentration of a chemical in soil to decrease to half of the starting concentration.

HEP **Habitat evaluation procedure**

HQ **Hazard quotient**

Hydric soils
Wetland soils that may spend a substantial portion of the year submerged under water.

Hysteresis
When equilibrium desorption of a chemical is vastly different from sorption; may be so slow that it may be termed "irreversible," implying that the chemical may be permanently sorbed.

I

Industrial fill
Fill used at most manufacturing or industrial sites.

Intensity
Equilibrium concentration of a dissolved free metal ion in the solution phase.

Indirect measure of bioavailability
Quantification of the response of an organism to chemical exposure, implicitly assuming that the observed response is due to the bioavailability of the chemical in the medium, which may or may not be quantified.

ISM **Integrated soil microcosm**

IVG **In vitro gastrointestinal**

K

K_d

Sorption distribution coefficient for partitioning of a chemical between soil and water, expressed as (moles/kg)/(moles/L).

K_{oc}

Partitioning factor between soil organic carbon and water, expressed as (mg/L water)/(mg/kg organic carbon).

K_{ow}

Octanol-water partitioning factor, expressed as (mg/L water)/(mg/L octanol).

L

Leaching

Advective and dispersive transport of chemicals in an aqueous phase through soil; transport of soluble chemical species with the flow of subsurface (soil) water.

LC50 **Median lethal concentration**

The concentration (mg/kg soil) or proportion (% by dry weight) of material in the soil that is estimated to cause lethality in 50% of the test organisms; in most instances, the LC50 (together with its 95% fiducial limits) is statistically derived by analysis of observed lethal responses for various exposure concentrations or dilutions, after a fixed period of exposure.

LD50 **Median lethal dose**

The total amount of contaminant ingested by an organism (e.g., concentration (mg/kg soil) X soil mass (kg)) that is estimated to cause lethality in 50% of the test organisms; in most instances, the LD50 (together with its 95% fiducial limits) is statistically derived by analysis of observed lethal responses for various exposure concentrations or dilutions, after a fixed period of exposure.

LEC **Lowest-effect concentration**

LFER **Linear free energy relationship**

LOAEC (LOAEL) **Lowest-observed-adverse-effect concentration (level)**

Lowest concentration of a test material to which organisms are exposed, that causes adverse effects on the organisms; detected by the observer.

Local population

Operational definition reflecting a combination of ecological considerations and risk management judgments (see Chapter 2) defined as

- Geographic boundary based on the site
- Geographic boundary based on contiguous or nearby habitat
- Ecological boundary.

LOEC (LOEL) **Lowest-observed-effect concentration (level)**
lowest concentration of a test material to which organisms are exposed that causes an effect on the organisms that is significantly different from those in the control treatment; detected by the observer.

M

"Manufactured" soils
Blends of natural and anthropogenic materials specifically designed for use as topsoils or agricultural soils.

Measurement endpoints
Readily measurable phenomena that are appropriate for the exposure pathways, temporal dynamics of exposure, and scale of the site being evaluated; must correspond to or predict assessment.

MDD **Mean daily dose**

MDL **Method detection limit**

MFO **Mixed-function oxidase**

Mine tailings
Finely ground mineral material resulting from the grinding of ores during the extraction of metal in mining operations.

Mineralization
Biological transformation of a compound to inorganic forms, such as conversion of benzene to carbon dioxide and water.

Mor
Type of forest soil in which there is little mixing of surface organic matter with mineral soil, with a sharp transition between the surface organic horizon and the underlying mineral horizon.

Mull
Type of forest soil in which the surface horizon consists of organic matter and mineral soil thoroughly mixed together, with a gradual transition to the underlying horizon.

MOS **Margin of safety**

MPN **Most probable number**

N

NAPL **Nonaqueous-phase liquid**
May alter soil-chemical interactions or be a source of contaminants themselves; commonly encountered in soil systems include petroleum fuels (gasoline, diesel, heating oil, jet fuel), industrial oils and lubricants, transformer oils (some including PCBs), and chlorinated solvents (e.g., trichloroethylene, perchloroethylene).

Native topsoils
Natural soils containing no amendments which may be in an undisturbed (virgin) or disturbed (physically manipulated, e.g., tilled) state.

NEC **No effect concentration**

NOAEC (NOAEL) **No-observed-adverse-effect concentration (level)**

Highest concentration (or lowest dilution) of a test material to which organisms are exposed that does not cause any observed and statistically significant adverse effect on the organisms.

NOEC (NOEL) **No-observed-effect concentration (level)**

Highest concentration (or lowest dilution) of a test material to which organisms are exposed that does not cause any observed and statistically significant effect on the organisms.

NRC **National Research Council**

NSP **Nonsettling particle**

Colloid size particles in soil solution made up of humic substances often associated with inorganic materials.

NSQG **Numerical soil-quality guideline**

Substance-specific benchmarks intended to indicate a particular level of environmental protection, risk, or suitability for use.

NURP **National Urban Runoff Program**

O

OECD **Organization for Economic Cooperation and Development**

OMAF **Ontario Ministry of Agriculture and Food**

OMEE **Ontario Ministry of Environment and Energy**

P

PAH **Polyaromatic hydrocarbon**

Partitioning

Distribution of a chemical between two phases.

PBET **Physiologically based extraction technique**

PCB **Polychlorinated biphenyl**

PCE **Perchloroethylene**

Phytomass

Quantity (e.g., mass) of plant material in a particular test unit that is either above (shoots and leaves) or below (roots) ground or both (whole-plant phytomass).

PICT **Pollution-induced community tolerance**

PLFA **Phospholipid fatty acid analysis**

POM **Particulate organic matter**

PRG **Preliminary remediation goal**

Preferential flow

Rapid movement of solutes through fractures, root holes, earthworm burrows, and other heterogeneities in soil at rates much greater than expected from consideration of the porous medium as a whole.

p-CSM	**Preliminary conceptual site model**

Q

QA–QC	**Quality assurance–quality control**

R

RA–RM	**Risk assessment–risk management**
RAGS	**Risk assessment guidance for Superfund**
RAMAS®	**Risk assessment risk managers audit systems**
RBCA	**Risk-based corrective action**
RBSL	**Risk-based screening level**
RCF	**Root concentration factor**

Reference soil

Sample of whole soil collected from a site within the general vicinity of a test soil (e.g., contaminated site soil); can be used in toxicity tests as an indicator of localized conditions exclusive of the specific contaminants of concern that may be present in the site soils; frequently selected for biological testing because of its physicochemical similarity (e.g., texture and fertility) to the site soils; can be used as a diluent of the contaminated site soil in order to expose the test organisms to a range of exposure dilutions.

Reference test

Whole-soil toxicity test in which a control soil (e.g. artificial or reference soil) is spiked with a reference toxicant (e.g., a seedling emergence toxicity test performed with boric acid as the toxicant).

Reference toxicant

Standard chemical used to assess the sensitivity of organisms to establish confidence in the toxicity data obtained for a test material. In most instances, a toxicity test with a reference toxicant is performed to assess the sensitivity of the organisms at the time the test material is evaluated, and to assess the precision of results obtained by the laboratory over time.

Replicate

Test units within a treatment that contain a test soil with the same concentration (or dilution) and moisture content, and the same number of seeds from the same batch for that species; replicates within treatments are treated in a manner identical to that applied to replicates in other treatments to estimate the experimental error derived from the variability among each of the measurements within a treatment.

Residential fill

Fill or topsoil imported for use at a residential plot.

RfD **Risk reference dose**

S

SDMP **Scientific decision management point**

Sequestered compound

Abiotic—compound that is unavailable for biological uptake or removal by mild solvent extractant yet remains chemically unchanged. Biotic—metals that have been absorbed by organisms and incorporated into an inorganic matrix, rendering them unavailable for further biological reactions.

Sequestration

Abiotic—the process of a chemical becoming unavailable for biological uptake; sometimes referred to as poorly reversible or irreversible sorption. Biotic—the incorporation of metals taken up by organisms into inorganic granules where they are unavailable for further biological interactions and are often excreted in this form.

SETAC **Society of Environmental Toxicology and Chemistry**

SIR **Substrate-induced respiration**

Site

From a regulatory perspective, wherever contamination resides or could come to reside; indicates both current and potential future conditions.

Soil amendments

Materials added to soils to adjust the soil properties to enhance plant growth or stabilize chemicals (manure, composted sewage sludge, composted or ground vegetation, peat moss, vermiculite, and chemicals intended to adjust soil properties to enhance plant growth [e.g., lime]); any substance that may be added to soil.

Soils

Any unconsolidated surface material, substantially of natural mineral origin, other than aquatic sediments.

SOM **Soil organic matter**

Sorption

The association of a chemical in one phase (e.g., water) with a second phase (e.g., soil) organic matter; sorption is used to imply a general process of association with a stationary phase, where the precise physical-chemical mechanism of interaction (e.g., adsorption, absorption, ion exchange, precipitation, etc.) is either not specified or is not known.

Spiked control soil

Control soil spiked with a specific amount of reference toxicant to achieve a specific concentration of reference toxicant in the soil; also serves as a positive control to determine whether the test organisms respond consistently over time to a specific concentration of a reference toxicant.

Spiked soil

Any whole soil to which a test material such as a chemical, a mixture of chemicals, or contaminated material has been added, and mixed thoroughly, for experimental purposes.

SPE **Solid-phase extraction**

SPMD **Semipermeable membrane device**

SPME **Solid-phase microextraction**

SPLP **Synthetic precipitation leaching procedure**

SQG **Soil-quality guideline**

SSL **Soil screening level**

SSM **Sequestered source model**

SSO **Site-specific objective**

SSTL **Site-specific target level**

STA **Site of toxic action**

T

TCE **Trichloroethylene**

TCLP **Toxicity characteristic leaching procedure**

Test unit

Test container and its contents; in the seedling emergence test, refers to the test container and cover, the test soil, and the organisms (seeds) in the soil.

TME **Terrestrial model ecosystem**

TNRCC **Texas Natural Resources Conservation Commission**

Toxicity

Inherent potential or capacity of a material to cause adverse effects toward the exposed organisms.

TPH **Total petroleum hydrocarbon**

TQ **Toxicity quotient**

Treatment

Particular set of conditions and manipulations that are applied to replicate test units; in seedling emergence tests described herein, treatment refers to the exposure concentration of boric acid and the control. Therefore, the 6 treatments in a test had concentrations of boric acid of 0 (control), 160, 240, 320, 480, and 640 mg/kg soil d.w.

Treatment

Particular set of experimental conditions that is applied to a test unit or number of test units (e.g., replicates).

TRV **Toxicity reference value**

Dose below which no chronic or acute effects are expected.

TSCF **Transpiration steam concentration factor**

U

UCL — **Upper confidence limit**

Urban fill

Fill present in essentially all cities for developed land.

USEPA — **U.S. Environmental Protection Agency**

USFDA — **U.S. Food and Drug Administration**

USLE — **Universal soil loss equation**

V

Vernal pools

Geographically small areas of great ecological significance.

Vicinal water layers

Water microlayers found at the interface of a solid particle and water phase; near surface water layers.

Volatilization

Diffusive transport of chemicals through soil air.

VOC — **Volatile organic compound**

W

Weathering

Decrease in the concentration of a chemical in soil and pore water over time due to fate and transport processes such as volatilization, biodegradation, hydrolysis, leaching, etc.

Index

Contaminated Soils: From Soil–Chemical Interactions to Ecosytem Management. Roman P. Lanno, editor.

D

E

F

O

P

S

X

Y

Z

Other Titles from SETAC

Environmental Impacts of Pulp and Paper Waste Streams
Stuthridge, van den Heuvel, Marvin, Slade, Gifford, editors
2003

Porewater Toxicity Testing: Biological, Chemical, and Ecological Considerations
Carr and Nipper, editors
2003

Reevaluation of the State of the Science for Water-Quality Criteria Development
Reiley, Stubblefield, Adams, Di Toro, Erickson, Hodson, Keating Jr, editors
2003

Bioavailability of Metals in Terrestrial Ecosystems: Importance of Partitioning for Bioavailability to Invertebrates, Microbes, and Plants
Allen, editor
2003

Community-Level Aquatic System Studies—Interpretation Criteria (CLASSIC)
Giddings, Brock, Heger, Heimbach, Maund, Norman, Ratte, Schäfers, Streloke, editors
2002

Interconnections between Human Health and Ecological Variability
Di Giulio and Benson, editors
2002

Silver in the Environment: Transport, Fate, and Effects
Andren and Bober, editors
2002

Test Methods to Determine Hazards for Sparingly Soluble Metal Compounds in Soils
Fairbrother, Glazebrook, van Straalen, Tararzona, editors
2002

Avian Effects Assessment: A Framework for Contaminants Studies
Hart, Balluff, Barfknecht, Chapman, Hawkes, Joermann, Leopold, Luttik, editors
2001

Ecological Variability: Separating Natural from Anthropogenic Causes of Ecosystem Impairment
Baird and Burton, editors
2001

Impacts of Low-Dose, High-Potency Herbicides on Nontarget and Unintended Plant Species
Ferenc, editor
2001

Risk Management: Ecological Risk-Based Decision-Making
Stahl, Bachman, Barton, Clark, deFur, Ells, Pittinger, Slimak, Wentsel, editors
2001

SETAC

A Professional Society for Environmental Scientists and Engineers and Related Disciplines Concerned with Environmental Quality

The Society of Environmental Toxicology and Chemistry (SETAC), with offices currently in North America and Europe, is a nonprofit, professional society established to provide a forum for individuals and institutions engaged in the study of environmental problems, management and regulation of natural resources, education, research and development, and manufacturing and distribution.

Specific goals of the society are:

- Promote research, education, and training in the environmental sciences.
- Promote the systematic application of all relevant scientific disciplines to the evaluation of chemical hazards.
- Participate in the scientific interpretation of issues concerned with hazard assessment and risk analysis.
- Support the development of ecologically acceptable practices and principles.
- Provide a forum (meetings and publications) for communication among professionals in government, business, academia, and other segments of society involved in the use, protection, and management of our environment.

These goals are pursued through the conduct of numerous activities, which include:

- Hold annual meetings with study and workshop sessions, platform and poster papers, and achievement and merit awards.
- Sponsor a monthly scientific journal, a newsletter, and special technical publications.
- Provide funds for education and training through the SETAC Scholarship/Fellowship Program.
- Organize and sponsor chapters to provide a forum for the presentation of scientific data and for the interchange and study of information about local concerns.
- Provide advice and counsel to technical and nontechnical persons through a number of standing and ad hoc committees.

SETAC membership currently is composed of more than 5,000 individuals from government, academia, business, and public-interest groups with technical backgrounds in chemistry, toxicology, biology, ecology, atmospheric sciences, health sciences, earth sciences, and engineering.

If you have training in these or related disciplines and are engaged in the study, use, or management of environmental resources, SETAC can fulfill your professional affiliation needs.

All members receive a newsletter highlighting environmental topics and SETAC activities, and reduced fees for the Annual Meeting and SETAC special publications.

All members except Students and Senior Active Members receive monthly issues of *Environmental Toxicology and Chemistry* (ET&C), a peer-reviewed journal of the Society. Student and Senior Active Members may subscribe to the journal. Members may hold office and, with the Emeritus Members, constitute the voting membership.

If you desire further information, contact the appropriate SETAC Office.

SETAC North America
1010 North 12th Avenue
Pensacola, Florida 32501-3367 USA
T 850 469 1500 F 850 469 9778
E setac@setac.org

SETAC Europe
Avenue de la Toison d'Or 67
B-1060 Brussels, Belgium
T 32 2 772 72 81 F 32 2 770 53 83
E setac@setaceu.org

www.setac.org

Environmental Quality Through Science®